RF POWER AMPLIFIER BEHAVIORAL MODELING

If you are an engineer or RF designer working with wireless transmitter power amplifier models, this comprehensive and up-to-date exposition of nonlinear power amplifier behavioral modeling theory and techniques is an absolute must-have. Including a detailed treatment of nonlinear impairments, as well as chapters on memory effects, simulation aspects for implementation in commercial system and circuit simulators, and model validation, this one-stop reference makes power amplifier modeling more accessible by connecting the mathematics with the practicalities of RF power amplifier design. Uniquely, the book explains how systematically to evaluate a model's accuracy and validity, compares model types, and offers recommendations as to which model to use in which situation.

DOMINIQUE SCHREURS is Associate Professor in the ESAT-TELEMIC Division, Department of Electrical Engineering, Katholieke Universiteit Leuven, where she also gained her Ph.D. in Electrical Engineering in 1997. She is a Senior Member of the IEEE and was Chair of the IEEE MTT-11 technical committee on microwave measurements. She was a steering committee member of TARGET (Top Amplifier Research Groups in European Team).

MÁIRTÍN O'DROMA is Director of the Telecommunications Research Centre, and Senior Lecturer in the Department of Electronic and Computer Engineering at the University of Limerick. A Fellow of the IET and Senior Member of the IEEE, he was a founding partner and steering committee member of TARGET and a section head of the RF power linearization and amplifier modeling research strand.

ANTHONY A. GOACHER is Research Projects Manager of the Telecommunications Research Centre, University of Limerick. He has an MBA, is a Member of the IET, an Associate Member of the Institute of Physics, and has held a senior management position in the electronics industry for 20 years.

MICHAEL GADRINGER is a Research Assistant in the Institute of Electrical Measurements and Circuit Design, Vienna University of Technology. He is currently involved with power amplifier modeling, linearization and device characterization.

The Cambridge RF and Microwave Engineering Series

Series Editor,
Steve C. Cripps

Peter Aaen, Jaime A. Plá, and John Wood, *Modeling and Characterization of RF and Microwave Power FETs*
Enrico Rubiola, *Phase Noise and Frequency Stability in Oscillators*
Dominique Schreurs, Máirtín O'Droma, Anthony A. Goacher, and Michael Gadringer, *RF Amplifier Behavioral Modeling*
Fan Yang and Yahya Rahmat-Samii, *Electromagnetic Band Gap Structures in Antenna Engineering*

Forthcoming
Sorin Voinigescu and Timothy Dickson, *High-Frequency Integrated Circuits*
Debabani Choudhury, *Millimeter Waves for Commercial Applications*
J. Stephenson Kenney, *RF Power Amplifier Design and Linearization*
David B. Leeson, *Microwave Systems and Engineering*
Stepan Lucyszyn, *Advanced RF MEMS*
Earl McCune, *Practical Digital Wireless Communications Signals*
Allen Podell and Sudipto Chakraborty, *Practical Radio Design Techniques*
Patrick Roblin, *Nonlinear RF Circuits and the Large-Signal Network Analyzer*
Dominique Schreurs, *Microwave Techniques for Microelectronics*
John L. B. Walker, *Handbook of RF and Microwave Solid-State Power Amplifiers*

RF POWER AMPLIFIER BEHAVIORAL MODELING

DOMINIQUE SCHREURS
Katholieke Universiteit Leuven

MÁIRTÍN O'DROMA
University of Limerick

ANTHONY A. GOACHER
University of Limerick

MICHAEL GADRINGER
Vienna University of Technology

CAMBRIDGE
UNIVERSITY PRESS

CAMBRIDGE UNIVERSITY PRESS
Cambridge, New York, Melbourne, Madrid, Cape Town, Singapore, São Paulo, Delhi

Cambridge University Press
The Edinburgh Building, Cambridge CB2 8RU, UK

Published in the United States of America by Cambridge University Press, New York

www.cambridge.org
Information on this title: www.cambridge.org/9780521881739

First published 2009

Printed in the United Kingdom at the University Press, Cambridge

A catalogue record for this publication is available from the British Library

Library of Congress Cataloguing in Publication data
Schreurs, Dominique.
RF power amplifier behavioral modeling / Dominique Schreurs . . . [et al.].
p. cm.
Includes bibliographical references and index.
ISBN 978-0-521-88173-9 (hardback)
1. Power amplifiers. 2. Amplifiers, Radio frequency. I. Goacher, Anthony A.
II. Gadringer, Michael. III. Title.
TK7871.58.P6S36 2008
621.384′12 – dc22 2008019173

ISBN 978-0-521-88173-9 hardback

Contents

Notation

$f(\cdot)$	general nonlinear function
$f_{\mathrm{R}}(\cdot)$	nonlinear extension of an IIR digital filter
$f_{\mathrm{D}}(\cdot)$	nonlinear extension of an FIR digital filter
t	continuous-time variable
s	discrete-time variable
T_{s}	sampling time
x	RF input signal or general input signal
y	RF output signal or general output signal
$\tilde{x}$	low-pass complex-envelope input signal
$\tilde{y}$	low-pass complex-envelope output signal
r	envelope amplitude
P	in-phase component of the envelope amplitude
Q	quadrature component of the envelope amplitude
ϕ	envelope phase
ω	angular frequency variable
f	frequency variable
ω_0	carrier angular frequency
f_0	carrier frequency
$G(\cdot)$	RF memoryless nonlinearity
$g(\cdot)$	AM–AM nonlinearity
$\Phi(\cdot)$	AM–PM nonlinearity
Re	real-part operator
Im	imaginary-part operator
$*$	convolution operator

Abbreviations

ACEPR	adjacent-channel error power ratio
ACI	adjacent-channel interference
ACLR	adjacent-channel leakage ratio
ACPR	adjacent-channel power ratio
ADC	analogue-to-digital converter
ADS	advanced design system
AM–AM/AM–PM	AM–AM and AM–PM model or characterisation
ANN	artificial neural network
APK	amplitude phase-shift keying
ARMA	autoregressive moving average
AWG	arbitrary waveform generator
AWR-MO	Applied Microwave Research's Microwave office
AWR-VSS	Applied Microwave Research's Visual system simulator
BBACS	broadband amplifier characterisation setup
BER	bit error rate
BF	Bessel–Fourier
BPSK	binary phase-shift keying
C/I	carrier-to-intermodulation ratio
C3IM	carrier-to-third-order intermodulation product ratio
CAD	computer-aided design
CCDF	complementary cumulative density function
CDMA	code-division multiple access

CW	continuous-wave
DAC	digital-to-analogue converter
DIDO	dual-input–dual-output
DSO	digital storage oscilloscope
DSP	digital signal processing
DTD	direct-time domain
DUT	device under test
EIRP	equivalent isotropic radiated power
ESDA	electronic system design automatisation
EVM	error-vector magnitude
FCC	Federal Communications Commission
FDMA	frequency-division multiple access
FET	field-effect transistor
FFT	fast Fourier transform
FIR	finite impulse response
FOBF	Fourier-series-optimised Bessel–Fourier
FOM	figure of merit
FPGA	field-programmable gate array
GSM	global system for mobile communications
HB	harmonic balance
HEMT	high-electron-mobility transistor
IBO	input power backoff
IC	integrated circuit
IF	intermediate-frequency
IFFT	inverse fast Fourier transform
IIR	infinite impulse response
IMD	intermodulation
IMP	intermodulation product
IP	intercept point
IRF	impulse response function

I–Q	in-phase–quadrature
IS-95	Interim Standard 95, also known as cdmaOne
LDMOS	laterally diffused metal oxide semiconductor
LF	low-frequency
LNA	low-noise amplifier
LS	least-squares
LSNA	large-signal network analyser
LTI	linear time-invariant
MDL	minimum description length
MESFET	metal–semiconductor field-effect transistor
MFTD	mixed frequency- and time-domain signal representation
MIMO	multiple-input–multiple-output
MLP	multilayer perceptron
MS	modified Saleh
NARMA	nonlinear autoregressive moving-average
NARMAX	nonlinear autoregressive moving-average with exogenous input
NIM	nonlinear integral model
NL	nonlinear
NMSE	normalised mean-square error
NOCEM	non-constant envelope modulated
NPR	noise power ratio
NVNA	nonlinear vector network analyser
OBO	output power backoff
ODE	ordinary differential equation
OFDM	orthogonal frequency-division multiplexing
OOK	on/off keying
PA	power amplifier
PAE	power-added efficiency
PAPR	peak to average power ratio
PCB	printed circuit board

PDF	probability density function
PL	percentage linearisation
PSB	Poza–Sarkozy–Berger
PSD	power spectral density
QAM	quadrature amplitude modulation
QPSK	quadrature phase-shift keying
RF	radio-frequency
RMS	root-mean-square
RRC	root-raised cosine
SAW	surface acoustic wave
SER	symbol-error rate
SISO	single-input–single-output
SNR	signal-to-noise ratio
SSPA	solid-state power amplifier
SVD	singular-value decomposition
SWANS	scalable wireless ad hoc network simulator
TDNN	time-delay neural network
TWTA	travelling-wave tube amplifier
UML	universal modelling language
VAF	variance accounted for
VCO	voltage-controlled oscillator
VDHL	very high speed IC hardware description language
VIOMAP	Volterra input–output map
VNA	vector network analyser
VSA	vector signal analyser
VSG	vector signal generator
WCDMA	wideband code-division multiple access

Preface

This book provides a comprehensive treatment of radio-frequency (RF) nonlinear power amplifier behavioural modelling, from the fundamental concepts and principles through to the range of classical and, especially, current modelling techniques.

The continuing rapid growth of wireless communications and radio transmission systems, with their ever increasing sophistication, complexity and range of application, has been paralleled by a similar growth in research into all aspects of electronic components, systems and subsystems. This has given rise to a great variety of new and advanced technologies catering for the breadth of frequencies, bandwidths and powers expected in new and existing air interfaces and in the mobile wireless world, for the ever increasing integration of widely differing interfaces into single devices, with the future likelihood that these devices will be active on two or more interfaces simultaneously. For radio communications, or simply radio transmission systems, from the high-frequency (HF) band to the microwave and millimetrewave bands, the transmitter power amplifier (PA) is a pivotal enabling component. This is especially apparent when setting and satisfying air-interface specifications, the correct transmitted signal power levels and tolerable levels of inband and out-of-band signal impairment. The reason for this high-profile role of the PA is that it is the major source of signal distortion and spurious signal generation, harmonics and intermodulation products. Further, it is by far the greatest energy-consuming component in the radio transmission path. Depending on the class of amplifier and the operating conditions dictated by the complexity of the signals to be amplified, its DC to RF power-conversion efficiency is generally poor, resulting in power wastage. As this wastage occurs mostly through heat dissipation, in many situations active extraction of this heat through cooling systems is necessitated, which in turn leads to further energy costs. Hence, in all applications, a reduction in energy consumption and heat dissipation through improved efficiency of the PA is a desired goal.

Technically, this increase in PA efficiency is usually achieved at the expense of increased nonlinear distortion effects. Predicting, assessing and quantifying the impact of these detrimental effects on the transmitted signals and on the radio environment requires accurate behavioural models of power amplifiers on the one hand and a detailed knowledge of the radio characteristics of the environment on the other. Accurate behavioural models are also required to support research into nonlinear impairment-reduction techniques (such as power amplifier linearisation), efficiency-improvement techniques, full transmitter and communications-link system design, investigation into new wireless communications systems and concepts (with new signal-modulation techniques and multiple-access techniques) and so forth. For such reasons RF nonlinear PA behavioural modelling has grown to become a topic of great interest for all those involved in radio communications engineering.

It is hoped that this book will provide RF research engineers in industry, research institutes and centres and also students and academics with a comprehensive resource covering this major area of wireless communications research and development engineering. Although there is an abundant literature covering different PA behavioural modelling approaches, mainly comprising specialist journals and books of international conference proceedings, there are few works dedicated to their comprehensive treatment, analysis and comparison. This work seeks to fill this gap, bringing together much of the classical treatment and modern conceptual, theoretical and algorithmic developments.

The theoretical foundations for PA behavioural modelling are presented in Chapter 1. This is a systematic overview and comparative assessment of the various approaches to RF power amplifier modelling that have received widespread attention by the scientific community. The chapter is organised into three sections, on power amplifier modelling basics, system-level power amplifier models and circuit-level power amplifier models. In the first section, a theoretical foundation to support the subsequent PA model classification and analysis is set out. The approach is to address the physical and behavioural modelling strategies and then to classify behavioural models as either static or dynamic with varying levels of complexity. Then a distinction is made between heuristic and systematic approaches, hence creating a theoretical framework for comparing different behavioural model formats with respect to their formulation, extraction and, in most cases, predictive capabilities.

Approaches to PA representations for use in system-level simulators are treated in the second section, on system-level power amplifier models. These are analytic signal- or complex-envelope-based techniques, leading to single-input–single-output (SISO) low-pass equivalent models, whose input and output are the complex functions needed to represent the bidimensional nature of amplitude and phase modulation.

The final section of this chapter, on circuit-level power amplifier models, provides an overview of behavioural models intended for use in conventional PA circuit simulators. Representing the voltage, current or power-wave signals as real entities, these models handle the complete, and computationally demanding – because of the different RF and envelope signal time scales involved – input and output signal dynamics. This includes taking into account the signals' harmonic content and, possibly, the input and output mismatches and other physical circuit features.

Having introduced and classified models according to their mathematical structures, in Chapter 2 we address other important properties and classifications of PA behavioural models that, in one way or another, are not directly related to the models themselves. These arise out of experimentally observed PA characteristics and may be grouped into those properties derived from the model structure, those introduced by the PA modelling application and those reflecting the behaviour of the observed amplifier under a specific excitation. Some of these properties may describe the same model characteristic but from different perspectives. This is borne in mind in the approach to their treatment here, where the aim is to provide an integrated and complete overview of behavioural models based on their properties.

Models extracted from amplifier measurements are optimised to mimic the behaviour seen in these measurements, i.e. to produce identical or near identical behavioural results. Such models, therefore, reflect influences of the particular amplifier characterisation technique. Hence an overview of typical amplifier measurement setups together with a compilation of models extracted by these means completes the treatment of PA properties in Chapter 2. The next three chapters deal with memoryless models, models with linear memory and models with nonlinear memory.

In Chapter 3 memoryless nonlinear PA behavioural models are considered; the most popular models presented and investigated are the complex power series expansion, the Saleh model in both polar and quadrature forms and the Bessel–Fourier model. Other models considered are the Fourier model, the Hetrakul and Taylor model, the Berman and Mahle model and the Wiener-based polynomial models. Static envelope characteristics, i.e. the static AM–AM and/or AM–PM characteristics, are taken as the basis for defining a behavioural model as memoryless. Some of these models are well established, though new developments and new insights keep occurring. An example of the latter, included in this chapter, is the new modified Saleh model, developed to overcome some particular weaknesses of the original Saleh model. Generally, a comparative approach is taken in parallel with the exposition of the models. All are applied to a particular memoryless-equivalent AM–AM and AM–PM (AM–AM/AM–PM) characterisation of an LDMOS amplifier amplifying a WCDMA signal, the predicted results being set against actual PA measurements. Other model aspects addressed comparatively include implementation and complexity, intermodulation product decomposition and harmonic handling capacity.

These conventional nonlinear memoryless models, based on static AM–AM and AM–PM representations, are frequency independent and can represent with reasonable accuracy the characteristics of various amplifiers driven by narrowband input signals. However, if an attempt is made to amplify 'wideband' signals, where the bandwidth of the signal is comparable with the inherent bandwidth of the amplifier, a frequency-dependent behaviour will be encountered. This phenomenon is described as a memory effect. The range of memory effects found in modern PA systems, especially higher-power solid state PAs, may be classified as linear or nonlinear or as short or long term. Knowing when and where these arise and how they contribute to system impairment is important to designers and researchers. Approaches to behavioural modelling that take account of both nonlinearities and memory effect phenomena is the theme of Chapters 4 and 5.

In Chapter 4 we focus on investigating those nonlinear models that handle memory effects (i.e. frequency-dependent behaviour) using linear filters. The models described in this chapter are structurally categorised into two-box, three-box or parallel-cascade structure. The two-box models presented are the Wiener and Hammerstein topologies, while the three-box models include the Poza–Sarkozy–Berger (PSB) model and the frequency-dependent Saleh model. These models represent some first attempts at extending the nonlinear static AM–AM models and AM–PM models to cover frequency-dependent effects. The parallel-cascade models presented in this chapter are the Abuelma'atti and polyspectral models. In these a parallel branch structure is used to describe linear memory.

Significantly more challenging is the behavioural modelling of nonlinear PAs that exhibit nonlinear memory effects. Chapter 5 contains a comprehensive overview of this topic and addresses memory polynomial models, the time-delay neural network (TDNN) model, the nonlinear autoregressive moving-average (NARMA) model, the parallel-cascade Wiener model, Volterra-series-based models and the state-space-based model. The simplest modelling approach is the memory polynomial. The introduction of non-uniform time-delay tabs yields better results. The TDNN and NARMA approaches are strongly related to the memory polynomial. In the TDNN model the memoryless nonlinear network is described by an artificial neural network. In the case of the NARMA model, the output depends not only on past values of the input but also on past values of the output. As stability may be an issue, criteria are derived to check for this.

Another way to model nonlinear PAs with nonlinear memory effects is by an extension of the Wiener modelling approach. By introducing parallel branches consisting of a linear time-invariant system followed by a memoryless nonlinear system, nonlinear memory effects can be modelled adequately. The Volterra-series-based models form a large class of models with nonlinear memory. The difficulty in computation and optimisation of the fitting parameters, e.g. the Volterra kernels, of the analytical functions for dynamic (envelope-frequency-dependent) input–output measured data is addressed. It is notable how the complexity of the model increases with increasing memory and nonlinearity order, requiring the extraction of an ever larger number of coefficients to achieve an adequate approximation. A number of extended approaches have been developed to overcome this intrinsic disadvantage of Volterra series models. A parallel FIR-based model has a reduced computational complexity. The Laguerre–Volterra modelling approach yields a reduction in the number of model parameters. The modified or dynamic Volterra model aims to handle higher levels of nonlinearity. Finally, a relationship between Volterra models and TDNN models is presented.

Memory polynomial models and Volterra-series-based models of a lower degree are only really efficient for systems with memory but which are weakly nonlinear. However, state-space-based behavioural models are not so restricted. The dynamics of the PA are determined directly from time-series data, resulting in a compact, accurate and transportable model. Ways in which multisine excitations can render model development more efficient are also presented.

In Chapter 6 PA model validation and comparison are addressed. As PAs are in general complex dynamic systems that combine both short- and long-term memory effects with nonlinear phenomena, there is quite a variety of ways to approach questions of validation and comparison. In contrast with linear systems with memory, where superposition holds, nonlinear dynamical systems must be 'locally' modelled and validated. Therefore, test signals and model comparison criteria must be carefully chosen to suit a particular set of typical operating conditions. In this chapter we set down suitable figures or characteristics of merit (metrics) for model performance comparison in different telecommunication-application contexts. Our overall goal is to present concepts in ways that will help a reader to formulate suitable figure(s) of merit for his or her application.

A two-part approach is taken. General figures of merit (FOMs) are presented first and the main concepts regarding their applicability are explained. Although most of the proposed metrics can be generalised for sampled and or stochastic signals, only deterministic continuous-time signals are considered here. Starting from a general time-domain metric, several variants are proposed each of which is specially suitable for a certain measurement setup. Then more realistic applications are considered. Here most of the proposed figures of merit are formulated for sampled (i.e. discrete-time) signals and in terms of statistical measures such as the covariance and the power spectral density. The stochastic-process approach is seen as potentially useful for modern measurement instruments and system simulators, where complex telecommunication standards test signals are usually characterised statistically. This part of Chapter 6 includes an application example in which different figures of merit are compared.

Simulation tools are widely used for designing and analysing complex communications systems. In Chapter 7, the final chapter, an overview of aspects of system simulation is provided with a view to the integration of RF power amplifier behavioural models into such simulations. Generally communications simulation tools seek to describe the operating characteristics and performances of a complete communications link, whether simple or complex, and to mimic through mathematical models all the analogue and digital signal-processing activities that occur in the real system, whether at baseband, intermediate or radio frequencies. Here distinctions between the different forms of simulation encountered in the telecommunication field are made and examples of the associated software products, mainly commercial ones, are presented. In this way the kind of full-system simulations that are relevant to the behavioural modelling of RF power amplifiers is highlighted.

Following this, in Section 7.3, a general overview of analogue signal behavioural simulators for wireless communication systems, together with figure of merit considerations in behavioural simulations, is presented. First, an explanation of some relevant simulation terminology is given. In this overview distinctions are made between circuit-level and system-level simulations, both of which are closely allied in RF PA behavioural modelling. For the former, harmonic-balance simulation, circuit-envelope simulation and mixed-signal high-frequency IC circuit-level simulation are briefly described and the respective contexts of their application set out.

As this book's focus is on system-level simulation, the latter part of Chapter 7 is concerned mainly with aspects relevant to the theme of the book. These include analogue signal representation, sampling and processing considerations, sampling rate issues – including multirate sampling – and signal decomposability. Continuous-in-time and finite-time-window time-domain simulation modes are also considered. An example of a general schema for the computation flow and execution of a communications-link simulation at system level is also given. This could be considered to be a heterogeneous simulation, or in this case 'co-simulation', as it integrates digital-logic and analogue signal system-level models of computation.

This book is the product of a significant integrated collaborative effort by many researchers from a wide range of research centres and universities across Europe. This was possible because of the proactive infrastructural support provided under TARGET (Top Amplifier Research Groups in a European Team), one of the European Networks of Excellence 2004–2007 (www.target-org.net), headed by Professor Gottfried Magerl of the Vienna University of Technology. Naturally, the book editors and all the contributors acknowledge this invaluable support. Full details of all authors are listed at the end of the book. The editors would like to express their thanks to the book reviewers, and to all the authors for their patient detailed revision of texts and other contributions.

entities, at two different time scales, one, very fast, for the RF carrier and another,

1 Overview of power amplifier modelling

1.1 Introduction

This chapter presents an overview and comparative assessment of the various approaches to RF power amplifier (PA) modelling that have received widespread attention by the scientific community. The chapter is organised into three sections: power amplifier modelling basics, system-level power amplifier models and circuit-level power amplifier models.

Section 1.2 on power amplifier modelling basics provides the basic knowledge to support the subsequent PA model classification and analysis. First, physical and behavioural modelling strategies are addressed and then behavioural models are classified as either static or dynamic with varying levels of complexity. Then, a distinction is made between the heuristic and systematic approaches, hence creating a theoretical framework for comparing different behavioural model formats with respect to their formulation, extraction and, in most cases, predictive capabilities.

In Section 1.3, dedicated to system-level power amplifier models, PA representations intended to be used in system-level simulators are considered. These are analytic signal- or complex-envelope-based techniques; they do not represent the RF carrier directly and RF effects are not specifically included. They are single-input–single-output (SISO) low-pass equivalent models, whose input and output constitute the complex functions needed to represent the bidimensional nature of amplitude and phase modulation.

The final section, on circuit-level power amplifier models, provides an overview of behavioural models intended for use in conventional PA circuit simulators. These models handle the complete input and output RF modulated signals, which are real entities, at two different time scales, one, very fast, for the RF carrier and another, much slower, for the modulating envelope. So, in contrast with system-level models, they also take into account the signals' harmonic content and, possibly, the input and output mismatches. For that, they need to represent the voltage and current, or incident and reflected power waves, of the PA input and output ports, thus becoming two-input–two-output model structures.

Although there is an abundant literature on the various different PA behavioural modelling approaches, there are only a few works dedicated to their analysis and comparison. A widely known reference in this field is the book of Jeruchim *et al.* [1]. More recently, a book edited by Wood and Root [2] and the papers of Isaksson *et al.* [3] and of Pedro and Maas [4] have appeared. This introduction draws from all these four references but follows the last most closely.

1.2 Power amplifier modelling basics

Power amplifiers have a major effect on the fidelity of wireless communications systems, which justifies the large number of studies undertaken to understand their limitations and then to optimise their performance. Although some earlier studies simply consisted of empirical observations of PA input–output behaviour, later works have applied scientific theories to account for the observed behaviour and, hence, to justify the resulting PA models [1–8]. Seen from the more general context of system identification, PA models can be divided into two major groups according to the type of data needed for their extraction: physical models and empirical models [9].

Physical models require knowledge of the electronic elements that constitute the PA, their relationships and the theoretical rules describing their interactions. They use nonlinear models of the PA active device and of the other, passive, components (these models may themselves be of a physical or empirical nature) to form a set of nonlinear equations relating the terminal voltages and currents. Using an equivalent-circuit description (typically having an empirical nature) of the PA, these models are appropriate to circuit-level simulation and provide a result accuracy that is, nowadays, limited almost only by the quality of the active device model. Unfortunately, such precision has a high price in simulation time and the need for a detailed description of the PA internal structure.

When such a PA equivalent circuit is not available, or whenever a complete system-level simulation is desired, PA behavioural models are preferred. Since they are solely based on input–output (behavioural) observations, their accuracy is highly sensitive to the adopted model structure and the parameter extraction procedure. So, it is no surprise that distinct model topologies and different observation data sets may lead to a large disparity in model applicability and simulation results. In fact, though such a behavioural-modelling approach may guarantee the accurate reproduction of the data set used for its extraction, or, possibly, of some other set pertaining to the same excitation class, it is not obvious that it will also produce useful results for a different data set, a different PA of the same family or a PA based on a completely different technology. That is, in contrast with the physical-modelling alternative, the generalisation of the predictive capability of a behavioural model should always be viewed with circumspection.

1.2.1 Nonlinear system identification background

In order to establish a theoretical framework with which to analyse the various approaches to PA behavioural modelling, it is convenient to recall some basic results of system identification theory.

In that framework, our power amplifier is described either by a nonlinear function or a system operator; it is assumed to be either static or dynamic respectively. In the static case its output $y(t)$ can be uniquely defined as a function of the instantaneous input $x(t)$, and the model reduces to

$$y(t) = f(x(t)) \tag{1.1}$$

or

$$y = f(x), \tag{1.2}$$

since the dependence with time is, in this case, immaterial.

When the PA presents memory effects to either the modulated RF signal or the modulating envelope, it is said to be dynamic. The output can no longer be uniquely determined from the instantaneous input. It now depends also on the input past and/or the system state. The relation between $y(t)$ and $x(t)$ cannot be modelled simply by a function but becomes an operator that maps a function of time $x(t)$ onto another function of time $y(t)$. Thus the input–output mapping of our PA is represented by a forced nonlinear differential equation,

$$f\left(y(t), \frac{d\,y(t)}{d\,t}, \ldots, \frac{d^p\,y(t)}{d\,t^p}, x(t), \frac{d\,x(t)}{d\,t}, \ldots, \frac{d^r\,x(t)}{d\,t^r}\right) = 0. \tag{1.3}$$

This states that the output and its time derivatives (in general, the system state) may be nonlinearly related to the input and its time derivatives. Since our PA behavioural models have to be evaluated in a digital computer, i.e. a finite-state machine, it is convenient to adopt a discrete-time environment, in which the time variable becomes a succession of uniform time samples of convenient sampling period T_{s}; thus the time and the continuous time signals may be translated as $t \to sT_{\mathrm{s}}$, $x(t) \to x(s)$ and $y(t) \to y(s)$, $s \in Z$. In this way, the solution of the nonlinear differential equation in Equation (1.3) can be expressed in the following recursive form [10]:

$$y(s) = f_{\mathrm{R}}(y(s-1), \ldots, y(s-Q_1), x(s), x(s-1), \ldots, x(s-Q_2)). \tag{1.4}$$

Here $y(s)$, the present output at time instant sT_{s}, depends in a nonlinear way, dictated by f_{R}, the nonlinear function, on the system state (herein expressed by $y(s-q), q = 1, \ldots, Q_1$), the present input $x(s)$ and its past values, $x(s-q)$. This nonlinear extension of infinite impulse response digital filters [10] (nonlinear IIR) is assumed to be the general form for recursive PA behavioural models.

System identification results have shown that, under a broad range of conditions [10–12] (basically operator causality, stability, continuity and fading memory), such a system can also be represented with any desirable small error by a non-recursive, or direct, form, where the relevant input past is restricted to $q \in \{0, 1, 2, \ldots, Q\}$, the so-called system memory span [10]:

$$y(s) = f_{\mathrm{D}}(x(s), x(s-1), \ldots, x(s-Q)) \tag{1.5}$$

in which $f_{\mathrm{D}}(\cdot)$ is again a multidimensional nonlinear function of its arguments. This nonlinear extension of finite impulse response digital filters [10] (nonlinear FIR), is again the general form that a direct, or feedforward, behavioural model should obey.

Various forms have been adopted for the multidimensional functions $f_{\mathrm{R}}(\cdot)$ and $f_{\mathrm{D}}(\cdot)$, although two of these have received particular attention in nonlinear system identification. This is due to their formal mathematical support and because they lead directly to a canonical realisation and so to a certain model topology. These two forms are polynomial filters [10–14] and artificial neural networks (ANNs) [15–17].

In the first case, $f_{\mathrm{D}}(\cdot)$ is replaced by a multidimensional polynomial approximation, so that Equation (1.5) takes the form

$$\begin{aligned} y(s) &= P_{\mathrm{D}}\left(x(s), x(s-1), \ldots, x(s-Q)\right) \\ &= \sum_{q=0}^{Q} a_1(q)x(s-q) + \sum_{q_1=0}^{Q}\sum_{q_2=0}^{Q} a_2(q_1,q_2)x(s-q_1)x(s-q_2) + \cdots \\ &\quad + \sum_{q_1=0}^{Q}\cdots\sum_{q_N=0}^{Q} a_N(q_1,\ldots,q_N)x(s-q_1)\cdots x(s-q_N). \end{aligned} \tag{1.6}$$

This form shows that the nonlinear system is approximated by a series of multilinear terms. Although simple in concept, this 'polynomial FIR' model architecture is known for its large number of parameters.

The function $f_{\mathrm{R}}(\cdot)$ can also be replaced by a multidimensional polynomial leading to recursive polynomial IIR structures. These provide similar approximation capabilities for many fewer parameters than the direct topology. However, the polynomial IIR is significantly more difficult to extract than the direct topology; this has impeded its application in the PA modelling field.

Indeed, the comparative ease of extraction of the polynomial FIR, in comparison with other PA models, provides its particular and attractive advantage. Since the output is linear in respect of the model parameters, i.e. the kernels $a_n(q_1,\ldots,q_n)$, and dependent only on multilinear functions of the delayed versions of the input, it can be extracted in a systematic way using conventional linear identification procedures.

If $f_{\mathrm{D}}(\cdot)$ or $P_{\mathrm{D}}(\cdot)$ is approximated by a Taylor series then this FIR filter is known as a Volterra series or Volterra filter [10–14]. This Volterra series approximation is particularly interesting as it produces an optimal approximation (in a uniform-error sense) near the point where it is expanded. Therefore it shows good modelling properties in the small-signal, or mildly nonlinear, regimes. However, it shows catastrophic degradation under strong nonlinear operation.

In fact, $f_{\mathrm{D}}(\cdot)$ can be replaced by any other multidimensional polynomial. For example, the Wiener series is orthogonal for white Gaussian noise as an excitation signal [13, 14]; other orthogonal polynomials have been proposed for other excitations [10, 13, 18, 19]. In these cases, the respective series produce results that are optimal (in a mean-square-error sense) in the vicinity of the power level used and for the particular type of input used in the model extraction. These representations are, therefore, amenable to the modelling of strong nonlinear systems when the excitation bandwidth and statistics can be considered close to those used in extraction experiments. A presentation of Wiener series expansions and their orthogonality under white Gaussian noise excitation is given in Section 3.11.

Such polynomial FIR filters can be realised in the form indicated in Figure 1.1. The multiplicity of nth-order cross products between all delayed inputs may be noted; it is to these that the nonlinear filter owes its notoriously complex, although general, form. In a similar way, polynomial IIR filters can be realised. A bilinear, recursive, nonlinear IIR filter implementation is shown in Figure 1.2 [4].

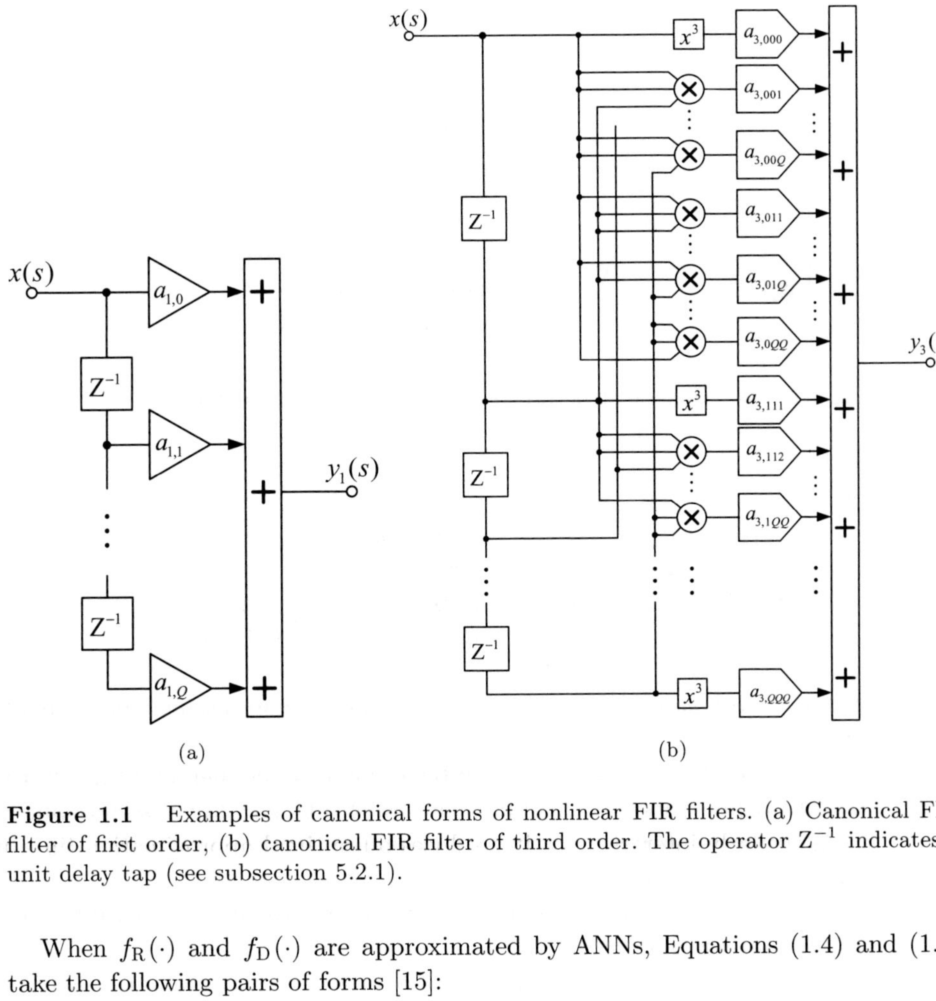

(a) (b)

Figure 1.1 Examples of canonical forms of nonlinear FIR filters. (a) Canonical FIR filter of first order, (b) canonical FIR filter of third order. The operator Z^{-1} indicates a unit delay tap (see subsection 5.2.1).

When $f_{\mathrm{R}}(\cdot)$ and $f_{\mathrm{D}}(\cdot)$ are approximated by ANNs, Equations (1.4) and (1.5) take the following pairs of forms [15]:

$$\begin{aligned} u_k(s) &= \sum_{q=1}^{Q_1} wy_k(q)y(s-q) + \sum_{q=0}^{Q_2} wx_k(q)x(s-q) + b_k, \\ y(s) &= b_{\mathrm{o}} + \sum_{k=1}^{K} wy_{\mathrm{o}}(k) f(u_k(s)) \end{aligned} \tag{1.7}$$

followed by a memoryless nonlinearity, as shown in Figure 1.5.

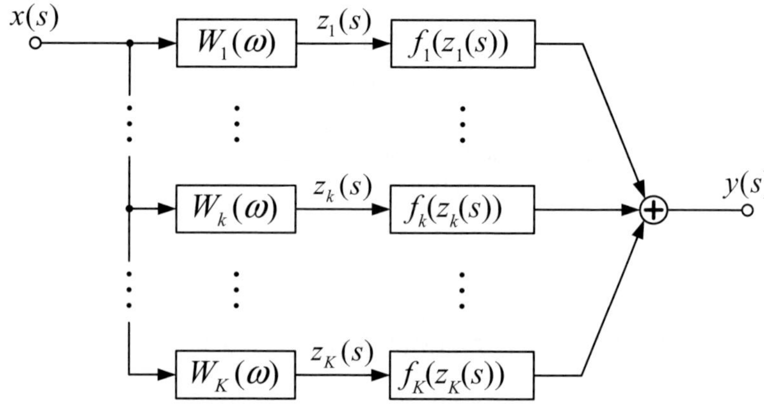

Figure 1.5 Equivalent structure of a feedforward single-hidden-layer perceptron ANN. Note that here the combinations of the branch biases b_k, the activation functions $f(u_k(s))$, the branch gains $w_o(k)$, and the final bias b_o are here represented by different branch memoryless nonlinearities $f_k(z_k(s))$.

If the branch memoryless nonlinearities were now approximated by polynomial functions we would end up again with a polynomial filter. This shows that there is essentially no distinction between a feedforward time-delay ANN and a non-recursive polynomial filter. They simply constitute two alternative ways of approximating the multidimensional function $f_D(\cdot)$, of Equation (1.5). There are, however, some slight differences in these two approaches that will be addressed below. These are worth mentioning because of their impact on PA behavioural modelling activities.

The series form of polynomial filters enables certain output properties to be related to each polynomial degree, and this can be used to guide the parameter extraction procedure. This is especially true if the polynomial series is orthogonal for the input used in the model identification process. For example, the relationship between the intermodulation content of the system's response to a multisine (a signal consisting of several sinusoidal tones) and the coefficients of an appropriate multidimensional orthogonal polynomial have recently been found [18, 19] (the structure and design of multisine signals will be discussed in subsection 2.5.6). However, since in an ANN all memoryless nonlinearities share a common form, there is no way to identify such relationships. Consequently, while polynomial filters can be extracted in a direct way, ANN parameters can be obtained only from some nonlinear optimisation scheme.

Moreover, despite the universal approximation properties of ANNs, there is no way of knowing *a priori* how many hidden neurons are needed to represent a specific system, nor is there any way of predicting the modelling improvement gained when this number is increased. It cannot even be ensured that the extracted ANN is unique or that it is optimal for a certain number of neurons. This can obviously pose some potential problems for the ANN's predictability, especially for inputs outside the signal class used for the identification, i.e. the ANN training process.

However, in contrast with the intrinsically local approximating properties of polynomials, ANNs behave as global approximates, an important advantage when one is modelling strongly nonlinear systems. Also, since the sigmoidal functions used in ANNs are bounded in output amplitude, ANNs are, in principle, better than polynomials at extrapolating beyond the zone where the system was operated during parameter extraction.

1.2.2 *Nonlinear dynamic properties of microwave PAs*

We now turn our attention to some typical nonlinear effects presented by practical microwave and wireless PAs. Considering the variety of available PA technologies, it is not easy to give a completely comprehensive view. Nevertheless, the technical literature in this subject indicates that a few effects at least are commonly observed in a fairly wide range of devices.

Both solid-state PAs (SSPAs) and travelling-wave tube PAs (TWTAs) have been frequently represented by cascade combinations of linear filters and a memoryless nonlinearity [20–22], the so-called two-box and three-box models. These structures introduce linear memory effects at the input and output that can be physically related to the PA's input and output tuned networks.

Beyond these linear memory effects, there are also some dynamic effects that show up only in the presence of nonlinear regimes. This is the case for the so-called long-term memory effects commonly attributed to the active device's low-frequency dispersion and electrothermal interactions and the interactions of the active device with the bias circuitry [23–29] (compare also subsection 2.4.1). Described by the dynamic interaction of two or more nonlinearities through a dynamic network, these long-term memory effects manifest nonlinear dynamics that cannot be modelled by any non-interacting linear filter and memoryless nonlinearity box models. Indeed, Pedro *et al.* [26] showed that such effects can be represented by a memoryless nonlinearity and a filter in a feedback path, as depicted in Figure 1.6, while Vuolevi *et al.* [25] and Vuolevi and Rahkonen [27] used a cascade connection of two nonlinearities with a linear filter in between.

As a common basis for the following behavioural-model discussion, we will assume that a general PA has the form shown in Figure 1.6. Through $H(\omega)$ and $O(\omega)$, this feedback model can account for linear memory effects not only in the carrier but also in the information envelope; these occur whenever the PA characteristic is not flat within the operating signal's bandwidth. In addition, the model is also capable of describing nonlinear memory effects in the carrier (AM–PM) and/or the envelope whenever the feedback filter $F(\omega)$ exhibits dynamic behaviour at the carrier frequency, the carrier harmonics frequencies or the demodulated envelope frequency [4, 26, 27].

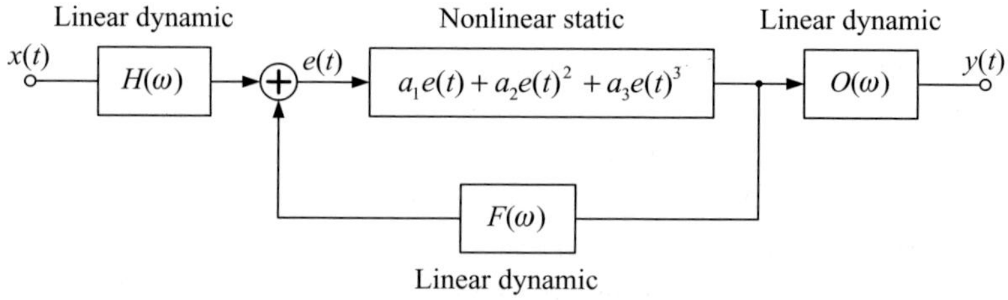

Figure 1.6 Typical nonlinear feedback structure of a microwave PA. Note the presence of the filters $H(\omega)$ and $O(\omega)$ representing linear memory effects related to the input and output matching networks; the feedback path represents nonlinear memory effects attributed to electrothermal and/or bias circuitry dynamics.

For reference, the first- and third-order Volterra nonlinear transfer functions of the dynamic feedback model of Figure 1.6 are [4, 26]:

$$S_1(\omega) = H(\omega)\frac{a_1}{D(\omega)}O(\omega) \tag{1.9}$$

and

$$\begin{aligned} S_3(\omega_1,\omega_2,\omega_3) = & \frac{H(\omega_1)\,H(\omega_2)\,H(\omega_3)}{D(\omega_1)\,D(\omega_2)\,D(\omega_3)}\frac{O(\omega_1+\omega_2+\omega_3)}{D(\omega_1+\omega_2+\omega_3)} \\ & \times\left\{a_3+\frac{2}{3}a_2^2\left[\frac{F(\omega_1+\omega_2)}{D(\omega_1+\omega_2)}+\frac{F(\omega_1+\omega_3)}{D(\omega_1+\omega_3)}+\frac{F(\omega_2+\omega_3)}{D(\omega_2+\omega_3)}\right]\right\}, \end{aligned} \tag{1.10}$$

where $D(\omega) = 1 - a_1F(\omega)$.

On expanding Equation (1.10), input and output linear memory effects are described by the terms $H(\omega_1)H(\omega_2)H(\omega_3)$, $O(\omega_1+\omega_2+\omega_3)$, $F(\omega_1)F(\omega_2)F(\omega_3)$ and $F(\omega_1+\omega_2+\omega_3)$, while nonlinear memory can be seen to arise from the harmonics $F(\omega_j+\omega_k)$, $j,k = 1,2,3$ and the envelope dynamics $F(\omega_j-\omega_k)$.

1.3 System-level power amplifier models

System-level PA behavioural modelling employs low-pass equivalent PA models and thus processes only the complex-envelope information signal. Any specific effects related to or arising from the carrier frequency used must be individually incorporated. This distinguishes such models from circuit-level PA models, which maintain the full RF circuit's band-pass nature and information and work with the actual RF signal.

The RF signal may be written [1, 30]:

$$s(t) = \mathrm{Re}\left\{r(t)e^{j[\omega_0 t+\phi(t)]}\right\} = r(t)\,\cos[\omega_0 t+\phi(t)], \tag{1.11}$$

where an RF carrier of frequency ω_0 is modulated by the complex envelope:

$$\tilde{s}(t) = r(t)e^{j\phi(t)}. \tag{1.12}$$

However, since only the envelope carries useful information, a PA may be thought of as an envelope-processing device.

Although we can also find system-level band-pass behavioural-model representations, i.e. operators $x(t) \to y(t)$, the majority of published PA behavioural models are low-pass complex-envelope equivalents in which $\tilde{x}(t)$ is directly mapped onto $\tilde{y}(t)$. So, and unless otherwise stated, the models we will consider here are of this low-pass equivalent system-level type.

1.3.1 Memoryless PA models

System-level memoryless behavioural models are those in which the output envelope reacts instantaneously to variations in the input envelope. Therefore, they can be represented by two algebraic functions of the instantaneous input envelope's amplitude $r_x(t)$: these are the real and imaginary output envelope components or, as is more common, the output envelope's amplitude $r_y(t)$ and phase $\phi_y(t)$.

Two commonly used examples of low-pass equivalent memoryless models are: (i) a polynomial with complex coefficients a_{2n+1} (see, for example [6]),

$$\tilde{y}(t) = f(r_x(t)) = \sum_{n=0}^{N-1} a_{2n+1} \left[r_x(t)\right]^{2n+1}; \tag{1.13}$$

(ii) the widely used Saleh model [7]:

$$r_y(r_x(t)) = \frac{\alpha_r r_x(t)}{1 + \beta_r [r_x(t)]^2}, \tag{1.14}$$

$$\phi_y(r_x(t)) = \frac{\alpha_\phi r_x(t)^2}{1 + \beta_\phi \left[r_x(t)\right]^2} \tag{1.15}$$

where α_r, β_r, α_ϕ and β_ϕ are fitting parameters for the measured PA's AM–AM characteristics $r_y(r_x(t))$ and AM–PM characteristics $\phi_y(r_x(t))$. The memoryless amplitude and phase nonlinearities can be represented by the model of Figure 1.7, which is given in terms of in-phase and quadrature nonlinearities.

To be accurately described by such a memoryless model, the earlier, more general, model given by Equations (1.9) and (1.10) must obey very restrictive conditions. First, its input and output filters $H(\omega)$ and $O(\omega)$ need to have a bandwidth much larger than the excitation bandwidth, so that they can be seen as flat filters by the band-pass signal. In this case, their complex-envelope low-pass equivalents [1, 30] can be considered approximately as all-pass networks and thus be neglected.

Second, the active device should not be able to produce any odd-order dynamic distortion components from even-degree nonlinearities (in our case $S_3(\omega_1, \omega_2, \omega_3)$ does not involve terms including a_2). This condition is fulfilled if the system can be modelled by a nonlinearity of pure odd symmetry or if $F(\omega)$ is memoryless even for out-of-band components. It should be noted that if the in-band behaviour of $F(\omega)$ were also memoryless, $S_3(\omega, \omega, -\omega)$ would have the same phase as $S_1(\omega)$ and

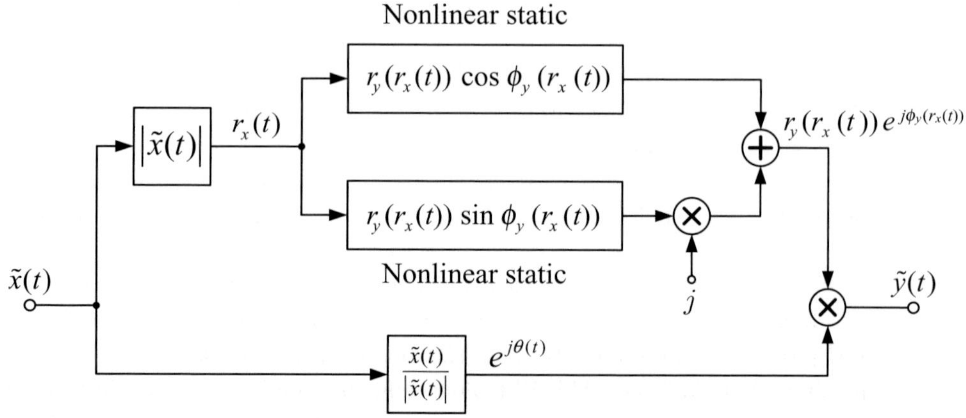

Figure 1.7 Memoryless behavioural model that features AM–AM and AM–PM, nonlinearities, $r_y(r_x(t))$ and $\phi_y(r_x(t))$ respectively.

our PA could be described by a simpler amplitude-only nonlinearity (it would not present any AM–PM conversion).

1.3.2 PA models for addressing linear memory

When considering wide-bandwidth signals, the memoryless narrowband approximation assumed by the low-pass equivalent AM–AM and AM–PM memoryless models referred to above is deficient. Many models have been conceived to address the memory effects arising from the band-pass-system bandwidth limitations observed in power amplifiers (most of such models address the TWTA characteristics) driven by wide-bandwidth signals. The idea behind these models is that the bandwidth of the input is no longer assumed to be so small, compared with that of the system, that a CW signal ceases to be a reasonable representation [1]. Recognising that the variation in the PA's so-called memoryless characteristics as a function of frequency over the PA bandwidth is the problem, the solution naturally consists in sampling that bandwidth at all possible frequency points. This leads to models that are typically AM–AM and AM–PM memoryless models parametrised in the frequency [7]. These models simply achieve the following extension:

$$r_y(r_x(t)) \to r_y(r_x(t), \omega_{\mathrm{RF}}) \tag{1.16}$$

and

$$\phi_y(r_x(t)) \to \phi_y(r_x(t), \omega_{\mathrm{RF}}), \tag{1.17}$$

in which ω_{RF} stands for the angular frequency of the RF carrier at which the frequency-dependent AM–AM and AM–PM conversions are extracted. If we recognise that a CW test of amplitude A and frequency ω_{RF}, displaced from ω_0 by

$\omega_{\rm m} = \omega_{\rm RF} - \omega_0$, corresponds to a complex sinusoidal envelope

$$\tilde{x}(t) = A\, e^{j(\omega_{\rm RF} - \omega_0)t} = A\, e^{j\omega_{\rm m} t} = A(\cos\omega_{\rm m} t + j\sin\omega_{\rm m} t) \tag{1.18}$$

then an $\omega_{\rm RF}$ sweep, as shown in Equations (1.16) and (1.17), describes a memory effect that can be understood as due to the variation in the input envelope frequency. Subsection 2.5.3 and Figure 2.14 present some sample experimental results of this. So, this effect can be represented by either an input, $\tilde{H}(\omega_{\rm m})$, or output, $\tilde{O}(\omega_{\rm m})$, linear low-pass equivalent filter (e.g. as represented in Figure 4.6).

If the required filter precedes the memoryless block (a structure known as the two-box Wiener model), it can model horizontal shifts in the AM–AM and AM–PM plots by an amount $|\tilde{H}(\omega_{\rm m})|$. That is, when variations in $\omega_{\rm RF}$, and thus in $\omega_{\rm m}$, produce AM–AM and AM–PM plots that are similar in shape but are shifted in input envelope amplitude, this frequency-dependent effect can be modelled by the magnitude of an input transfer function $\tilde{H}(\omega_{\rm m})$ (e.g. as represented in Figures 4.10 and 4.11 in relation to the Poza–Sarkozy–Berger model). If the filter is placed after the memoryless block (an arrangement known as the two-box Hammerstein model) then it can model an $|\tilde{O}(\omega_{\rm m})|$ vertical shift in the AM–AM curves and a $\angle\tilde{O}(\omega_{\rm m})$ vertical shift in the AM–PM curves [1]. The simultaneous fulfilment of these two effects usually requires prefiltering, $\tilde{H}(\omega_{\rm m})$, and post-filtering, $\tilde{O}(\omega_{\rm m})$ and for this reason these models tend to share a linear-filter–memoryless-nonlinearity–linear-filter three-box structure (known as the three-box Wiener–Hammerstein model).

Examples of such models are the Poza–Sarkozy–Berger model [5], dealt with in some detail in subsection 4.3.2, the Saleh model [7], discussed in subsection 4.3.3 and the Abuelma'atti model [8], presented in subsection 4.4.1. All these models are based on the heuristic principles explained above, although they use different alternatives for fitting the frequency-dependent AM–AM and AM–PM curves and thus for building the required filters.

It should be noted, however, that the Abuelma'atti model does not share the Wiener–Hammerstein structure but uses a series of parallel Hammerstein branches for the in-phase and quadrature nonlinearities [1, 8], acquiring in this way a much greater flexibility in its modelling capabilities. As in [5, 7], the memoryless nonlinearities (Bessel series) of the Abuelma'atti model are extracted from swept-tone AM–AM and AM–PM (e.g. single-tone) measurements. This approach leaves the model open to the limitation that it can only predict the response to narrowband input signals. Applying a broadband excitation signal to a model extracted in such a way is equivalent to assuming that the superposition principle is valid for this nonlinear system, which is of course wrong.

Other models in this group do not try to deduce any 'convenient' model architecture. Instead they extract a certain set of parameters, assuming one of the model structures already described above. Models of this type use either filter–memoryless-nonlinearity cascades [31], memoryless-nonlinearity–filter cascades [21] or even filter–memoryless-nonlinearity–filter cascades [20, 21], in which the memoryless nonlinearity consists of the AM–AM, AM–PM curves measured at the centre frequency ω_0 (or $\omega_{\rm m} = 0$).

1.3.3 Polyspectral PA models addressing linear memory

Polyspectral filter–memoryless-nonlinearity and memoryless-nonlinearity–filter models are now being given a theoretical basis [31–33]. When their memoryless nonlinearities are implemented as polynomials, polyspectral models become one-dimensional polynomial filters whose higher-order frequency-domain nonlinear transfer functions can be obtained from one-dimensional higher-order input–output cross correlations [33].

Basically, there are two relevant simplifying assumptions involved in these polyspectral models. Each simplification leads to a benefit but has drawbacks.

First, it is assumed that the nonlinear FIR kernels can be represented by one-dimensional systems. So, according to Equation (1.10), they can represent only the cascade of the input or output filters plus the nonlinearity, thus acquiring a filter–memoryless-nonlinearity or memoryless-nonlinearity–filter two-box topology, where the nonlinearity consists of the measured AM–AM/AM–PM behaviour [21, 22, 31, 32]. Although this restriction allows model extraction using a CW carrier test followed by another one-dimensional envelope test, it also brings an inherent incapability, that of describing nonlinear memory. In fact, if we imagine a PA whose nonlinearity arises from the envelope dynamics (for example, a significant reduction in the instantaneous applied supply voltage caused by a deficient bias network design at the envelope-frequency components), such an AM–AM/AM–PM extraction procedure would lead to a linear model.

The second assumption involves the concept of an arbitrary memoryless nonlinearity. If the nonlinearity can be any arbitrary nonlinear function, it does not need to be expanded in a series. Hence, it can represent strong nonlinear effects with many fewer parameters than would be required by a polynomial FIR filter.

The particular polyspectral models used to represent the dynamic characteristics of TWTAs and SSPAs share the format shown in Figure 1.8 [22, 32]; this consists of an extension of the previously proposed three-box model of Silva *et al.* [21].

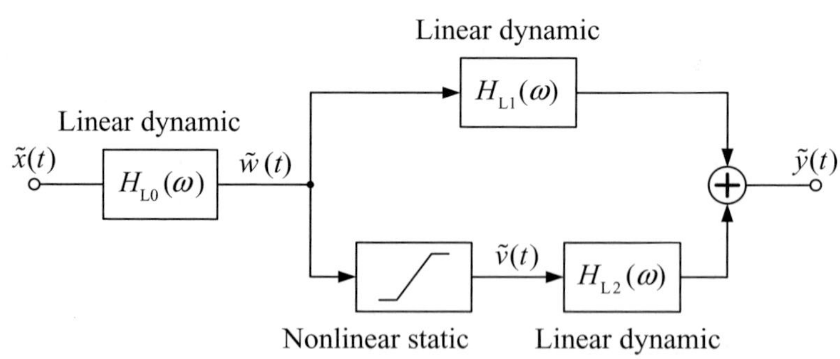

Figure 1.8 Polyspectral model of the memoryless-nonlinearity–filter type, used as a behavioural model of TWTAs and SSPAs.

Although the canonical form of these memoryless-nonlinearity–filter polyspectral models does not include any prefilter [33, 34], the inclusion of the filter $H_{L0}(\omega)$ was empirically justified by its resemblance to the physical operation of the active device [22] and by the superior predictive fidelity obtained by its inclusion [32]. So it was selected as the small-signal transfer function of the SSPA or a corrected version of the former (due to power saturation) in the case of the TWTA. In both cases, the memoryless nonlinearity was arbitrarily selected as a Bessel series fitting of the centreband measured AM–AM and AM–PM conversions.

The final elements in the model, the filters $H_{L1}(\omega)$ and $H_{L2}(\omega)$, were then extracted by minimising the error between the predicted and observed system responses for a predefined excitation. However, since the responses of the upper and lower model branches are, in general, correlated and the upper branch is purely linear, $H_{L2}(\omega)$ may be determined by observing that its branch is the only one responsible for representing the signal's uncorrelated response (the stochastic nonlinear distortion); $H_{L1}(\omega)$ may then be extracted by minimising the residual error in such a way as to obtain the best linear approximation to the real system.

The model proposed by Ibnkahla *et al.* [20] is another three-box model intended to represent linear memory. However, not only is its memoryless nonlinearity represented by two gain (AM–AM) and phase (AM–PM) ANNs, but it is extracted in a completely different way from the Bessel series fitting used for the polyspectral models mentioned above. First, all three blocks are extracted simultaneously, training the neural networks and the linear FIR filters' parameters for a certain optimised error using actual device input and output signal measurements. Second, the selected training data of Ibnkahla *et al.* was uniformly distributed white noise for TWT models and an equally separated multisine signal for SSPA models. Therefore, despite the emphasis still being put on the memory effects arising from the PA's bandwidth limitations, this model is capable of addressing some nonlinear memory effects.

1.3.4 PA models for addressing nonlinear memory

In some situations it is necessary to consider behavioural models addressing nonlinear memory effects. Such effects are seen for frequency-dependent two-tone intermodulation (IMD) responses [35–37], so-called nonlinear impulse responses [38, 39] and even digital random modulation responses [37, 40]; these effects are all observed even under the narrowband approximation. Indeed, memory effects observed in common wireless PAs driven by signals whose relative bandwidth are 0.25% or 0.01% (e.g. in wideband code-division multiple access (WCDMA) [41] or GSM-1800 [42] PAs, where carriers near 2 GHz are modulated with bandwidths as small as 5 MHz or 200 kHz respectively) can hardly be attributed to bandwidth limitations of the PA's matching networks.

An heuristic parametric approach to the PA behavioural-modelling problem was recently proposed by Asbeck *et al.* [28] and then complemented by Draxler *et al.* [29]. The idea was to extend the memoryless AM–AM/AM–PM characterisation by postulating that, in a PA showing long-term memory effects, its gain and phase characteristics will no longer depend only on the instantaneous envelope amplitude

$r_x(t)$ but also on a parameter $\tilde{z}(t)$. This is then used to model dynamic effects such as those caused by self-heating of the active device or by a varying power supply [28]. Consequently, the PA output is expressed as a dynamically varying nonlinear complex gain function:

$$\tilde{y}(t) = f(r_x(t), \tilde{z}(t))r_x(t)e^{j\theta(t)}, \tag{1.19}$$

which is then approximated by a first-order Taylor series as

$$f(r_x(t), \tilde{z}(t)) \approx f_0(r_x(t))[1 + h_z(t) * r_x(t)], \tag{1.20}$$

where $f_0(r_x(t))$ is the measured memoryless AM–AM/AM–PM conversion, $h_z(t)$ is the impulse response of an arbitrary linear filter and $*$ is the convolution operator. According to Draxler *et al.* [29], this filter can be extracted by fitting Equation (1.20) to the measured PA response to a modulated RF stimulus whose envelope is a step function.

Therefore, the amplifier becomes modelled, as shown in Figure 1.9, by a complex gain function that depends in a nonlinear way on the instantaneous amplitude envelope and on a parameter $\tilde{z}(t)$ obtained from the input amplitude envelope by linear filtering.

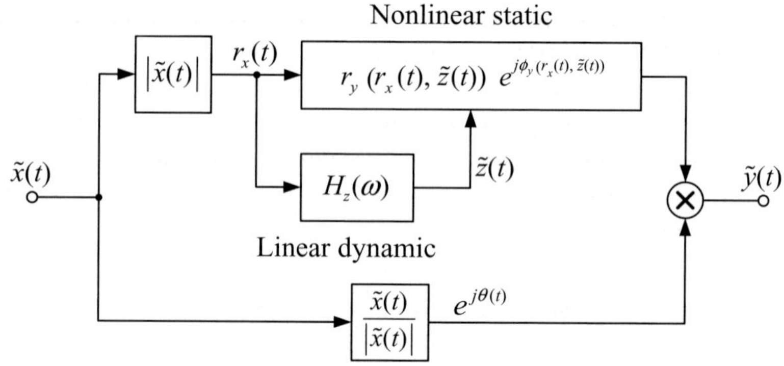

Figure 1.9 Parametric PA nonlinear behavioural model in which the gain is nonlinearly dependent on the instantaneous envelope amplitude and on a linear dynamic parameter.

In the approach followed by Ku *et al.* [35] the objective was to model the memory effects observed in power amplifiers excited by a two-tone RF signal whose amplitudes are both A and whose frequencies are ω_{RF1} and ω_{RF2}. Imagining these frequencies as being symmetrically located about a non-existent carrier ω_0, this two-tone signal corresponds to a sinusoidal complex envelope (in this case purely real) of frequency ω_{m}. So, to incorporate memory into their model, Ku and his co-authors [35] began with the AM–AM/AM–PM memoryless odd-order polynomial representation of complex coefficients, Equation (1.13), and then supposed that the polynomial complex coefficients would now vary with the frequency of the envelope stimulus, i.e. would have the form $a_{2n+1}(\omega_{\mathrm{m}})$. Hence, as these turn into filters, the model adopts the topology of Figure 1.10.

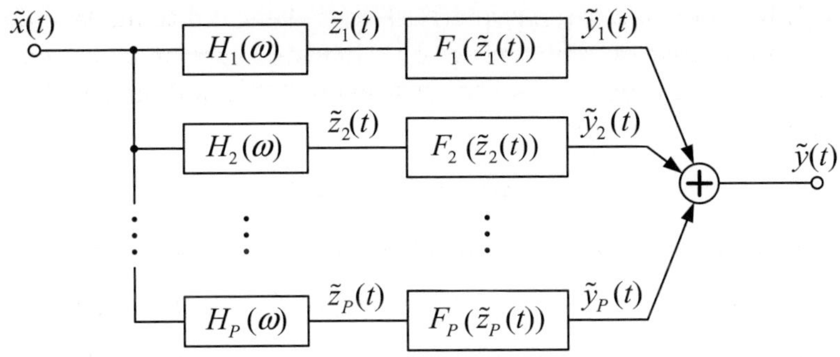

Figure 1.10 The parallel linear-filter–memoryless-nonlinearity cascade model structure adopted in the model of Ku *et al.* [35].

As discussed above, the parallel connection of an arbitrary large, but finite, number of branches each composed of a linear filter followed by a memoryless nonlinearity is equivalent to a non-recursive ANN and is known for its universal modelling capabilities [43–45].

In a later development of their work, Ku and Kenney [36] proposed a behavioural model capable of also accommodating the amplitude- and frequency-dependent asymmetric distortion responses of PAs excited by two-tone inputs, the so-called intermodulation (IMD) asymmetries [46]. Assuming that the eventual asymmetric $(2n+1)$th-order IMD component (where $2n+1$ can now be positive or negative since the envelope response may be asymmetric) has an amplitude and phase represented by the sum of a series of $2N$ complex polynomials $f_{2n+1}(A_\mathrm{i}, \omega_\mathrm{m})$ dependent on the input envelope amplitudes A_i and frequencies ω_m, the output envelope $\tilde{y}(t)$ may be expressed as

$$\tilde{y}(t, A_\mathrm{i}, \omega_\mathrm{m}) = \sum_{n=-N}^{N-1} f_{2n+1}(A_\mathrm{i}, \omega_\mathrm{m}) e^{j(2n+1)\omega_\mathrm{m} t}, \tag{1.21}$$

which, in a discrete-time domain of length $Q+1$ (in an FIR realisation), can be written as

$$\tilde{y}(s, A_\mathrm{i}, \omega_\mathrm{m}) = \sum_{n=0}^{N-1} \sum_{q=0}^{Q} a_{2n+1,q} \tilde{x}(s-q, A_\mathrm{i}, \omega_\mathrm{m}) \left|\tilde{x}(s-q, A_\mathrm{i}, \omega_\mathrm{m})\right|^{2n}, \tag{1.22}$$

where s is the sample instant at which the output is calculated, $a_{2n+1,q}$ is the coefficient multiplying power degree $2n+1$ at the instant $(s-q)T_\mathrm{s}$ of the input envelope $\tilde{x}(q, A_\mathrm{i}, \omega_\mathrm{m})$. This model can be synthesised by adding N structures in parallel, as shown in Figure 1.11.

If the summations in the delay q and in the nonlinear order n of Equation (1.22) are interchanged, then we obtain another, equivalent, structure, which can be synthesised as the FIR filter of Figure 1.12. This form of nonlinear moving-average filter (a generalisation of the linear moving-average model [1]) was previously used by Heutmaker *et al.* [47].

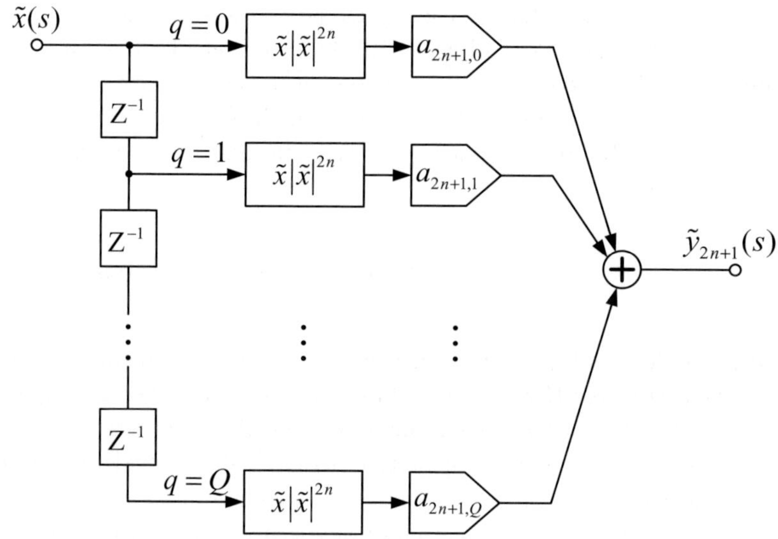

Figure 1.11 Example of $(2n+1)$th-order section of the nonlinear FIR filter model structure adopted by Ku and Kenney [36] for representing frequency-dependent asymmetric IMD behaviour.

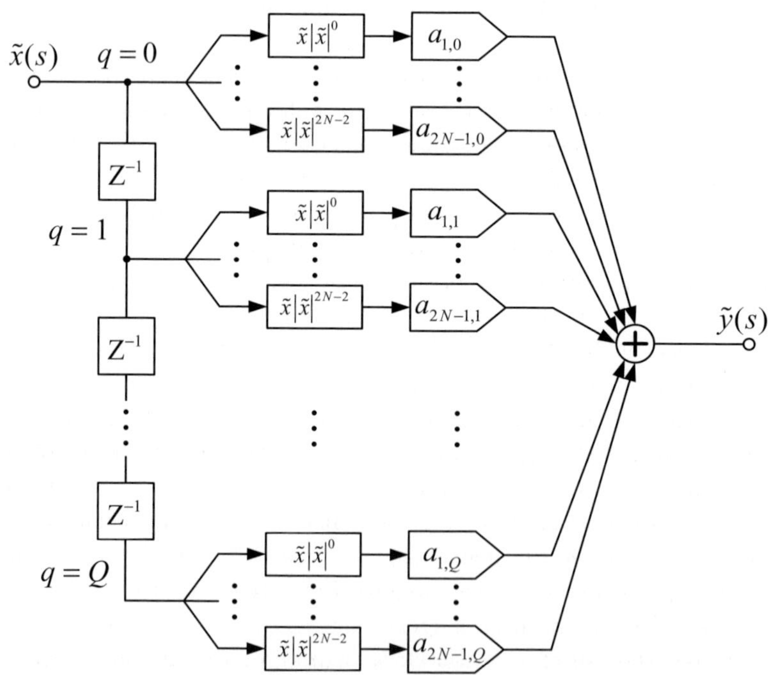

Figure 1.12 One-dimensional nonlinear FIR filter model structure adopted in the model of Heutmaker *et al.* [47], which is equivalent to that shown in Figure 1.11.

It may be noted that neither of the structures shown in Figures 1.11 and 1.12 are as general as the nonlinear FIR filter presented in Figure 1.1, but they are simple one-dimensional approximations because they do not involve cross products between the various delayed versions of the inputs.

Moreover, because Equation (1.22) can also be expressed as a sum of N convolutions,

$$\tilde{y}(s) = \sum_{n=0}^{N-1} \sum_{q=0}^{Q} \tilde{h}_{2n+1}(q)\tilde{x}(s-q)|\tilde{x}(s-q)|^{2n}, \tag{1.23}$$

we conclude that Equation (1.22) can also be interpreted as a parallel connection of nonlinearity–filter Hammerstein branches, in which the memoryless nonlinearities are simply $(2n+1)$th-order monomials. As discussed above, this is the structure already used in the Abuelma'atti model [8] although extracted in a completely different way.

In conclusion, these models share a similar topology with the Silva model [22, 32] although they are more general in the sense that they do not use a single linear filter after a memoryless nonlinearity (the conventional Hammerstein structure), but distinct filters for different nonlinear orders. Furthermore, because they extract the filters and memoryless nonlinearities from the same input–output modulated data, they do not suffer from the limitations imposed by the AM–AM/AM–PM nonlinearity.

In a more formal approach, Brazil and co-workers have tried extracting complete Volterra series models for PAs. In one of their earlier works, Wang and Brazil [48] proposed the extraction of an RF band-pass model for a PA using envelope transient harmonic-balance simulation data gathered from a circuit-level model of the device. A least-squares error-extraction procedure of a hybrid time-and-frequency-domain Volterra formulation allowed the accurate prediction of fundamental and two-tone IMD up to fifth order and also of the spectral regrowth caused by one or two IS-95 CDMA carriers at the input of the amplifier.

In a recent development, Zhu *et al.* [40] proposed a low-pass equivalent model based on a discrete-time Volterra series. Although eventually limited in scope by the mildly nonlinear restrictions imposed by the Volterra series, this model has the advantages that it is applicable to a much broader behavioural representation than the previous models and that it is founded in the solid theoretical ground of Volterra series. In fact, it is no longer a one-dimensional approximation (or a parallel connection of such branches), but a true multidimensional nonlinear dynamic representation of the system. Envisaging future nonlinearity-compensation schemes, the authors used a special arrangement of the Volterra terms [10, 49] and suggested the use of an adaptive learning process [40, 50] as opposed to a simpler and more obvious direct extraction.

Also using a formal, but less general, approach to the behavioural modelling problem, is the work of Mirri *et al.* [38], which was followed by that of Ngoya *et al.* [2, 51] and Soury *et al.* [37, 39]. Starting with a general nonlinear FIR model, the authors state that it can be approximated by a first-order Taylor series expansion around a predetermined nonlinear memoryless operation state $x_0(t)$:

$$y(t) \approx \sum_{q=1}^{Q} f_q(x_0(t), \tau_q)x(t - \tau_q), \tag{1.24}$$

thus defining a so-called nonlinear impulse response $h(x_0(t), \omega)$ and a nonlinear convolution representation.

Such a representation is an ingenious application of the nonlinear integral model (NIM) of Filicori *et al.* [52] to low-pass equivalent nonlinearities and is based on the assumption that, while the signal may be nonlinearly processed in a memoryless way, the dynamic effects are linear. In fact it, once again, is a one-dimensional approximation.

A curious aside on this analysis is that it can be shown [4] that the Mirri *et al.* and the Soury *et al.* models can be implemented as a parallel Hammerstein branch topology and are thus identical to the Ku and Kenney [36] and Heutmaker *et al.* [47] models.

The two models investigated by Fang *et al.* [53] are of the recursive neural network type. They were intended to model a 1 GHz band-pass amplifier by using one recursive ANN for the transient regime and another for the steady-state regime.

1.4 Circuit-level power amplifier models

An alternative is to model the PA as a circuit-level entity. Since they handle the true RF modulated signal, such models are conceived to process real excitations accounting for the possible harmonic content and the distinct time scales of the RF carrier and the baseband information. They may also include possible interstage mismatching in the input and output PA ports.

1.4.1 Equivalent circuit models

The most common PA model for circuit-level simulation uses the equivalent circuit approach. Its topology is usually derived from direct inspection of the real PA, while the value of its parameters (the equivalent-circuit elements) can be derived by either inspection or measurement. This is particularly evident if we note the way in which we model the distributed element-matching networks on the one hand and the nonlinear elements of the active device on the other. As a result, these equivalent-circuit PA models are the consequence of a more or less conscious combination of behavioural and physical modelling approaches. Having much higher detail than common behavioural models, they naturally produce a higher level of accuracy. However, this benefit comes at the high cost of computational efficiency, which justifies the demand for more compact models.

1.4.2 *Circuit-level behavioural models*

If behavioural models are expected to predict port mismatches and harmonic distortion content, they must be able to fully represent in the time domain the ports' voltage and current or incident and reflected power wave relationships. They must therefore be of a double-input–double-output nature. This is a generalisation of the single-input–single-output recursive or direct models of Equations (1.4) and (1.5) and so takes the form

$$\begin{cases} y_1(s) = f_{\mathrm{R}_1}(y_1(s-p), \ldots, y_1(s-1), x_1(s-r), \ldots, x_1(s-1), x_1(s); \\ \qquad\qquad y_2(s-p), \ldots, y_2(s-1), x_2(s-r), \ldots, x_2(s-1), x_2(s)), \\ y_2(s) = f_{\mathrm{R}_2}(y_1(s-p), \ldots, y_1(s-1), x_1(s-r), \ldots, x_1(s-1), x_1(s); \\ \qquad\qquad y_2(s-p), \ldots, y_2(s-1), x_2(s-r), \ldots, x_2(s-1), x_2(s)) \end{cases} \tag{1.25}$$

and

$$\begin{cases} y_1(s) = f_{\mathrm{D}_1}(x_1(s), \ldots, x_1(s-Q); x_2(s), \ldots, x_2(s-Q)), \\ y_2(s) = f_{\mathrm{D}_2}(x_1(s), \ldots, x_1(s-Q); x_2(s), \ldots, x_2(s-Q)). \end{cases} \tag{1.26}$$

If $x_1(t), x_2(t)$ and $y_1(t), y_2(t)$ are the components of the incident and reflected power waves respectively, then Equations (1.25) and (1.26) represent a nonlinear generalisation of the linear scattering matrix. If they are stated as voltages and currents, we end up with a nonlinear generalisation of the linear admittance or impedance matrix formulations.

A rigorous Volterra series formulation of general multi-input–multi-output mildly nonlinear networks was presented by Saleh [54], and Weiner and Naditch [55] developed an appropriate particularisation for describing a nonlinear microwave network using scattering variables. More recently, Verbeyst and Vanden Bossche developed the so-called Volterra input–output map (VIOMAP) [56], which was later shown to be a simplification of the complete double Volterra model [9]. This modelling strategy was implemented in the discrete-time and frequency domains by Wang and Brazil [57].

In an attempt to overcome the mild nonlinearity limitations of the double Volterra model, Schreurs *et al.* [58] used general polynomials while Rizzoli *et al.* [59], Xu *et al.* [60] and Wood and Root [61] used ANNs to approximate the required multidimensional recursive or direct dynamic functions.

Rizzoli and his co-workers [59] used a non-recursive multilayer perceptron ANN in the frequency domain, which was extracted from harmonic-balance simulation results under CW excitation. This ANN used the port voltages, specifically the voltage magnitudes, frequencies and phase differences from the driving signal, as inputs. The outputs of the ANN were composed of the amplitude and phase of the corresponding in-band port currents [59].

The ANN of Xu *et al.* [60] is of the recursive type and expresses the amplifier input current and output voltage as dynamic functions of the amplifier input voltage and output current. In contrast with the recursive ANN shown in Figure 1.3, here the ANN deals, in the continuous-time domain, with the time derivatives of the input and output variables.

Since they were trained by a CW stimulus of varying amplitude and frequency (a static envelope), we can anticipate that these ANNs will be appropriate to fit a PA's frequency-dependent AM–AM/AM–PM characteristics, but it is unlikely that they will predict accurately the PA's nonlinear memory effects (although the two-tone tests performed in [60] are somewhat encouraging).

In an attempt to reduce the number of ad hoc assumptions, first Schreurs *et al.* [58] and then Wood and Root [61] based their work on the mathematical field of nonlinear time series analysis [62]. They proposed nonlinear device models whose multidimensional functions are approximated by general polynomials [58] or by a recursive multilayer perceptron ANN [61] involving the two port input voltages, output currents and their continuous time derivatives. The stimulus used to train the recursive multilayer perceptron was composed of a two-tone signal driving the PA input and another tone at the output.

An interesting advantage of this ANN approach is that, irrespective of the number of hidden layers and neurons, the system's dimension, i.e. the number of embedded input variables [62], can be determined in a systematic way. Furthermore, since this ANN is trained from data obtained using a two-tone signal (at the PA input), i.e. a CW envelope stimulus, it is expected to give better results than the previous ANN models [59, 60] in reproducing a PA's envelope dynamics. In fact, while the models of Rizzoli *et al.* [59] and Xu *et al.* [60] would eventually lead to a memoryless (although carrier-frequency-dependent) low-pass equivalent, the ANN of Wood and Root [61] leads, at least, to a one-dimensional nonlinear dynamic low-pass equivalent model.

In a more recent circuit-level model, first Verspecht *et al.* [63] and then Root *et al.* [64] conceived a mathematically founded extension of the linear S-parameter matrix to the nonlinear case. To achieve this they assumed that the amplifier could be described accurately through its linear response to a small-signal perturbation about a dynamic quiescent point. So the model represents the response of the PA to a small-signal CW excitation of an offset frequency, injected into either the input or output, when the circuit is operating at a dynamic (time-varying) quiescent point determined by a large-signal CW excitation at the nominal centre frequency, injected at the PA input. As the model was conceived to describe the PA behaviour around each harmonic of the large-signal CW, or pumping, quiescent-point signal, the authors called it the polyharmonic distortion model.

Finally, the so-called two-slice model proposed by Walker *et al.* [65] is a circuit-level behavioural model that is of the single-input–single-output form. Although it handles the modulated RF signal in the same manner as all the other models of this section, it does not predict any port mismatches or even any output components other than the ones at, or close to, the fundamental band. Composed of two branches, called slices, the model uses one branch for predicting the odd-order fundamental and intermodulation components while the other branch introduces the intermodulation asymmetry caused by second-order effects. The branch responsible for the fundamental output has a Wiener–Hammerstein topology using a memoryless nonlinearity with an AM–AM characteristic. The second branch uses a memoryless nonlinearity as a detector to extract the envelope signal, which can be afterwards filtered and then relocated in the original frequency band by multiplication with the input RF modulated signal.

References

[1] M. C. Jeruchim, P. Balaban and K. S. Shanmugan, *Simulation of Communication Systems, Modeling, Methodology, and Techniques,* second edition, Kluwer/Plenum, 2001.

[2] J. Wood and D. Root, *Fundamentals of Nonlinear Behavioral Modeling for RF and Microwave Design,* Artech House, 2005.

[3] M. Isaksson, D. Wisell and D. Rönnow, "A comparative analysis of behavioral models for RF power amplifiers," *IEEE Trans. Microw. Theory Tech.*, vol. 54, no. 1, pp. 348–359, January 2006.

[4] J. Pedro and S. Maas, "A comparative overview of microwave and wireless power-amplifier behavioral modeling approaches," *IEEE Trans. Microw. Theory Tech.*, vol. 53, no. 4, pp. 1150–1163, April 2005.

[5] H. Poza, Z. Sarkozy and H. Berger, "A wideband data link computer simulation model," in *Proc. NAECON Conf.*, June 1975, pp. 71–78.

[6] K. G. Gard, H. M. Gutierrez, and M. B. Steer, "Characterization of spectral regrowth in microwave amplifiers based on the nonlinear transformation of a complex Gaussian process," *IEEE Trans. Microw. Theory Tech.*, vol. 47, no. 7, pp. 1059–1069, July 1999.

[7] A. Saleh, "Frequency-independent and frequency-dependent nonlinear models of TWT amplifiers," *IEEE Trans. Communications*, vol. 29, no. 11, pp. 1715–1720, November 1981.

[8] M. Abuelma'atti, "Frequency-dependent nonlinear quadrature model for TWT amplifiers," *IEEE Trans. Communications*, vol. 32, no. 8, pp. 982–986, August 1984.

[9] J. Pedro, J. Madaleno and J. Garcia, "Theoretical basis for the extraction of mildly nonlinear behavioral models," *Int. J. RF and Microwave CAE*, vol. 13, no. 1, pp. 40–53, January 2003.

[10] V. Mathews and G. Sicuranza, *Polynomial Signal Processing,* John Wiley & Sons, 2000.

[11] W. J. Rugh, *Nonlinear System Theory, the Volterra–Wiener Approach,* Johns Hopkins University Press, 1981.

[12] S. Boyd and L. Chua, "Fading memory and the problem of approximating nonlinear operators with Volterra series," *IEEE Trans. Circuits and Systems*, vol. CAS-32, pp. 1150–1161, November 1985.

[13] M. Schetzen, *The Volterra and Wiener Theories of Nonlinear Systems,* John Wiley & Sons, 1980.

[14] M. Schetzen, "Nonlinear system modeling based on the Wiener theory," *Proc. IEEE*, vol. 69, no. 12, pp. 1557–1573, December 1981.

[15] Q. J. Zhang and K. C. Gupta, *Neural Networks for RF and Microwave Design,* Artech House, 2000.

[16] G. Cybenko, "Approximation by superpositions of a sigmoidal function," *Math. Control, Signals, and Systems*, vol. 2, pp. 303–314, December 1989.

[17] K. Hornik, M. Stinchcombe and H. White, "Multilayer feedforward networks are universal approximators," *Neural Networks*, vol. 2, pp. 359–366, June 1989.

[18] P. Lavrador, J. Pedro and N. Carvalho, "A new Volterra series based orthogonal behavioral model for power amplifiers," in *Proc. Asia Pacific Microwave Conf. Dig.*, December 2005, pp. 4–7.
[19] J. Pedro, P. Lavrador and N. Carvalho, "A formal procedure for microwave power amplifier behavioral modeling," in *IEEE MTT-S Int. Microwave Symp. Dig.*, June 2006, pp. 848–851.
[20] M. Ibnkahla, N. J. Bershad, J. Sombrin and F. Castanié, "Neural network modeling and identification of nonlinear channels with memory: algorithms, applications, and analytic models," *IEEE Trans. Signal Processing*, vol. SP-46, no. 5, pp. 1208–1220, May 1998.
[21] C. P. Silva, C. J. Clark, A. A. Moulthrop and M. S. Muha, "Optimal-filter approach for nonlinear power amplifier modeling and equalization," in *IEEE MTT-S Int. Microwave Symp. Dig.*, June 2000, pp. 437–440.
[22] C. Silva, "Time-domain measurement and modeling techniques for wideband communication components and systems," *Int. J. RF and Microwave CAE*, vol. 13, no. 1, pp. 5–31, January 2003.
[23] J. C. Pedro and N. B. Carvalho, *Intermodulation Distortion in Microwave and Wireless Circuits,* Artech House, 2003.
[24] W. Bösch and G. Gatti, "Measurement and simulation of memory effects in predistortion linearizers," *IEEE Trans. Microw. Theory Tech.*, vol. 37, no. 12, pp. 1885–1890, December 1989.
[25] J. Vuolevi, T. Rahkonen and J. Manninen, "Measurement technique for characterizing memory effects in RF power amplifiers," *IEEE Trans. Microwave Theory Tech.*, vol. 49, no. 8, pp. 1383–1389, August 2001.
[26] J. Pedro, N. Carvalho and P. Lavrador, "Modeling nonlinear behavior of band-pass memoryless and dynamic systems," in *IEEE MTT-S Int. Microwave Symp. Dig.*, June 2003, pp. 2133–2136.
[27] J. Vuolevi and T. Rahkonen, *Distortion in RF Power Amplifiers,* Artech House, 2003.
[28] P. Asbeck, H. Kobayashi, M. Iwamoto, G. Nanington, S. Nam and L. E. Larson, "Augmented behavioral characterization for modeling the nonlinear response of power amplifiers," in *IEEE MTT-S Int. Microwave Symp. Dig.*, June 2002, pp. 135–138.
[29] P. Draxler, I. Langmore, T. P. Hung and P. M. Asbeck, "Time domain characterization of power amplifiers with memory effects," in *IEEE MTT-S Int. Microwave Symp. Dig.*, June 2003, pp. 803–806.
[30] S. Benedetto and E. Biglieri, *Principles of Digital Transmission with Wireless Applications,* Kluwer, 1999.
[31] G. Chrisikos, C. Clark, A. A. Moulthrop, M. Muha and C. Silva, "A nonlinear ARMA model for simulating power amplifiers," in *IEEE MTT-S Int. Microwave Symp. Dig.*, June 1998, pp. 733–736.
[32] C. Silva, A. A. Moulthrop and M. Muha, "Introduction to polyspectral modeling and compensation techniques for wideband communications systems," in *ARFTG Conf. Dig.*, November 2001, pp. 1–15.
[33] J. Bendat and A. Piersol, *Engineering Applications of Correlation and Spectral Analysis,* John Wiley & Sons, 1993.
[34] J. Bendat, *Nonlinear Systems: Techniques and Applications,* John Wiley & Sons, 1998.
[35] H. Ku, M. Mckinley and J. Kenney, "Quantifying memory effects in RF power amplifiers," *IEEE Trans. Microw. Theory Tech.*, vol. 50, no. 12, pp. 2843–2849, December 2002.

[36] H. Ku and J. Kenney, "Behavioral modeling of RF power amplifiers considering IMD and spectral regrowth asymmetries," in *IEEE MTT-S Int. Microwave Symposium Dig.*, June 2003, pp. 799–802.

[37] A. Soury, E. Ngoya and J. Nebus, "A new behavioral model taking into account nonlinear memory effects and transient behaviors in wideband SSPAs," in *IEEE MTT-S Int. Microwave Symp. Dig.*, June 2002, pp. 853–856.

[38] D. Mirri, F. Filicori, G. Iuculano, and G. Pasini, "A non-linear dynamic model for performance analysis of large-signal amplifiers in communication systems," in *IMTC/99 IEEE Instrumentation and Measurement Technology Conf. Dig.*, May 1999, pp. 193–197.

[39] A. Soury, E. Ngoya, J. Nebus and T. Reveyrand, "Measurement based modeling of power amplifiers for reliable design of modern communication systems," in *IEEE MTT-S Int. Microwave Symp. Dig.*, June 2003, pp. 795–798.

[40] A. Zhu, M. Wren and T. Brazil, "An efficient Volterra-based behavioral model for wideband RF power amplifiers," in *IEEE MTT-S Int. Microwave Symp. Dig.*, June 2003, pp. 787–790.

[41] H. Holma and A. Toskala, *WCDMA for UMTS: Radio Access for Third Generation Mobile Communications,* Artech House, 2000.

[42] S. Redl, M. Weber and M. Oliphant, *GSM and Personal Communications Handbook,* Artech House, 1998.

[43] G. Palm, "On the representation and approximation of nonlinear systems. part II: discrete time," *Biol. Cybern.*, vol. 34, pp. 49–52, September 1979.

[44] M. J. Korenberg, "Parallel cascade identification and kernel estimation for nonlinear systems," *Ann. Biomedical Eng.*, vol. 19, pp. 429–455, July 1991.

[45] H.-W. Chen, "Modeling and identification of parallel nonlinear systems: structural classification and parameter estimation methods," *Proc. IEEE*, vol. 83, no. 1, pp. 39–66, January 1995.

[46] N. B. Carvalho and J. C. Pedro, "A comprehensive explanation of distortion sideband asymmetries," *IEEE Trans. Microw. Theory Tech.*, vol. 50, no. 9, pp. 2090–2101, September 2002.

[47] M. Heutmaker, E. Wu and J. Welch, "Envelope distortion models with memory improve the prediction of spectral regrowth for some RF amplifiers," in *ARFTG Conf. Dig.*, December 1996, pp. 10–15.

[48] T. Wang and T. Brazil, "Volterra-mapping-based behavioral modeling of nonlinear circuits and systems for high frequencies," *IEEE Trans. Microw. Theory Tech.*, vol. 51, no. 5, pp. 1433–1440, May 2003.

[49] V. J. Mathews, "Adaptive polynomial filters," *IEEE Signal Processing Mag.*, pp. 10–26, July 1991.

[50] A. Zhu and T. Brazil, "An adaptive Volterra predistorter for the linearization of RF high power amplifiers," in *IEEE MTT-S Int. Microwave Symp. Dig.*, June 2002, pp. 461–464.

[51] E. Ngoya, N. Le Gallou, J. M. Nébus, H. Burêt and P. Reig, "Accurate RF and microwave system level modeling of wideband nonlinear circuits," in *IEEE MTT-S Int. Microwave Symp. Dig.*, June 2000, pp. 79–82.

[52] F. Filicori, G. Vannini and V. Monaco, "A nonlinear integral model of electron devices for HB circuit analysis," *IEEE Trans. Microw. Theory Tech.*, vol. 40, no. 7, pp. 1456–1465, July 1992.

[53] Y. Fang, M. C. Yagoub, F. Wang, and Q. J. Zhang, "A new macromodeling approach for nonlinear microwave circuits based on recurrent neural networks," *IEEE Trans. Microw. Theory Tech.*, vol. 48, no. 12, pp. 2335–2344, December 2000.

[54] A. Saleh, "Matrix analysis of mildly nonlinear multiple-input, multiple-output systems with memory," *Bell System Tech. J.*, vol. 61, pp. 2221–2243, November 1982.

[55] D. Weiner and G. Naditch, "A scattering variable approach to the Volterra analysis of nonlinear systems," *IEEE Trans. Microw. Theory Tech.*, vol. 24, no. 7, pp. 422–433, July 1976.

[56] F. Verbeyst and M. Vanden Bossche, "VIOMAP, the S-parameter equivalent for weakly nonlinear RF and microwave devices," *IEEE Trans. Microw. Theory Tech.*, vol. 42, no. 12, pp. 2531–2533, December 1994.

[57] T. H. Wang and T. J. Brazil, "A Volterra mapping-based S-parameter behavioral model for nonlinear RF and microwave circuits and systems," in *IEEE MTT-S Int. Microwave Symp. Dig.*, June 1999, pp. 783–786.

[58] D. Schreurs, N. Tufillaro, J. Wood, D. Usikov, L. Barford and D. E. Root, "Development of time domain behavioural non-linear models for microwave devices and ICs from vectorial large-signal measurements and simulations," in *Gallium Arsenide Applications Symp. Dig.*, October 2000, pp. 236–239.

[59] V. Rizzoli, A. Neri, D. Masotti and A. Lipparini, "A new family of neural network-based bidirectional and dispersive behavioral models for nonlinear RF/microwave subsystems," *Int. J. RF and Microwave CAE*, vol. 12, no. 1, pp. 51–70, January 2002.

[60] J. Xu, M. Yagoub, R. Ding and Q. J. Zhang, "Neural-based dynamic modeling of nonlinear microwave circuits," *IEEE Trans. Microw. Theory Tech.*, vol. 50, no. 12, pp. 2769–2780, December 2002.

[61] J. Wood and D. Root, "The behavioral modeling of microwave/RF ICs using non-linear time series analysis," in *IEEE MTT-S Int. Microwave Symp. Dig.*, June 2003, pp. 791–794.

[62] H. Kantz and T. Schreiber, *Nonlinear Time Series Analysis,* Cambridge University Press, 1997.

[63] J. Verspecht, D. E. Root, J. Wood and A. Cognata, "Broad-band multi-harmonic frequency domain behavioral models from automated large signal vectorial network measurements," in *IEEE MTT-S Int. Microwave Symp. Dig.*, June 2005, pp. 1975–1978.

[64] D. E. Root, J. Verspecht, D. Sharrit, J. Wood and A. Cognata, "Broad-band poly-harmonic distortion (PHD) behavioral models from fast automated simulations and large-signal vectorial network measurements," *IEEE Trans. Microw. Theory Tech.*, vol. 53, no. 11, pp. 3656–3664, November 2005.

[65] A. Walker, M. Steer, K. Gard and K. Gharaibeh, "Multi-slice behavioral model of RF systems and devices," in *Radio and Wireless Conf. Dig.*, September 2004, pp. 71–74.

2 Properties of behavioural models

2.1 Introduction

In Chapter 1 we introduced and classified models according to the mathematical structure of the model. However, several other properties and classifications of PA behavioural models are possible that are not directly related to the model itself. They are a result of the observed PA characteristics, which are reflected by the model.

Here we will highlight aspects and properties of behavioural models and will also discuss important issues concerning the characterisation of microwave amplifiers. The model properties and classifications used in this context can be divided into the following major groups:

- properties derived from the model structure;
- properties introduced by the PA modelling application;
- properties reflecting the behaviour of the observed amplifier under a specific excitation.

The most important of these properties are summarised in Figure 2.1.

Obviously, a model can represent several of these characteristics simultaneously. Some properties listed in the figure describe the same model characteristic from different points of view. For example, one may classify models according to the presence of linear or nonlinear memory effects using the structure presented in Figure 1.6, whereas another possibility is to assume a specific structure for the observed amplifier and then divide the dynamic nonlinear behaviour into short- and long-term memory effects.

In Section 2.2, model- and model-structure-based properties of behavioural models are summarised for completeness. Most of these properties and classifications were introduced in Chapter 1.

In Section 2.3 we briefly present the differences between models classified as circuit-level and system-level. The relationship between the mathematical descriptions of a system in the band-pass channel and in the low-pass equivalent channel is exemplified by the use of a Volterra series for nonlinear dynamic models and the Chebyshev transform for nonlinear static models.

Amplifier-based model properties are discussed in Section 2.4. At the beginning of this section a typical solid-state power amplifier (SSPA) transistor is represented by an equivalent-circuit model. The impedances presented at the ports of this model at the baseband, fundamental and harmonic frequencies have a significant influence

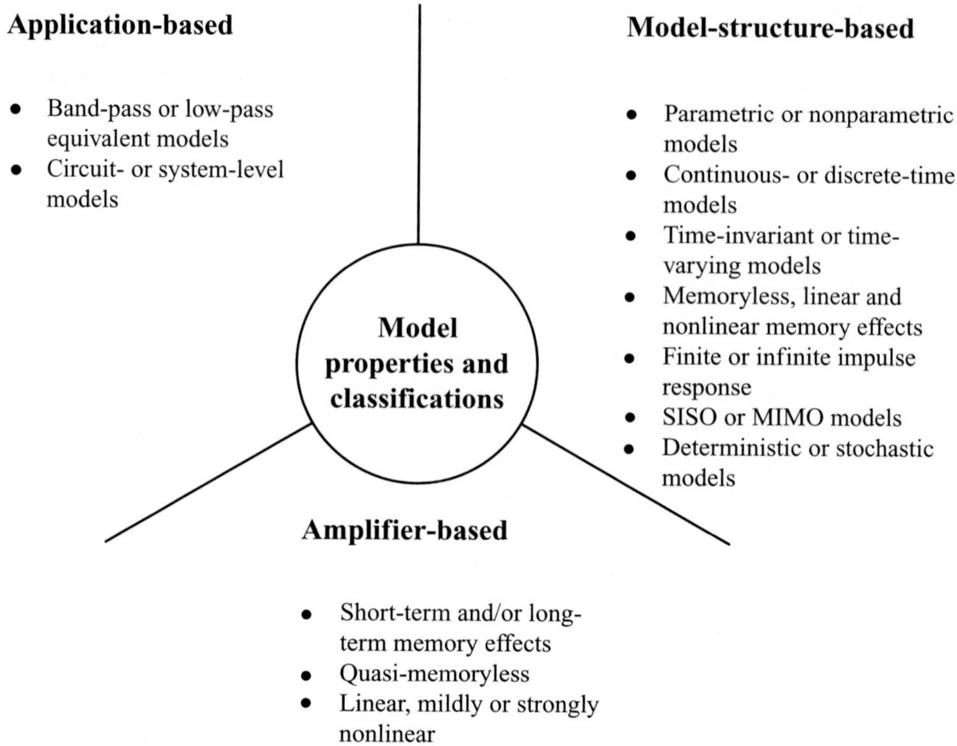

Figure 2.1 Important properties of behavioural models grouped according to their source of origin.

on the dynamic behaviour of the corresponding amplifier and, of course, on the model. On the basis of this equivalent circuit model the classification of memory effects in the short and long term will be explained. Then the impact of thermal memory on the long-term memory effects is considered. Thereafter, the conditions necessary to correctly represent a dynamic nonlinear system using a memoryless nonlinear description are explored. This simplified representation is usually named quasi-memoryless. Finally, the classification of a static nonlinear amplifier according to the magnitude of the input modulation, as linear, mildly nonlinear or strongly nonlinear is presented.

Section 2.5 covers aspects of the characterisation of microwave power amplifiers. Power amplifier behavioural models are discussed as a part of a system-identification process in subsection 2.5.1. The close relation between the model and the underlying amplifier-characterisation process will be highlighted. Afterwards, factors in the selection and design of an excitation signal are considered. Single-tone, two-tone, broadband and multisine amplifier characterisation setups are presented. A process to create a periodic multisine excitation signal which can substitute for other more complex input signals is described, together with multisine measurement setups.

2.2 Model-structure-based properties of behavioural models

The properties presented in Figure 2.1 are related to the mathematical description of behavioural models. There are many possibilities for describing the input–output relationship of a system. Within the scope of this book the most relevant properties in regard to PA behavioural models are treated. A summary of these properties is as follows.

Parametric or nonparametric models According to [1], nonparametric models are described by a curve or function or in a look-up table. Parametric models are characterised by a matrix which parametrises the interrelationship between the system input and output. A frequency plot or an impulse response are examples of nonparametric models, while Volterra series or memory polynomials are parametric.

Continuous- or discrete-time models A continuous-time input–output PA mapping represented by a nonlinear differential equation is given in Equation (1.3). A related discrete-time description is given in Equation (1.4). Within the scope of this book, only equidistant sampling systems are considered.

Time-invariant or time-varying models A time-invariant model responds to an arbitrary time-shifted input signal with the same shift in the output signal, as long as the initial state is also shifted in the same way [2]. Time-varying models alter their system state with time. Models such as the parametric PA nonlinear behavioural model presented in Figure 1.9 can be interpreted as a time-varying memoryless nonlinearity in which the time-dependent nonlinear function is controlled by the linearly filtered magnitude of the input signal. If the filter providing the control of the memoryless function is static, the complete model behaves in a time-invariant way.

Memoryless, linear and nonlinear memory effects A presentation of the different classes of nonlinear dynamic systems was given in subsection 1.2.2. Through the use of Figure 1.6, conditions for representing each of the three types of memory effect were stated.

Finite or infinite impulse response A discrete-time nonlinear infinite impulse response (IIR) model was presented in Equation (1.4). The corresponding nonlinear finite impulse response (FIR) model is given in Equation (1.5). By the feedback of output signal samples, the IIR structure model can describe slowly decaying impulse response functions using only a small number of coefficients compared with the corresponding FIR description. This efficient modelling of slowly decaying impulse response functions is tempered by the potential instabilities of the feedback structure of IIR models. In subsection 5.4.2 a methodology for ensuring the stabilities of a nonlinear autoregressive moving-average (NARMA) model is presented.

SISO or MIMO models A single-input–single-output (SISO) model is typically used for system-level behavioural models. If further excitations of the amplifier (such as bias currents) are to be considered, additional inputs and outputs have to be added to the model [3]. An example of a multiple-input–multiple-output system (MIMO) is the circuit-level model of a mixer.

Deterministic or stochastic models The output of a deterministic model can be evaluated exactly if the input signal is known [1]. A stochastic model depends on random terms which circumvent the exact calculation of the output signal. Such stochastic terms can, for example, describe the various sources of noise added to the output signal.

2.3 Application-based model properties

Behavioural models are used in several different applications. Three major applications for these models are:

- solving ordinary differential equations;
- predicting the system response for a given input signal;
- characterising or classifying the behaviour of an observed system by parametrising a selected model structure on the basis of an observed system behaviour.

To use a behavioural model for the solution of ordinary differential equations (ODEs), a specific mathematical description of the input–output mapping is assumed. As presented in [4], using a Volterra series representation an improved analysis of ODEs is possible compared with solving the linearised differential equation system. The (time-invariant) Volterra series can be expressed as:

$$\begin{aligned} y(t) &= \sum_{n=1}^{\infty} \int_{-\infty}^{\infty} \int_{-\infty}^{\infty} \cdots \int_{-\infty}^{\infty} h_n(\tau_1, \tau_2, \ldots, \tau_n) d\tau_1 d\tau_2 \cdots d\tau_n \prod_{r=1}^{n} x(t - \tau_r) \\ &= \sum_{n=1}^{\infty} \mathrm{H}_n[x(t)], \end{aligned} \tag{2.1}$$

where $x(t)$ and $y(t)$ are the system input and output respectively, $h_n(\tau_1, \tau_2, \ldots, \tau_n)$ specifies the nth-order Volterra kernel and H_n is the corresponding nth-order Volterra operator. The idea is to present the input–output mapping given by the ODE as

$$\begin{aligned} y(t) &= \mathrm{H}[x(t)] = \mathrm{H}_1[x(t)] + \mathrm{H}_\mathrm{r}[x(t)], \\ \mathrm{H}_\mathrm{r}[x(t)] &= \sum_{n=2}^{\infty} \mathrm{H}_n[x(t)]. \end{aligned} \tag{2.2}$$

This separation of an ODE into a linear dynamic and a residual part is based on an important property of the Volterra series [4, 5]. Assuming an input signal $x'(t) = Cx(t)$, where C is an arbitrary constant, the Volterra series response will be

$$y'(t) = \sum_{n=1}^{\infty} \mathrm{H}_n[x'(t)] = \sum_{n=1}^{\infty} C^n \mathrm{H}_n[x(t)]. \tag{2.3}$$

Only coefficients of the same C^n power can contribute to the solution of the ODE for the nth-order Volterra operator. Hence, after inserting $x'(t) = Cx(t)$ and $y'(t) = \sum_{n=1}^{\infty} C^n \mathrm{H}_n[x(t)]$ into the ODE and equating the coefficients of the different powers of C (as indicated in Equation (2.2)), the linear operator H_1 can be calculated by selecting terms having the first power of C. Knowing H_1, the same procedure can be used to extract the higher-order Volterra operators. In [5] a similar approach, using the so-called nonlinear currents method, is discussed; this approximates an ODE for a specific excitation. A detailed discussion on Volterra series and their properties will be given in Section 5.6.

The two other behavioural-model applications mentioned above utilise the models in a system identification context. For both applications a model structure is selected and parametrised to represent the behaviour of an observed amplifier. The observed input and output signals are taken either from measurements (direct empirical modelling) or from simulations (indirect empirical modelling) of the amplifier under consideration. The extraction of behavioural models from simulations is often motivated by computational efficiency improvements. First, time-consuming physical or equivalent-circuit-level simulations are used to evaluate the behaviour of a system. After that, a behavioural model can be extracted and used to predict the response of the original system to other inputs.

When modelling microwave PAs it can often be assumed that the carrier frequency is significantly higher than the maximum envelope frequency (and the odd-order harmonics of the maximum envelope frequency) of the input signal:

$$\begin{aligned} x(t) &= \mathrm{Re}\{\tilde{x}(t)e^{j2\pi \mathrm{f}_0 t}\}, \\ \mathrm{f}_0 &\gg N\mathrm{f}_{\tilde{x},\max}, \end{aligned} \tag{2.4}$$

where the signal $x(t)$ is the RF input to the PA. The corresponding output signal will contain intermodulation and harmonic distortion. Assuming that distortion components up to Nth order are considered, Equation (2.4) is applicable for the actual input signal and the in-band intermodulation distortion. The validity of this relationship allows the use of the low-pass equivalent (or complex-envelope) representation of the signal $x(t)$ [6] (cf. Section 1.3 and Equations (1.11) and (1.12). An important result from Equation (2.4) is that a nonlinear system in the pass band can present harmonic and even-order intermodulation-distortion products while in the low-pass equivalent description it is assumed that the band-pass nonlinear system output is filtered to remove these distortion products. The suppression of the harmonics of a signal is called zonal filtering [7–10] and is depicted in Figure 2.2. After addition of the zonal filter, only odd-order distortion products can be recognised (since even-order distortion products are suppressed by the zonal filter). A pass-band nonlinear system that is able to represent both harmonic and intermodulation distortion is called an *instantaneous nonlinearity* while the corresponding low-pass equivalent system is an *envelope nonlinearity* [8]. The part $y(t)$ of the output signal that is located in the same band as the input $x(t)$ is often called the zonal-band output.

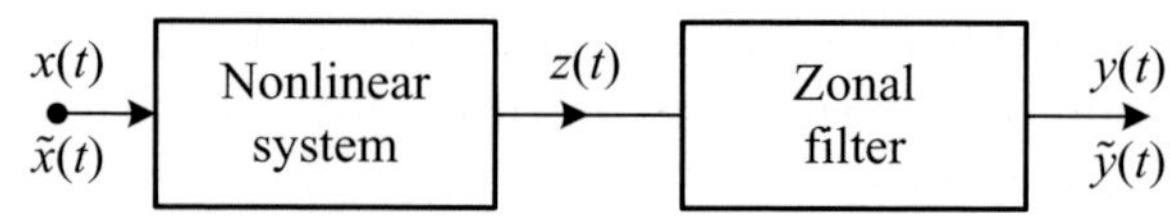

Figure 2.2 Zonal filtering [7–10].

The model presented in Figure 1.6 is a band-pass model while most of the models discussed in Section 1.3 employ the low-pass equivalent description. This is to be expected, as system-level models are typically low-pass equivalent models processing the input and output signal envelopes while circuit-level models handle the true RF modulated signals (see Section 1.4) and process the complete input and output ports' voltage and current (or incident and reflected power waves).

The mathematical description of a band-pass nonlinear system is different from the corresponding equivalent low-pass representation. This is a direct consequence of the zonal filtering of the band-pass output signal, as shown in Figure 2.2. In this section, the relationship between these two descriptions will be examined (in a similar way to that in [10, 11]). A Volterra series, as shown in Equation (2.1), is used to represent the dynamic nonlinear system in the passband. From this band-pass nonlinear dynamic system the corresponding low-pass equivalent description may be extracted. Without loss of generality, the kernels of the Volterra series can be assumed to be symmetric (see subsection 5.6.1); a procedure for symmetrising the kernels of a Volterra series is presented in [4]. Applying a modulated input signal,

$$\begin{aligned} x(t) &= \mathrm{Re}\{\tilde{x}(t)e^{j2\pi \mathrm{f}_0 t}\} \\ &= \tfrac{1}{2}[\tilde{x}(t)e^{j2\pi \mathrm{f}_0 t} + \tilde{x}^*(t)e^{-j2\pi \mathrm{f}_0 t}], \end{aligned} \tag{2.5}$$

to the input of the band-pass nonlinear system (as presented in Figure 2.2) results in

$$\begin{aligned} z(t) &= \sum_{n=1}^{N} z_n(t), \\ z_n(t) &= \frac{1}{2^n}\int_{-\infty}^{\infty}\int_{-\infty}^{\infty}\cdots\int_{-\infty}^{\infty} h_n(\tau_1,\tau_2,\ldots,\tau_n)d\tau_1 d\tau_2\cdots d\tau_n \\ &\times\prod_{r=1}^{n}\left[\tilde{x}(t-\tau_r)e^{j2\pi \mathrm{f}_0(t-\tau_r)} + \tilde{x}^*(t-\tau_r)e^{-j2\pi \mathrm{f}_0(t-\tau_r)}\right] \end{aligned} \tag{2.6}$$

where N represents the number of Volterra kernels. Assuming symmetric kernels, the expression for $z_n(t)$ can be rewritten as

$$\begin{aligned} z_n(t) &= \frac{1}{2^n}\int_{-\infty}^{\infty}\int_{-\infty}^{\infty}\cdots\int_{-\infty}^{\infty} h_n(\tau_1,\tau_2,\ldots,\tau_n)d\tau_1 d\tau_2\cdots d\tau_n \\ &\times\sum_{r=0}^{n}\binom{n}{r}e^{j2\pi(2r-n)\mathrm{f}_0 t}\prod_{o=0}^{r}\tilde{x}(t-\tau_o)e^{-j2\pi \mathrm{f}_0\tau_o}\prod_{p=r+1}^{n}\tilde{x}^*(t-\tau_p)e^{j2\pi \mathrm{f}_0\tau_p} \end{aligned} \tag{2.7}$$

For the zonal-band output, only terms located at the fundamental frequency can contribute (i.e. n must be odd and $2r - n = \pm 1$). The zonal-band output of Equation (2.7) is given by

$$\begin{aligned} y_n(t) &= \frac{1}{2^n} \int_{-\infty}^{\infty} \int_{-\infty}^{\infty} \cdots \int_{-\infty}^{\infty} h_n(\tau_1, \tau_2, \ldots, \tau_n) d\tau_1 d\tau_2 \cdots d\tau_n \\ &\times \binom{n}{(n-1)/2} \prod_{o=1}^{(n-1)/2} \tilde{x}(t - \tau_o) e^{-j2\pi f_0 \tau_o} \prod_{p=(n+1)/2}^{n-1} \tilde{x}^*(t - \tau_p) e^{j2\pi f_0 \tau_p} \\ &\times \Big[\tilde{x}(t - \tau_n) e^{j2\pi f_0 (t-\tau_n)} + \tilde{x}(t - \tau_n) e^{-j2\pi f_0 (t-\tau_n)} \Big]. \end{aligned} \tag{2.8}$$

Therefore, the complete first-zone filtered band-pass Volterra series can be written as

$$\begin{aligned} y(t) &= \sum_{m=1}^{(N+1)/2} y_m(t), \\ y_m(t) &= \frac{1}{2^{2m-1}} \int_{-\infty}^{\infty} \int_{-\infty}^{\infty} \cdots \int_{-\infty}^{\infty} h_{2m-1}(\tau_1, \tau_2, \ldots, \tau_{2m-1}) d\tau_1 d\tau_2 \cdots d\tau_{2m-1} \\ &\times \binom{2m-1}{m-1} \prod_{o=1}^{m-1} \tilde{x}(t - \tau_o) e^{-j2\pi f_0 \tau_o} \prod_{p=m}^{2(m-1)} \tilde{x}^*(t - \tau_p) e^{j2\pi f_0 \tau_p} \\ &\times \Big[\tilde{x}(t - \tau_{2m-1}) e^{j2\pi f_0 (t-\tau_{2m-1})} + \tilde{x}(t - \tau_{2m-1}) e^{-j2\pi f_0 (t-\tau_{2m-1})} \Big], \end{aligned} \tag{2.9}$$

and the corresponding low-pass equivalent representation is given by

$$\begin{aligned} \tilde{y}(t) &= \sum_{m=1}^{(N+1)/2} \int_{-\infty}^{\infty} \cdots \int_{-\infty}^{\infty} \tilde{h}_{2m-1}(\tau_1, \tau_2, \ldots, \tau_{2m-1}) d\tau_1 d\tau_2 \cdots d\tau_{2m-1} \\ &\times \prod_{o=1}^{m} \tilde{x}(t - \tau_o) \prod_{p=m+1}^{2m-1} \tilde{x}^*(t - \tau_p), \end{aligned} \tag{2.10}$$

where the low-pass equivalent kernels $\tilde{h}_{2m-1}$ are composed of scaled versions of the corresponding band-pass kernels and of the phase factors resulting from the time shift of the input signals:

$$\begin{aligned} \tilde{h}_{2m-1}(\tau_1, \tau_2 \ldots, \tau_{2m-1}) &= \frac{1}{2^{2(m-1)}} \binom{2m-1}{m-1} h_{2m-1}(\tau_1, \tau_2, \ldots, \tau_{2m-1}) \\ &\times \prod_{o=1}^{m} e^{-j2\pi f_0 \tau_o} \prod_{p=m+1}^{2m-1} e^{j2\pi f_0 \tau_p}. \end{aligned} \tag{2.11}$$

In the static nonlinear case ($\tau_i = 0$) the low-pass equivalent description reduces to

$$\begin{aligned} y(t) &= \mathrm{Re}\{\tilde{y}(t) e^{j2\pi f_0 t}\} \\ &= \mathrm{Re}\left\{ \sum_{m=1}^{(N+1)/2} \frac{a_{2m-1}}{2^{2(m-1)}} \binom{2m-1}{m-1} |\tilde{x}(t)|^{2(m-1)} \tilde{x}(t) e^{j2\pi f_0 t} \right\}, \end{aligned} \tag{2.12}$$

where $h_{2m-1}(0, 0, \ldots, 0) = a_{2m-1}$. Equation (2.12) shows that the band-pass and low-pass equivalent power series coefficients differ only by an order-dependent scaling factor. These scaling factors are not influenced by the even-order band-pass coefficients, and so to obtain the corresponding low-pass description it is sufficient to remove the even-order coefficients.

If the system considered is memoryless, the band-pass–low-pass relationship can be evaluated by the use of the Chebyshev transform [7, 12, 13]. The modulated input signal as in Equation (2.5),

$$\begin{aligned} x(t) &= \mathrm{Re}\{\tilde{x}(t)e^{j2\pi \mathrm{f}_0 t}\} \\ &= r(t)\cos[2\pi \mathrm{f}_0 t + \phi(t)], \end{aligned} \tag{2.13}$$

is fed into a memoryless nonlinear function $G(\cdot)$ (neglecting the time dependence):

$$y = G(r\cos\theta) \tag{2.14}$$

where $\theta = 2\pi \mathrm{f}_0 t + \phi$. Since the output y is a periodic function of θ it can be represented by a Fourier series in θ:

$$\begin{aligned} G(r\cos\theta) = &\tfrac{1}{2}a_0(r) + a_1(r)\cos\theta + b_1(r)\sin\theta \\ &+ a_2(r)\cos 2\theta + b_2(r)\cos 2\theta + \cdots. \end{aligned} \tag{2.15}$$

The coefficients of this Fourier expansion are given by

$$\begin{aligned} a_m(r) &= \tfrac{2}{\pi}\int_0^{\pi} G(r\cos\theta)\cos m\theta d\theta, \\ b_m(r) &= \tfrac{2}{\pi}\int_0^{\pi} G(r\cos\theta)\sin m\theta d\theta. \end{aligned} \tag{2.16}$$

This expansion is valid regardless of how r and θ vary with time. The first term of Equation (2.15) represents the 'baseband' output of the nonlinear system, acting as a detector. This 'baseband' output is caused by even-order mixing products of the nonlinear system and is, in general, different from the complex envelope of the pass-band output signal. For $m = 1$ the desired zonal-band output of the nonlinear system is selected. The terms $m > 1$ incorporate the higher-order harmonic output. Reinserting the time dependence for the zonal-band output yields

$$y(t) = a_1(r(t))\cos[2\pi \mathrm{f}_0 t + \phi(t)] + b_1(r(t))\sin[2\pi \mathrm{f}_0 t + \phi(t)] \tag{2.17}$$

and the equivalent low-pass representation of the RF signal $y(t)$ is therefore

$$\tilde{y}(t) = [a_1(r(t)) - jb_1(r(t))]e^{j\phi(t)} \tag{2.18}$$

According to [13] the complex function

$$g(r)e^{j\Phi(r)} = \frac{1}{r}[a_1(r) + jb_1(r)] \tag{2.19}$$

is known as the *describing function* in control-system literature. The existence of a close relationship between the describing function and the result presented in Equation (2.12) is confirmed by the fact that the powers of $r\cos\theta$ are the eigenfunctions of the Chebyshev transform ([13], Table 1 T5):

$$G(r\cos\theta) = (r\cos\theta)^n \leftrightarrow v_m(r)\cos m\theta = 2\binom{n}{\frac{1}{2}n - \frac{1}{2}m}\left(\frac{r}{2}\right)^n \cos m\theta, \qquad (2.20)$$

for $n = 0, 1, 2, \ldots$ and $m = n, n-2, n-4, \ldots$

2.4 Amplifier-based model properties

2.4.1 *Short- and long-term memory effects*

The difference between linear and nonlinear memory effects was discussed in subsection 1.2.2 on the basis of the model presented in Figure 1.6. These effects can also be classified as short- and long-term memory effects. This distinction is deduced from an equivalent-circuit-based description of a microwave transistor. The equivalent-circuit modelling of the nonlinear distortion effects appearing in band-pass systems with significant memory has received considerable attention [14, 15].

By strict definition, a memoryless circuit is one in which no charge or magnetic-flux storage elements (no capacitors or inductors) exist, so that the voltages and currents at any instant do not depend upon previous values of voltage or current [16]. Power amplifiers, however, as all electronic circuits, have a memory. In fact the familiar circuit topology of a power amplifier includes capacitors and inductors, not only within the transistors' equivalent schemes but also in the passive electrical networks (the matching networks and bias networks etc.). Nevertheless the memoryless assumption, so widely used in the past, remains valid for many purposes, particularly when a very high ratio between the centre operating frequency and the information bandwidth allows one to neglect the influence of those elements having a frequency-dependent contribution (see subsection 2.4.2 and Section 3.1). In Figure 2.3, a typical equivalent circuit for a power MESFET or HEMT is shown [17]. The reactive behaviour is associated not only with parasitic effects, which is the case for the extrinsic elements, but also with the device's intrinsic behaviour. As a result, electrical capacitance values of the electrodes in the femtofarad range may coexist with for instance thermal capacitance values several orders of magnitude larger. A microwave transistor shows, therefore, a frequency-dependent low-pass behaviour from DC up to its transition frequency f_T.

Memory effects in a power amplifier may have both short and long time constants compared with the period of the RF carrier signal or the slow variations in its complex envelope. The band-pass characteristics of the PA input and output matching networks as well as the low-pass characteristics of the transistor contribute to the short-term memory effects. They can be modelled by two filters with a memoryless transfer nonlinearity in between, as discussed in subsection 1.2.2.

Long-term memory effects are much more difficult to characterise and model.

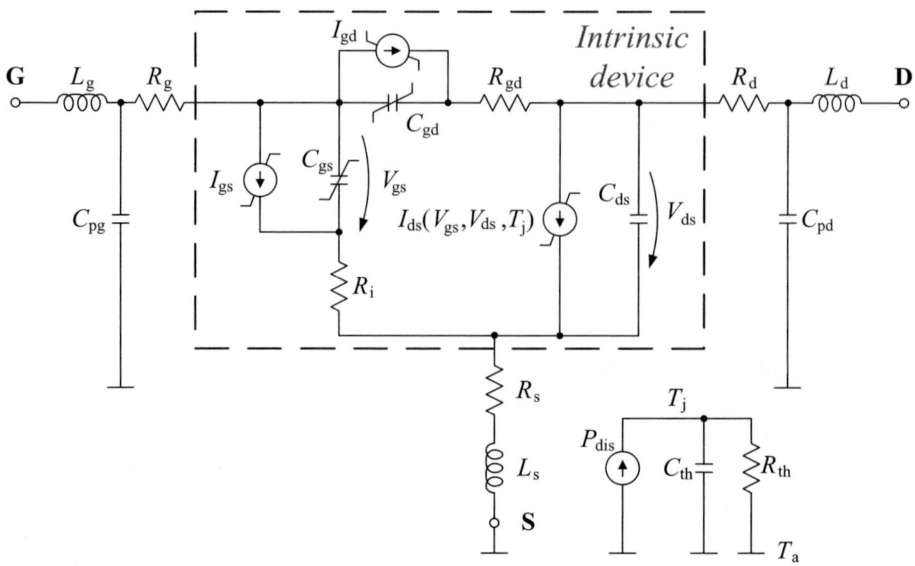

Figure 2.3 General electrothermal equivalent circuit for a power MESFET or HEMT: G, gate; D, drain; S, source.

They have been related to a variety of PA characteristics, including low-frequency dispersion due to trapped states (a physical effect) [18], self-heating [19] and envelope feedback paths as well as input and output bias circuits with long time constants [14, 15]. From this wide set of characteristics, drain or base circuitry-induced memory seem to be predominant for FET- or bipolar-based PAs respectively.

As it is usually designed to handle signals that are narrowband compared with its available bandwidth, a band-pass PA for modern wireless standards would be expected to be nearly memoryless. In that condition, as described above, short-term memory effects can be completely neglected. However, long-term effects are observed and become particularly critical when one is trying to fit the standard specification in terms of linearity. Low-frequency signal components, for which the circuit is no longer memoryless, must be remixed with the original RF signal to create such long-term memory effects. Since these low-frequency or envelope components can only be generated from the band-pass RF signal through a demodulation process, some form of low-frequency (LF) feedback should be available to explain the remixing of the components thus generated with the original signal in the nonlinearity. In the following, two important classes of LF feedback, bias-circuitry-based and self-heating-based, are discussed.

Bias-circuitry-based memory effects

The low-frequency feedback, as pointed out in [20], may be due to a physical path from the output to the input of the transistor. It may be created, in either a deliberate or an accidental way, when designing and implementing the PA. By way of illustration, envelope feedback can appear if the output and input biasing circuits are not appropriately isolated and if both biasing voltages are derived from the same power supply.

The main sources of LF feedback come from conceptual mechanisms, however. In an FET-based PA, the envelope-frequency components of the output current, generated from even-order terms in the $I_{ds}(V_{gs})$ transfer nonlinearity, are converted into drain-to-source voltage components when circulating along the load mesh. Owing to the load impedance values at the envelope frequencies determined by the bias tee, which is usually designed without too much attention being paid to its low-frequency response, envelope components of $V_{ds}(t)$ may be expected to suffer from strong frequency dispersion. Owing to the output $I_{ds}(V_{ds})$ nonlinearity, they are then remixed with the original RF signal, reproducing the LF feedback mechanism responsible for long-term memory effects as described before [14, 20].

In Figure 2.4(a), a typical biasing network is presented. The variation with frequency for the real and imaginary parts of the impedance, as seen from the drain side, is shown in Figure 2.4(b). The existence of a region where $Z_d(\omega_{LF})$ is highly reactive would probably give rise to strong asymmetries in the intermodulation-distortion sidebands.

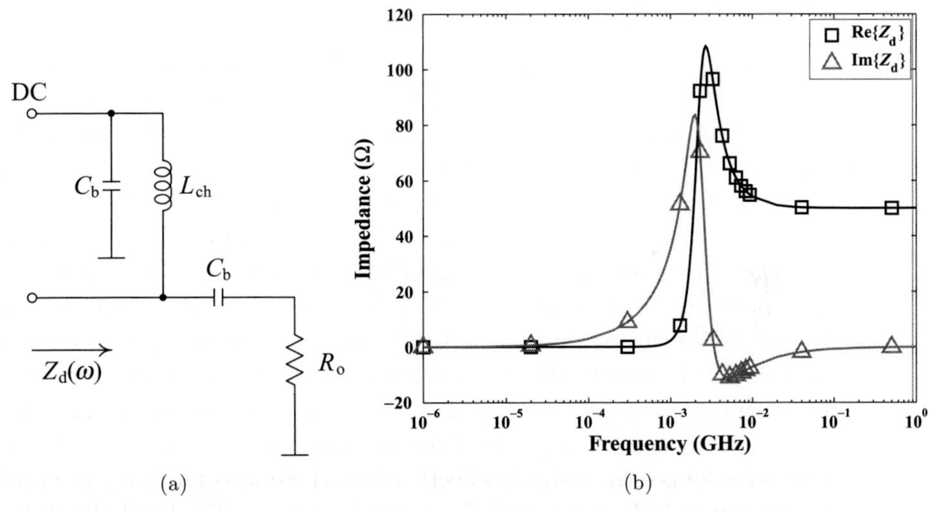

Figure 2.4 (a) Biasing network schematic and (b) drain-impedance frequency dispersion; $L_{ch} = 4.7\,\mu H$ and $C_b = 1\,nF$ as described in [15].

Additional conclusions were obtained in [14] regarding the appearance of asymmetries. An important factor is the second-harmonic band termination, which produces in-band distortion components through a remixing process. It should be pointed out that long-term memory effects are always important near the IMD sweet spots (an example of IMD sweet spots is given in subsection 2.5.3), where the IMD generated by the LF feedback mechanism is not masked by the direct-distortion contributions of the odd-order terms in the transfer nonlinearity.

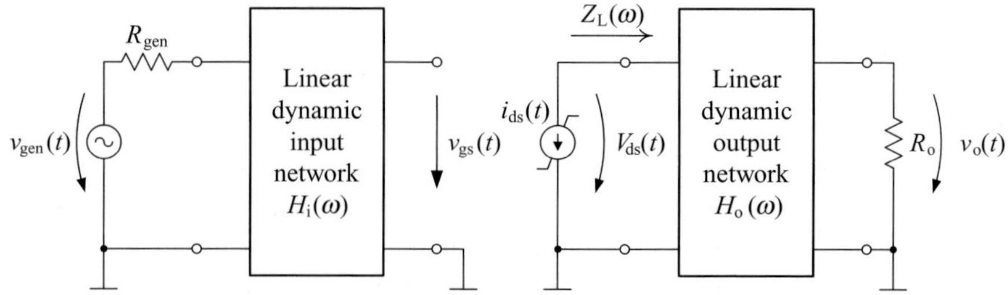

Figure 2.5 Simplified FET-based PA circuit.

With the aim of analysing the origin of bias-circuitry long-term memory effects, and as a way to establish a link with their modelling through nonlinear band-pass dynamic system approaches, already introduced in subsection 1.2.2, a simplified PA circuit is presented in Figure 2.5. This derivation is based on the work of Pedro *et al.* [20]. The input and output networks, $H_\mathrm{i}(\omega)$ and $H_\mathrm{o}(\omega)$, include not only matching circuits but also gate and drain elements of the device's equivalent circuitry.

The device was considered to be unilateral [35]. Besides avoiding unnecessary complexity in the analysis, this simplification allowed the existence of a bias-circuitry-based feedback path to be shown.

The PA model is driven by an input voltage $v_\mathrm{gen}(t)$ resulting in an output signal $v_\mathrm{o}(t)$. The Taylor series expansion of the main nonlinearity $I_\mathrm{ds}(V_\mathrm{gs}, V_\mathrm{ds})$ is given by

$$\begin{aligned} I_\mathrm{ds}\left(V_\mathrm{gs}, V_\mathrm{ds}\right) = {} & I_\mathrm{DS}\left(V_\mathrm{GS}, V_\mathrm{DS}\right) + G_\mathrm{m} v_\mathrm{gs} + G_\mathrm{ds} v_\mathrm{ds} \\ & + G_\mathrm{m2} v_\mathrm{gs}^2 + G_\mathrm{md} v_\mathrm{gs} v_\mathrm{ds} + G_\mathrm{d2} v_\mathrm{ds}^2 \\ & + G_\mathrm{m3} v_\mathrm{gs}^3 + G_\mathrm{m2d} v_\mathrm{gs}^2 v_\mathrm{ds} + G_\mathrm{md2} v_\mathrm{gs} v_\mathrm{ds}^2 + G_\mathrm{d3} v_\mathrm{ds}^3 + \cdots . \end{aligned} \tag{2.21}$$

Introducing the effects of the linear input-and-output matching networks into this Taylor series expansion, the first- and third-order nonlinear transfer functions can be identified as

$$S_1(\omega) = -H_\mathrm{i}(\omega) G_\mathrm{m} \frac{Z_\mathrm{L}(\omega)}{D(\omega)} H_\mathrm{o}(\omega) \tag{2.22}$$

and

$$\begin{aligned} & S_3(\omega_1, \omega_2, \omega_3) \\ & = -H_\mathrm{i}(\omega_1) H_\mathrm{i}(\omega_2) H_\mathrm{i}(\omega_3) \frac{Z_\mathrm{L}(\omega_1+\omega_2+\omega_3)}{D(\omega_1+\omega_2+\omega_3)} H_\mathrm{o}(\omega_1+\omega_2+\omega_3) \\ & \quad \times \left\{ \left(G_\mathrm{m3} + G_\mathrm{m2d} A_v + G_\mathrm{m2d} A_\mathrm{v}^2 + G_\mathrm{d3} A_\mathrm{v}^3\right) - \frac{1}{3}\left(G_\mathrm{md} + 2 G_\mathrm{d2} A_\mathrm{v}\right) \right. \\ & \quad \left. \times \left(G_\mathrm{m2} + G_\mathrm{md} A_\mathrm{v} + G_\mathrm{d2} A_\mathrm{v}^2\right) \left[\frac{Z_\mathrm{L}(\omega_1+\omega_2)}{D(\omega_1+\omega_2)} + \frac{Z_\mathrm{L}(\omega_1+\omega_3)}{D(\omega_1+\omega_3)} + \frac{Z_\mathrm{L}(\omega_2+\omega_3)}{D(\omega_2+\omega_3)} \right] \right\} . \end{aligned} \tag{2.23}$$

where

$$D(\omega) = 1 + G_{\rm ds} Z_{\rm L}(\omega), \qquad A_{\rm v}(\omega) = -G_{\rm m} \frac{Z_{\rm L}(\omega)}{D(\omega)} \tag{2.24}$$

are assumed to be constant within the signal bandwidth.

Assuming that the signal bandwidth is narrow, the impedances $H_{\rm i}(\omega)$, $H_{\rm o}(\omega)$ and $Z_{\rm L}(\omega)$ are approximately constant in the fundamental and higher-order-harmonic frequency bands. However, the LF feedback through $Z_{\rm L}(\omega_{\rm LF})$ cannot be considered narrowband and could produce nonlinear envelope effects.

Comparing Equations (2.22) and (2.23) with Equations (1.9) and (1.10), the equivalent structure introduced in subsection 1.2.2 can be fully understood.

Self-heating and trap-related mechanisms

The self-heating process causes temperature changes due to the dissipated power. This process shows a frequency-dependent behaviour determined by the physical structure of the transistor. In an FET, the power dissipation is a function of the drain current, which at the same time is influenced by the temperature dependence of the carrier mobility, the threshold voltage and the carrier saturation velocity [19, 21]. The dependence of the junction temperature on the dissipated power follows a low-frequency response, represented by the thermal network in the electrothermal model of Figure 2.3. Thus the heating process is also a low-frequency feedback mechanism, which in principle is expected to produce long-term memory effects [22].

A system-level representation of such a mechanism is shown in Figure 2.6(a), where the temperature dependence of the carrier mobility and the threshold voltage were considered to be dominated by the variation of the carrier saturation velocity [19, 21]. The isothermal current at ambient temperature is denoted by $I_{\rm d0}(T_{\rm a})$; the term δ is a function of both the thermal resistance and the temperature sensitivity of the drain current, $P_{\rm d}(t)$ is the dissipated power and the Fourier transform pair $h_{\rm Th}(t)$ and $H_{\rm Th}(\omega)$ represent the thermal impulse and frequency response respectively.

Trapping has also been considered as an additional source of PA long-term memory [22]. The influence of hole traps and impact ionisation phenomena on high-electron-mobility transistor (HEMT) distortion was studied in detail in [18]. Hole trapping was described as a low-frequency gate voltage shift controlled by the drain voltage, as presented in Figure 2.6(b). The envelope components appearing at the drain side owing to the even-order terms in the transfer nonlinearity are then fed back to the gate terminal. In this way remixing with the RF components takes place, in a low-frequency feedback mechanism similar to that previously discussed. The series generator at the gate terminal, which depends on the drain voltage, models the trap-induced difference between the effective control voltage $v_{\rm cs}(t)$ and $v_{\rm gs}(t)$ in the envelope frequency range.

Although these two sources of device frequency dispersion could certainly result

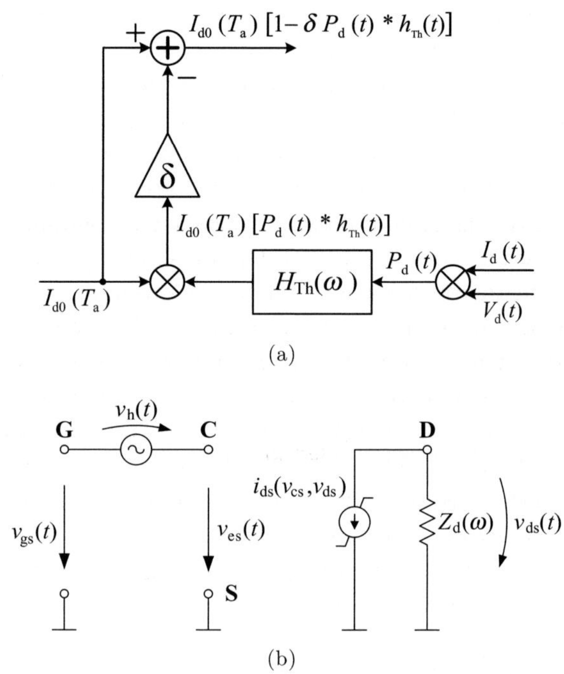

Figure 2.6 (a) Simplified schematic of the self-heating process, and (b) the low-frequency equivalent circuit for a high-electron-mobility transistor (HEMT) with trapping.

in long-term memory effects, most published measurement results show only a minor dependence (differences in the measurement values vary by a few dB) [18, 23]. Therefore, it seems that the electrical networks determining drain or base low-frequency impedance variations maintain a predominant role in their generation.

2.4.2 *Quasi-memoryless description of amplifiers*

Static, frequency-independent, AM–AM and AM–PM envelope characteristics are the basis for defining a behavioural model as memoryless. It is good practice to ensure through measurement that the characteristics are static over the input signal band. The qualifier 'quasi' arises in reference to the AM–PM characteristic. Some authors use this term if the model simply takes account of the AM–PM characteristic as well as the AM–AM characteristic. Perhaps this is in recognition of a power-dependent variable group delay that, nonetheless, is constant over the signal band. It could be argued that this is not strictly a memory effect and so the qualifier 'quasi' is unnecessary. However, there are two situations that may be considered to merit the term 'quasi-memoryless'. First, envelope-dependent non-negligible memory effects in the AM–AM and AM–PM characteristics may be present but removable by good biasing-circuit adjustments [24]. Second, a more subtle distinction is possible when one attempts to take account of the dynamic-transition process to a new phase corresponding to a new envelope power, i.e. modelling its delay and dynamic-transition characteristic. However, in reported measurement there is sparse reference to performance deterioration due to this dynamic aspect of the AM–PM effect, indicating that the time constants involved are much smaller than the reciprocal of the envelope bandwidth.

Where amplifier behaviour, through inherent characteristics combined with good matching and bias-circuit design, yields an instantaneous memoryless nonlinear AM–PM behaviour, the corresponding modelling approach may be properly referred to as memoryless without the qualification 'quasi'. This is the approach taken in this book. A final consideration in this context: it is worth noting that, in some SSPAs, AM–PM distortion is not significant, being typically in single-figure degrees over the PA's small-signal to large-signal dynamic range. Hence omitting the AM–PM nonlinearity in a modelling problem can at times be reasonably justified.

The use of the phrase 'quasi-memoryless system' can also be justified for cases in which a nonlinear dynamic system is driven by a sufficiently narrowband input signal. The dynamic nonlinear system can be represented by, for example, a Volterra series as introduced in Equation (2.1). The frequency-domain representation of the Volterra series is given by [4]

$$
\begin{aligned}
y(t) &= \frac{1}{2\pi}\sum_{n=1}^{\infty}\int_{-\infty}^{\infty} Y_n(j\omega)e^{j\omega t}\,d\omega, \\
Y_n(j\omega) &= \frac{1}{(2\pi)^{n-1}}\int_{-\infty}^{\infty}\int_{-\infty}^{\infty}\cdots\int_{-\infty}^{\infty} H_n(j\omega - j\mu_1, j\mu_1 - j\mu_2, \ldots, j\mu_{n-1}) \\
&\quad \times X(j\omega - j\mu_1)X(j\mu_1 - j\mu_2)\cdots X(j\mu_{n-1})d\mu_1 d\mu_2\cdots d\mu_{n-1},
\end{aligned}
\tag{2.25}
$$

where $H_n(j\omega_1, \ldots, j\omega_n)$ is the Fourier transform of the nth-order Volterra kernel:

$$
\begin{aligned}
H_n(j\omega_1, \ldots, j\omega_n) &= \int_{-\infty}^{\infty}\int_{-\infty}^{\infty}\cdots\int_{-\infty}^{\infty} d\omega_1 \ldots d\omega_n \\
&\quad \times h_n(\tau_1, \ldots, \tau_n)e^{-j\omega_1\tau_1}\ldots e^{-j\omega_n\tau_n}.
\end{aligned}
\tag{2.26}
$$

For an input signal bandlimited to a bandwidth $\mathrm{f}_0 \pm BW/2$ the integration limits for $Y_n(j\omega)$ in Equation (2.25) change as in the following:

$$
\begin{aligned}
Y_n(j\omega) &= \frac{1}{(2\pi)^{n-1}}\int_{-2\pi(\mathrm{f}_0+BW/2)}^{-2\pi(\mathrm{f}_0-BW/2)}\cdots\int_{-2\pi(\mathrm{f}_0+BW/2)}^{-2\pi(\mathrm{f}_0-BW/2)} d\mu_1 d\mu_2 \ldots d\mu_{n-1} \\
&\quad \times H_n(j\omega - j\mu_1, j\mu_1 - j\mu_2, \ldots, j\mu_{n-1})X(j\omega - j\mu_1) \\
&\quad \times X(j\mu_1 - j\mu_2)\cdots X(j\mu_{n-1}) \\
&\quad + \frac{1}{(2\pi)^{n-1}}\int_{2\pi(\mathrm{f}_0-BW/2)}^{2\pi(\mathrm{f}_0+BW/2)}\cdots\int_{2\pi(\mathrm{f}_0-BW/2)}^{2\pi(\mathrm{f}_0+BW/2)} d\mu_1 d\mu_2 \ldots d\mu_{n-1} \\
&\quad \times H_n(j\omega - j\mu_1, j\mu_1 - j\mu_2, \ldots, j\mu_{n-1})X(j\omega - j\mu_1) \\
&\quad \times X(j\mu_1 - j\mu_2)\cdots X(j\mu_{n-1}).
\end{aligned}
\tag{2.27}
$$

If this bandwidth BW is 'sufficiently' small then the Volterra kernels will not change within the integration limits and can be approximated by a constant factor $K_n \approx H_n(j\omega_1, \ldots, j\omega_n)$. Hence the input signal samples the Volterra kernels, and the nonlinear dynamic system reduces to a quasi-memoryless equivalent. The validity of this approximation depends on the properties of the nonlinear dynamic system, the bandwidth of the input signal and the desired modelling accuracy. The applicability of the quasi-memoryless approximation has to be determined in each case.

2.4.3 Classification of amplifier and model nonlinearities

The classification into linear, mildly nonlinear or strongly nonlinear is often used to express the degree of nonlinearity covered by an amplifier model. In subsection 1.2.1 the mildly nonlinear operation regime was mentioned as typical for polynomial filters under specific excitation conditions. To be classified in this way, an amplifier must be stable and also show a behaviour going from linear to mildly nonlinear and then to strongly nonlinear as the magnitude of the input signal increases. This is the usual behaviour for 'linear' amplifiers, in which there is a reduction in generated distortion as the input power is reduced. Such a characteristic is typical of class A or class AB amplifiers (an introduction to the different RF and microwave amplifier classes can be found in [5, 25, 26]). Class C and switching-mode amplifiers (designed for highly efficient operation) do not show this type of behaviour. Such amplifiers cannot be brought into linear operation by reducing the power of the input signal. A comparison of the input–output power characteristics of these two categories of amplifier is shown in Figure 2.7.

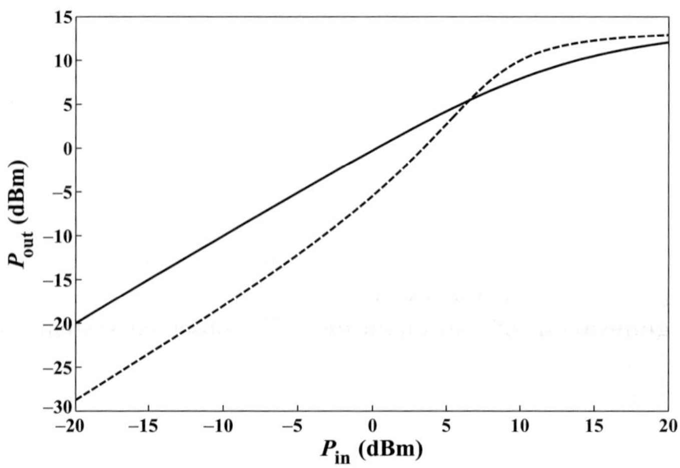

Figure 2.7 Input–output power characteristics for a class A amplifier (continuous line) and a class C amplifier (broken line). The former was set to a small-signal gain of 0 dB, while the latter shows its maximal gain at $P_{in} = 9.4$ dB m (it was also set to 0 dB).

In connection with behavioural amplifier models, the linear and mildly nonlinear classification implies the ability to describe the behaviour of the amplifier considered using a truncated power or Volterra series. Strongly nonlinear behaviour refers to amplifiers and models that cannot meet the mildly nonlinear requirements. To define the boundaries between these operating regimes, limits for the acceptable error between the approximating model and the actual amplifier behaviour must be defined. For the class A amplifier characteristic shown in Figure 2.7, linear and fifth-order power series approximations have been parametrised. In Figure 2.8(a) the operation-regime boundaries for the linear and the mildly nonlinear models are shown. The error between the amplifier behaviour and the approximating models together with the chosen error limits are presented in Figure 2.8(b). These figures highlight the fact that a mildly nonlinear amplifier model is only correctly specified if the corresponding boundaries for the input signal power are given. As mentioned in subsection 1.2.1 and also shown in Figure 2.8(b), evaluating a mildly nonlinear model outside the operation-regime boundaries is accompanied by an abrupt rise in the modelling error. This is typical for a truncated Volterra series.

Besides presenting definitions of linear and mildly and strongly nonlinear systems, Pearson [27] discusses the classification of a nonlinear model or a system regime from a process-control point of view. As a first step he defines the phenomena which can be present in the behaviour of dynamic nonlinear systems, as follows.

An asymmetric response to symmetric input changes An important property of linear time-invariant (LTI) systems is that they satisfy the homogeneity condition. For an input signal $x(t) \to \lambda x(t), \lambda \in \mathrm{R}$, such a system will respond with $y(t) \to \lambda y(t)$. The special case $\lambda = -1$ corresponds to the odd-symmetry requirement, and the violation of this requirement is a sufficient indication of the presence of nonlinearities in the behaviour of a system.

Input multiplicity This refers to a system where the steady-state output response $y_{\text{steady-state}}(t)$ corresponds to more than one steady-state input $x_{\text{steady-state},n}(t)$.

Output multiplicity This refers to a system where the steady-state input $x_{\text{steady-state}}(t)$ corresponds to more than one steady-state output response $y_{\text{steady-state},n}(t)$. The chosen steady-state response is a sensitive function of the system parameters and the input signal magnitude.

The generation of harmonics

The generation of subharmonics The observed system responds to an input signal of period T with an output signal that has a longer period nT, where n is an integer larger than 1.

Chaos The system reacts to a simple input signal such as a sinusoid, a step function, an impulse train and so on with a highly irregular response.

Input-dependent stability Some nonlinear systems show an unstable behaviour that depends on the magnitude of the input signal. For example, the first-order nonlinear dynamic model [27] $y(s) = ay(s-1) + by(s-1)\sin y(s-1) + x(s-1)$ for $a = 0.8$ and $b = -0.3141$ reacts to step responses of magnitudes -3 and $+1$ with an oscillating behaviour and for -1 and $+3$ with an oscillatory transient.

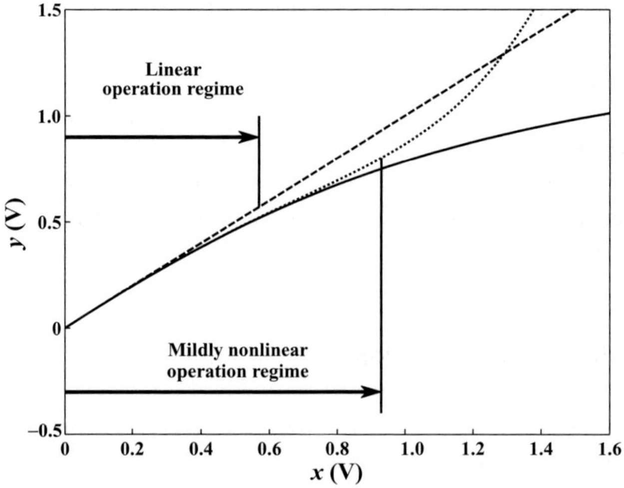

(a)

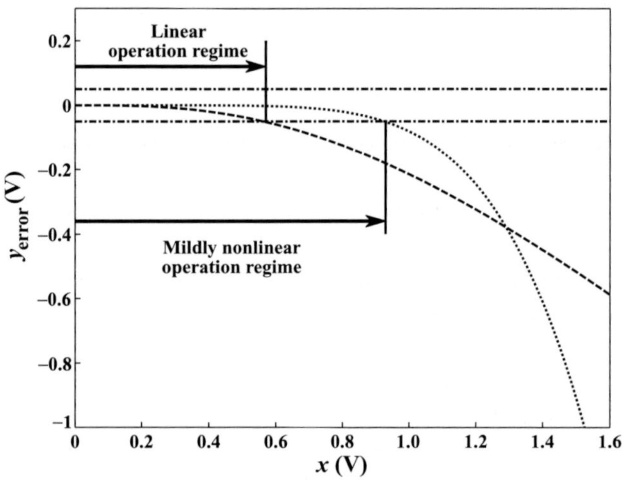

(b)

Figure 2.8 The different operation regimes for a class A amplifier. (a) The continuous line shows the class A amplifier transfer characteristic. The broken and the dotted lines present the linear and the mildly nonlinear approximations. Additionally, the boundaries for the linear and the mildly nonlinear operation regime, based on chosen error limits are shown. (b) The errors, between each approximating model and the amplifier characteristic. The upper and lower broken-and-dotted lines indicate the chosen limits for the acceptable error.

Pearson then classifies a system as mildly nonlinear when it shows at least one of the following features: asymmetric response to symmetric input changes, the generation of harmonics, input multiplicity. Strongly nonlinear systems are therefore characterised by the presence of output multiplicity and the generation of subharmonics and/or chaotic behaviour. An intermediate nonlinear system exhibits input-dependent stability and therefore cannot exhibit the full range of strongly nonlinear behaviour. On the basis of Pearson's formal classification of model nonlinearities, even the strongly nonlinear behaviour presented in Figure 2.8 can be categorised as mildly nonlinear.

In the following we will apply the Volterra-series-based definitions on model nonlinearity classification, as this approach is commonly used in conjunction with PA behavioural modelling.

2.5 Amplifier characterisation

Power amplifier characterisation provides the data needed for model extraction. Hence, the characterisation process has a significant influence on the resulting model and its properties. This close linkage is expressed in many articles on PA modelling, where amplifier characterisation, the PA model and the parameter extraction are discussed together (see for example [28–32]).

2.5.1 System identification background

Basically there are two alternative ways of constructing a mathematical model [1].

Mathematical modelling The basic laws of physics are used to describe the system's dynamic behaviour.

System identification A model is fitted to the data extracted from experimental measurements by assigning suitable numerical values to its parameters.

In many cases (such as multistage amplifiers, transmitter setups etc.) the systems are so complex that it is not possible to obtain reasonable models using only physical insight (see Section 1.2) and system identification techniques have to be used. Models obtained using such techniques have the following properties when compared to mathematical models [1]:

- limited validity (they are valid only for certain operation points, certain types of excitation signals etc.);
- little physical insight;
- ease of construction.

System identification does not provide foolproof methods that always and directly lead to the correct results. However, it may provide a number of theoretical results that are useful from a practical point of view.

A system identification procedure, as depicted in Figure 2.9, has a natural logical flow [1, 33, 36]. At the beginning, *a priori* knowledge concerning the amplifier and the desired model application is available. This could be, for example, a certain set of model structures that can be implemented in the chosen simulation environment.

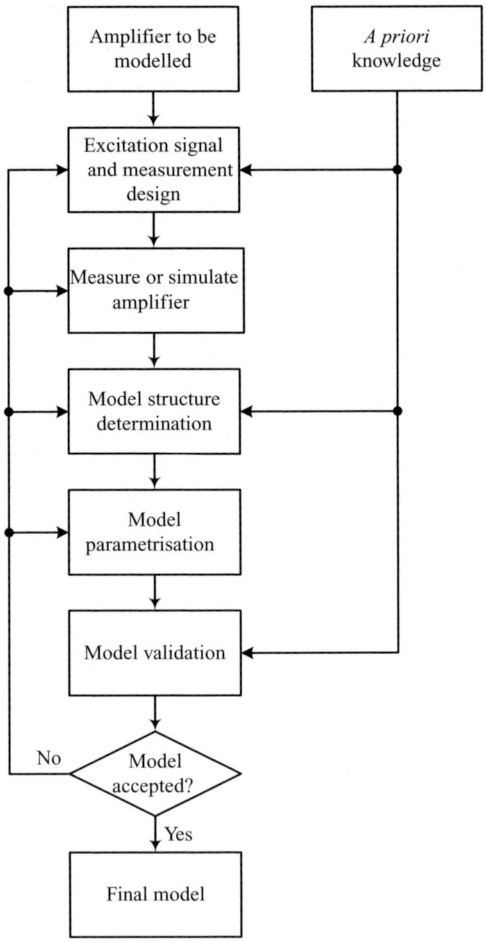

Figure 2.9 The system identification procedure, after [1] (© 1989 Prentice Hall).

On the basis of this knowledge the excitation signal is designed, the amplifier measurement setup is selected, the measurement procedure is specified and the operation range is defined. The design of the excitation signal and of the measurement procedure has a significant influence on the characteristics of the final model. If the measurements are badly planned the resulting data may not be very useful.

As behavioural modelling relies completely on the gathered measurement results, a good representation of the amplifier's behaviour is only possible if the measurements have excited all possible states of the amplifier. This property of the excitation signal and the measurements is called *persistent excitation*. As the acquired data can be generated by either measurements or simulations of the selected amplifier, the excitation-signal issue holds, in general, also for the simulation approach. After the measurements or simulations, have been performed the data can be analysed and a proper model structure selected. The selection of a suitable nonlinear model is complicated by the enormous structural heterogeneity of these models [27]. Together with this structural heterogeneity comes a vast array of specialised parametrisation algorithms, which exploit the different model details. Therefore a change in the model structure inevitably results in a change in the parametrisation algorithm. Ljung [33] summarises this model structure selection problem as follows: 'This is no doubt the most important and, at the same time, the most difficult choice [in] the system identification procedure. It is here that *a priori* knowledge and engineering intuition and insight have to be combined with the formal properties of the models'.

The selection of a particular model also involves considerations about the number of coefficients. The model must provide enough degrees of freedom to describe the full range of amplifier behaviour without using unnecessary parameters. A so-called *parsimony* model meets these demands. As an example, some detail on the proper specification of model size is given for two-box models in Section 4.2 and for state-space models in Section 5.7. Furthermore, the degree of complexity of the appropriate parameter-extraction and model-evaluation procedures can influence the model selection and the determination of model size.

After choosing the model and the corresponding identification algorithm, the model parameters can be extracted from the measurement data. Then the quality of the model has to be assessed. This step is known as model validation. It involves various procedures to evaluate how the model relates to the measurement results, to the *a priori* knowledge and to the intended usage [33]. Determining the correlation between the modelled results and the measurement results requires the selection of metrics to quantify the modelling accuracy. In order to be able to detect systematic errors in the applied extraction procedure, this evaluation should not be performed using the same data set as was used in the parameter-extraction process. If the model fails to meet the identified requirements, the model-identification procedure has to be restarted from an earlier step, as shown in Figure 2.9. Various factors can lead to a deficient model:

- the amplifier measurements do not provide sufficient information or are too noisy;
- the selected model is inappropriate to represent the behaviour of the amplifier;
- the chosen order of the model is too low or too high;
- the model-identification algorithm failed to extract all the parameters correctly.

A major part of the identification process consists of addressing such problems. In practice, this leads to an iterative estimation of the model structure and parameters, as discussed in [1].

2.5.2 *Excitation signal design*

The excitation signals have a significant influence on the success of the model-identification process. A proper design of the test signals is a requirement not only for model parametrisation but also for model validation. For microwave PAs these signals must be designed taking into consideration the bandlimited structure of the amplifiers. This leads to the frequent use of sine or multisine signals to excite the device under test (DUT). In addition to single-tone, two-tone and multisine signals, (bandlimited) noise signals and digitally modulated signals are commonly applied to test the amplifiers. These input signals are then swept over the input power range and frequency range of the DUT to capture its behaviour. Classical test signals in system identification, such as the step function or pseudo-random binary sequences, are rarely used to characterise RF and microwave amplifiers, as they must be low-pass filtered and then modulated onto a carrier signal.

Multisine signals are important in the discussion of excitation signals for PAs. They can be used to generate periodic input signals that can be optimised to substitute for other more complex or even non-periodic test signals. To do so they must provide certain degrees of freedom to allow a proper approximation of the reference signals. It is, therefore, assumed that a multisine test signal is composed of at least of four tones. This assumption distinguishes them from single- and two-tone input signals. A discussion of multisine signals and a presentation of the design procedures that enable them to substitute other input signals is given in subsection 2.5.6.

The question of how to design a suitable input signal to excite all states of an amplifier cannot be answered in general. In contrast with linear systems, the theory of dynamic nonlinear systems does not provide a closed solution to this problem. Two approaches to input signal design, one for the linear case and one for the Volterra series model, will be considered in this section.

For linear systems, correlation-based extraction of the coefficients of an FIR filter can be used to elaborate the concept of persistent excitation [1]. Starting from the low-pass equivalent discrete-time FIR model,

$$\tilde{y}(s) = \sum_{q=0}^{Q-1} \tilde{h}(q)\, \tilde{x}(s-q) + \tilde{v}(s) \tag{2.28}$$

where $\tilde{h}(q)$ are the FIR filter coefficients and $\tilde{v}(s)$ represents the measurement noise, assume that $\tilde{x}(s)$ is a zero-mean stationary random process and that $\tilde{v}(s)$ and $\tilde{x}(s)$ are statistically independent. Then the input–output covariance function is given by

$$R_{\tilde{y}\tilde{x}}(s) = \sum_{q=0}^{Q-1} \tilde{h}(q) R_{\tilde{x}\tilde{x}}(s-q), \tag{2.29}$$

where $R_{\tilde{y}\tilde{x}}(s) = \mathrm{E}\{\tilde{y}(p+s)\tilde{x}^*(p)\}$ and E denotes the expectation operator. If now the two covariance functions are estimated from the measurement data,

$$\hat{R}_{\tilde{y}\tilde{x}}(s) = \frac{1}{M}\sum_{p=1}^{M}\tilde{y}(p+s)\tilde{x}^*(p),$$
$$\hat{R}_{\tilde{x}\tilde{x}}(s) = \frac{1}{M}\sum_{p=1}^{M}\tilde{x}(p+s)\tilde{x}^*(p), \tag{2.30}$$

then an estimate for the FIR filter's coefficients $\hat{\tilde{h}}(q)$ can be found by solving

$$\hat{R}_{\tilde{y}\tilde{x}}(s) = \sum_{q=0}^{Q-1}\hat{\tilde{h}}(q)\hat{R}_{\tilde{x}\tilde{x}}(s-q). \tag{2.31}$$

This results in the following system of linear equations:

$$\begin{pmatrix} \hat{R}_{\tilde{x}\tilde{x}}(0) & \cdots & \hat{R}_{\tilde{x}\tilde{x}}(Q-1) \\ \vdots & \ddots & \vdots \\ \hat{R}_{\tilde{x}\tilde{x}}(Q-1) & \cdots & \hat{R}_{\tilde{x}\tilde{x}}(0) \end{pmatrix} \begin{pmatrix} \hat{\tilde{h}}(0) \\ \vdots \\ \hat{\tilde{h}}(Q-1) \end{pmatrix} = \begin{pmatrix} \hat{R}_{\tilde{y}\tilde{x}}(0) \\ \vdots \\ \hat{R}_{\tilde{y}\tilde{x}}(Q-1) \end{pmatrix}. \tag{2.32}$$

A solution is only possible if the matrix in Equation (2.32) is nonsingular. This leads to the definition of persistent excitation [1]: a signal $\tilde{x}(s)$ is said to be persistent of order Q if the matrix

$$\mathbf{R}_{\tilde{x}\tilde{x}}(Q) = \begin{pmatrix} R_{\tilde{x}\tilde{x}}(0) & \cdots & R_{\tilde{x}\tilde{x}}(Q-1) \\ \vdots & \ddots & \vdots \\ R_{\tilde{x}\tilde{x}}(Q-1) & \cdots & R_{\tilde{x}\tilde{x}}(0) \end{pmatrix} \tag{2.33}$$

is positive definite.

In [4], the conditions necessary to calculate a nth-order Volterra kernel are highlighted. It is shown that for the correct determination of an nth-order Volterra kernel the input signal should be the sum of at least n signals. For example, the response of the symmetric second-order Volterra operator to the sum of two input signals is given by

$$\begin{aligned} y(t) &= \mathrm{H}_2[x_1(t)+x_2(t)] \\ &= \mathrm{H}_2[x_1(t)] + 2\,\mathrm{H}_2\{x_1(t),x_2(t)\} + \mathrm{H}_2[x_2(t)], \end{aligned} \tag{2.34}$$

where H_2 is the second-order Volterra operator as introduced in Equation (2.1) and $\mathrm{H}_2\{x_1(t),x_2(t)\}$ specifies the bilinear Volterra operator:

$$\mathrm{H}_2\{x_1(t),x_2(t)\} = \int_{-\infty}^{\infty}\int_{-\infty}^{\infty} h_2(\tau_1,\tau_2)\,x_1(t-\tau_1)x_2(t-\tau_2)\,d\tau_1 d\tau_2. \tag{2.35}$$

An expression for this operator can be found by rearranging Equation (2.34), to give

$$2\,\mathrm{H}_2\{x_1(t), x_2(t)\} = \mathrm{H}_2[x_1(t) + x_2(t)] - \mathrm{H}_2[x_1(t)] - \mathrm{H}_2[x_2(t)]; \tag{2.36}$$

then substituting $x_1(t) = e^{j\omega_1 t}$ and $x_2(t) = e^{j\omega_2 t}$ into Equation (2.36) results in a frequency-domain representation of the bilinear Volterra operator:

$$2\,\mathrm{H}_2(\omega_1, \omega_2)e^{j(\omega_1+\omega_2)t} = \mathrm{H}_2(e^{j\omega_1 t} + e^{j\omega_2 t}) - \mathrm{H}_2(\omega_1, \omega_1)e^{j2\omega_1 t} - \mathrm{H}_2(\omega_2, \omega_2)e^{j2\omega_2 t}, \tag{2.37}$$

where the Fourier transform of the kernel $h_2(\tau_1, \tau_2)$ is given by

$$\mathrm{H}_2(\omega_1, \omega_2) = \int_{-\infty}^{\infty}\int_{-\infty}^{\infty} h_2(\tau_1, \tau_2)e^{-j(\omega_1\tau_1+\omega_2\tau_2)}d\tau_1 d\tau_2. \tag{2.38}$$

It should be noted that, for a general second-order Volterra operator composed of $\mathrm{H}_1[x(t)]$ and $\mathrm{H}_2[x(t)]$, the first-order operator will not influence the evaluation of the second-order bilinear Volterra operator, since

$$\mathrm{H}_1[x_1(t) + x_2(t)] - \mathrm{H}_1[x_1(t)] - \mathrm{H}_1[x_2(t)] = 0. \tag{2.39}$$

The generalisation of these results for the nth-order Volterra kernel is extremely labour intensive. The expression for the third-order trilinear Volterra operator is already significantly more complex:

$$\begin{aligned} 3!\,\mathrm{H}_3\{x_1(t), x_2(t), x_3(t)\} = {} & \mathrm{H}_3[x_1(t) + x_2(t) + x_3(t)] - \mathrm{H}_3[x_1(t) + x_2(t)] \\ & - \mathrm{H}_3[x_2(t) + x_3(t)] - \mathrm{H}_3[x_3(t) + x_1(t)] \\ & - \mathrm{H}_3[x_1(t)] - \mathrm{H}_3[x_2(t)] - \mathrm{H}_3[x_3(t)]. \end{aligned} \tag{2.40}$$

An expression for the nth-order Volterra kernel was derived in [4].

In conclusion, for the evaluation of an nth-order Volterra kernel in the frequency domain the input signal must be composed of at least n complex exponential functions and the extraction of this kernel will not be affected by the presence of the $n-1$ lower-order kernels.

2.5.3 *Amplifier response to typical excitation signals*

In this subsection the typical response of an amplifier to a single-tone signal, a two-tone signal and a wideband WiMax signal will be described. Details on the orthogonal frequency-division multiplex (OFDM) signal used in this section are given in Appendix A. In the first part of the subsection the responses of nonlinear static and nonlinear dynamic amplifier models are discussed and compared. The advantage of using models for this comparison is that it is possible to add nonlinear memory effects to the model behaviour without changing the nonlinear static characteristic. In the second part of the subsection the measurement results of a Freescale™ amplifier board using an MRF7S38010H LDMOS transistor are considered. This 10 W device has been optimised for European HiperMAN base-station applications [34] operating in the frequency band between 3.4 and 3.6 GHz.

This HiperMAN standard adapts the WiMax specification according to the needs of the European authorities. The amplifier board is characterised by a dominant static nonlinear behaviour. To allow comparison with a nonlinear dynamic amplifier, the drain bias circuitry was modified so that strong nonlinear memory effects were introduced into the PA characteristics. This amplifier allows a link to be established between typical (modelled) responses and the real-world measured results.

Nonlinear static and nonlinear dynamic amplifier model responses

The models used to generate typical responses represent amplifiers characterised by a gain of 20 dB and a 1 dB compression point of 9 dBm. They cover harmonic and intermodulation distortion products up to fifth order. The relationship between the desired output signals and the nonlinear distortion products is given by the intercept points (IP), which were chosen as follows:

- $\mathrm{IP}_2 = 36\,\mathrm{dBm}$
- $\mathrm{IP}_3 = 30\,\mathrm{dBm}$
- $\mathrm{IP}_4 = 32\,\mathrm{dBm}$
- $\mathrm{IP}_5 = 27\,\mathrm{dBm}$

(2.41)

The concept of the 1 dB compression and intercept points is well presented in the literature (see e.g. [25, 26]).

For single- and two-tone input signals all distortion products located in the DC, fundamental and second-order harmonic zones were considered. The response of the models to the WiMax signal was evaluated only in the zonal band. Figure 2.10 presents the input and output signals from the nonlinear static amplifier model for the three above-mentioned input signals, all at a level of $P_{\mathrm{in}} = -10\,\mathrm{dBm}$. All signal and distortion components in this figure are scaled to represent the correct power levels or the corresponding power spectral densities (PSDs) respectively. The different markers for the distortion products identify the nonlinear mechanism that caused this output. The arrow markers indicate the input and the (nonlinearly amplified) output tones. The circle markers indicate distortions that are harmonics of the input tones. The square markers identify intermodulation products (IMPs) located at frequencies generated by adding and subtracting the frequencies of the input sinusoids (e.g., $\mathrm{f}_2 - \mathrm{f}_1$, $2\mathrm{f}_2 - \mathrm{f}_1$, $\mathrm{f}_2 + \mathrm{f}_1$, ...).

In the same way, Figure 2.11 depicts the response of the nonlinear dynamic amplifier model. Its nonlinear dynamic behaviour is caused by frequency-dependent second-order harmonic products and is seen in the different lengths of the two circle markers in the second-harmonic zone of the two-tone response. This frequency dependence introduces an imbalance in the third- and fifth-order in-band IMPs of the two-tone response and a difference in the upper and lower adjacent channel distortions of the WiMax output signal.

In Figure 2.12 a two-tone power sweep of the same nonlinear static and nonlinear dynamic models is presented. Both models exhibit a sweet spot in the behaviour of the third-order IMPs. In the nonlinear dynamic case, imbalances between the IMPs

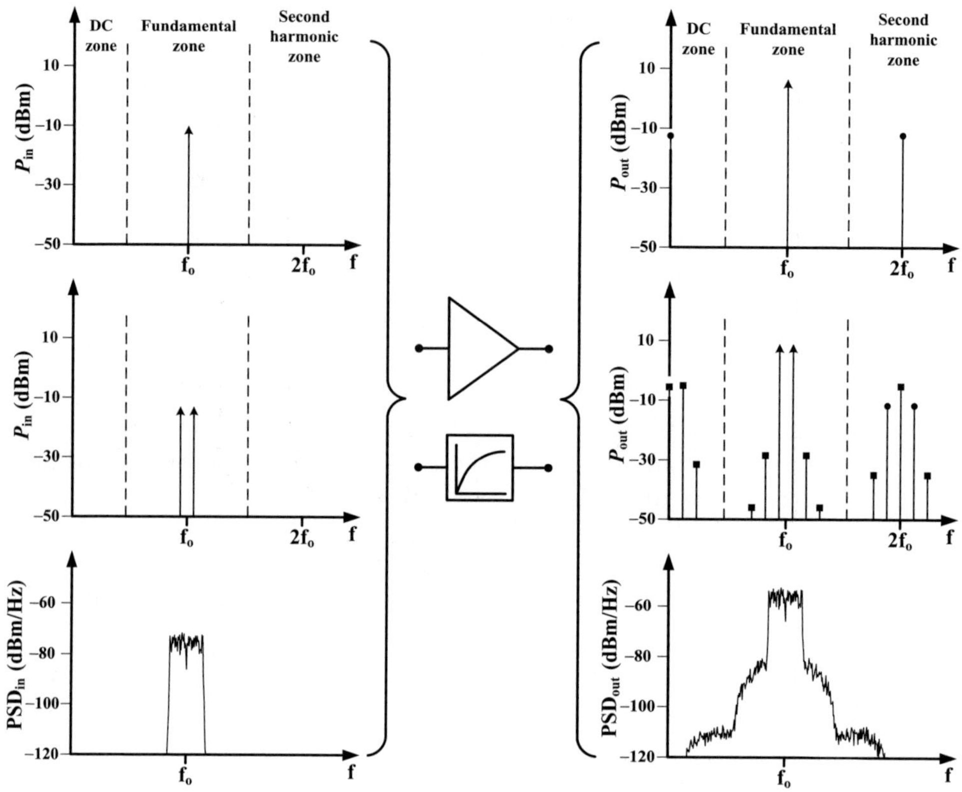

Figure 2.10 Responses of a nonlinear static amplifier to a single-tone, a two-tone and a WiMax input signal. The various signals are identified by markers (see the text). In contrast with the single-tone and two-tone responses, the broadband signal is shown only in the zonal band.

can be observed and the two sweet spots appear at different input power levels. The lack of sweet spots in the response of the fifth-order IMPs is caused by the neglect of higher-order distortion products.

The AM–AM conversion curves for the nonlinear static and nonlinear dynamic amplifiers with a WiMax input signal are given in Figure 2.13. Only in the memoryless case are the AM–AM conversion plot and the mean amplifier gain identical. The AM–AM conversion of the nonlinear dynamic amplifier is broadened by the presence of (nonlinear) memory effects. The depicted conversion plots coincide with the output signal spectra shown in Figures 2.10 and 2.11.

Amplifier measurement results

The amplifier board described above was biased for Class-AB operation. In this condition the amplifier exhibited a gain of 15 dB, a 1 dB compression point of 41 dBm and the following intercept points: $IP_3 = 51$ dBm and $IP_5 = 43$ dBm. In Figure 2.14 the magnitude and phase of the gain surface measured with a single-tone input are

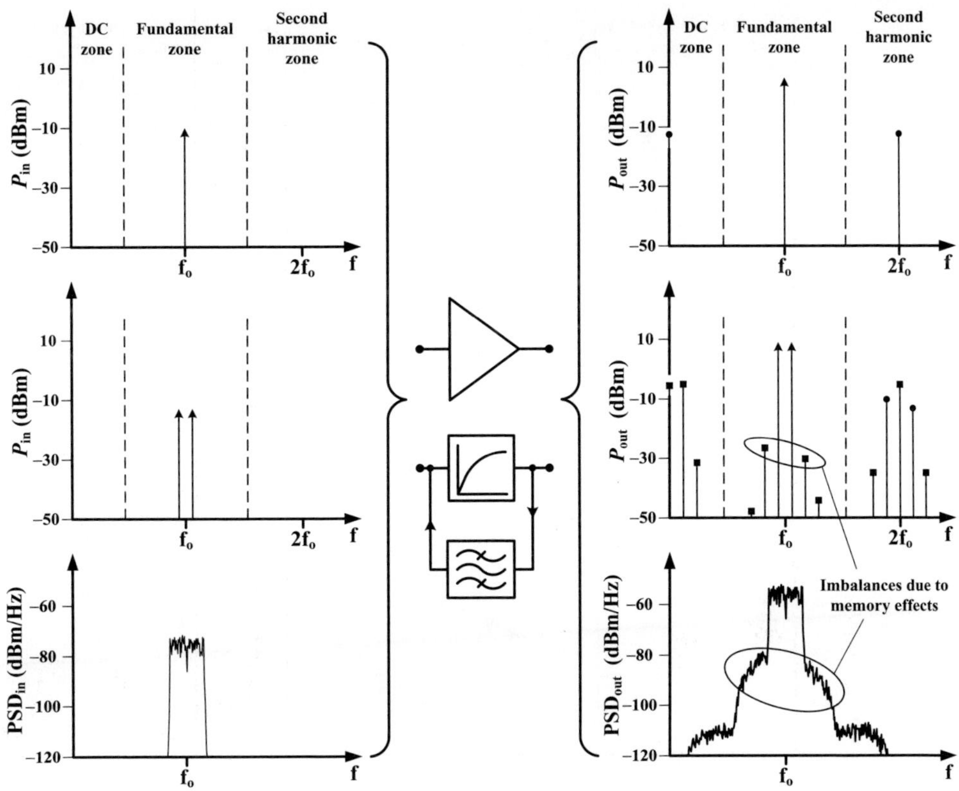

Figure 2.11 Responses of a dynamic nonlinear amplifier to a single-tone, a two-tone and a WiMax signal. The meaning of the different markers is explained in the text. The imbalances in the distortion products of the two-tone and the broadband signal were caused by nonlinear memory effects.

shown (swept-tone measurement was used). These results were achieved by the use of the measurement setup shown in Figure 2.21. To allow a better comparison of the frequency-dependent gain variation, the AM–AM and AM–PM conversion plots are depicted in Figure 2.15. Both plots have been normalised to the small-signal amplifier gain at the corresponding frequency, and the phase measurement has been normalised to remove the effects of amplifier delay; which was estimated to be 1.8 ns from the original AM–PM conversion.

The results of the two-tone measurements at $f_0 = 3.5$ GHz are shown in Figure 2.16. The IMPs were evaluated at tone spacings $\Delta f = 1$, 5 and 10 MHz. A measurement setup similar to that in Figure 2.23 was used to evaluate the power level of the different tones. For third- and fifth-order IMPs only a very low dependence of the generated distortion on the tone spacing can be detected. Also, no imbalance between the upper and the lower third-order IMPs is visible. Only at the 10 MHz tone spacing do the fifth-order IMPs show an imbalance of up to 5 dB.

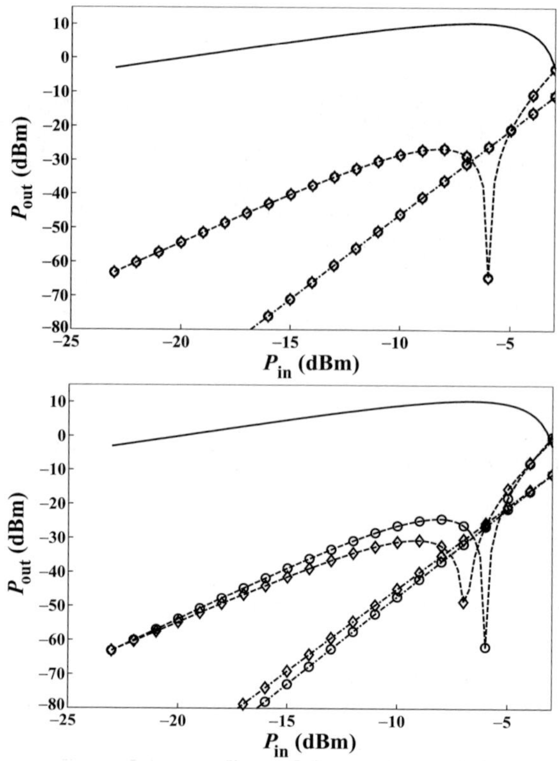

Figure 2.12 Input power sweep for the (a) nonlinear static and (b) nonlinear dynamic amplifier model under two-tone excitation. Power values for the third- and fifth-order intermodulation products (IMPs) are shown. (a) Solid line, amplifier output; broken line, IMP_3; broken-and-dotted line, IMP_5. (b) The same as (a) but now the curves for upper and lower distortion products are different: the circles correspond to the lower tone and the diamonds to the upper tone.

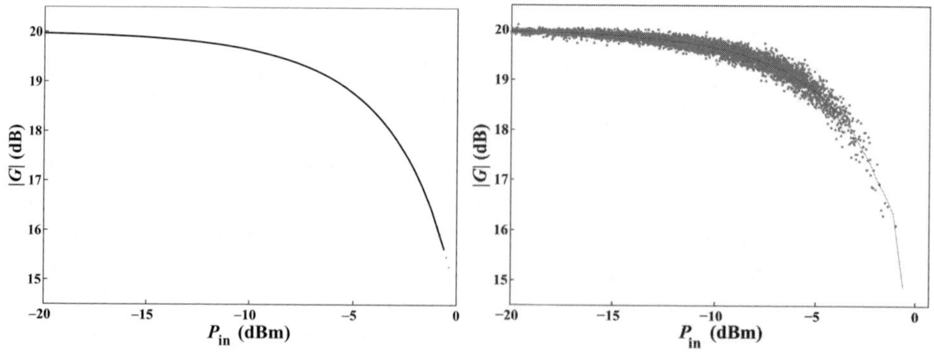

(a) (b)

Figure 2.13 The AM–AM conversion of a WiMax modulated input signal for (a) the nonlinear static model and (b) the nonlinear dynamic amplifier model. The grey dots depict the AM–AM conversion while the solid black line represents the mean amplifier gain.

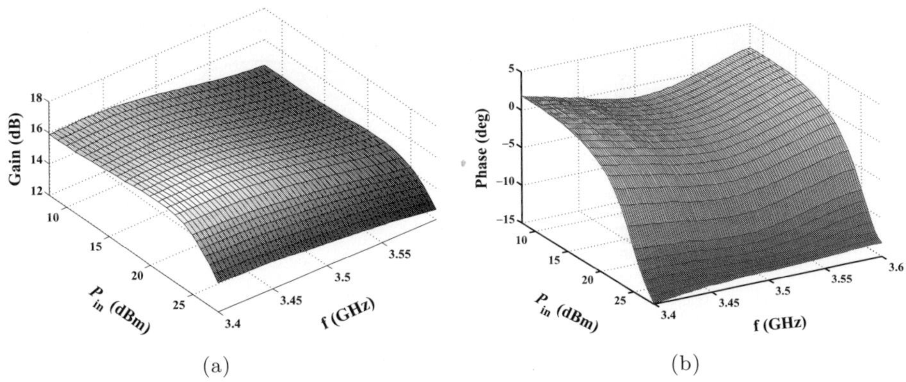

Figure 2.14 (a) Magnitude and (b) phase of the gain surface of the Freescale™ MRF7S38010H LDMOS amplifier board.

These results justify the classification of the amplifier behaviour as dominantly nonlinear static.

The response of the amplifier to a 15 dBm WiMax modulated input signal at a carrier frequency of 3.5 GHz is shown in Figure 2.17. This signal covers a bandwidth of 3.5 MHz and shows a peak to average power ratio (PAPR) of 9.8 dB [26].

The corresponding AM–AM and AM–PM conversion plots are presented in Figure 2.18. The deviation of the instantaneous gain from the nonlinear static characteristic is caused by measurement noise, and memory effects. By performing a measurement without the amplifier (i.e. connecting the generator of the measurement system directly to the receiver) the influence of the noise could be characterised. This measurement showed that noise affects the amplifier measurements up to an input power $P_{\text{in}} = 10$ dBm. For power levels below this point some memory effects add to the measurement noise, causing an expansion of the instantaneous AM–AM plot. At 10 dBm input power the gain variation introduced by the memory effects is about ± 0.2 dB. Above this level the impact of the memory effects reduces significantly. A broadband time-domain measurement system, as shown in Figure 2.27, was used to capture the input and output signals of the amplifier. A time alignment of the two signals was performed before the conversion plots were evaluated.

To emphasise the differences between a dominantly nonlinear static and a nonlinear dynamic amplifier, the drain bias-circuitry of the amplifier board was modified to introduce significant long-term memory effects. This modification resulted in the changes in the amplifier's characteristic values (at $f_0 = 3.5$ GHz) summarised in Table 2.1.

The wide range of values for the intercept point IP_3 of the modified amplifier is caused by a significant dependence on the tone spacing. Figure 2.19 shows the corresponding two-tone measurement results. In comparison with the dominantly nonlinear static two-tone response, the imbalance between the upper and lower

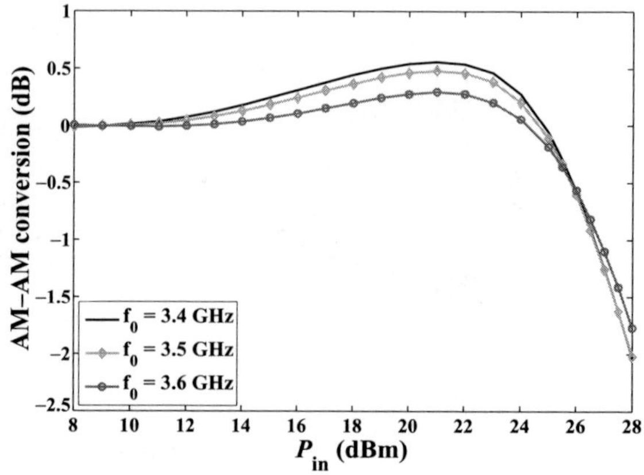

(a)

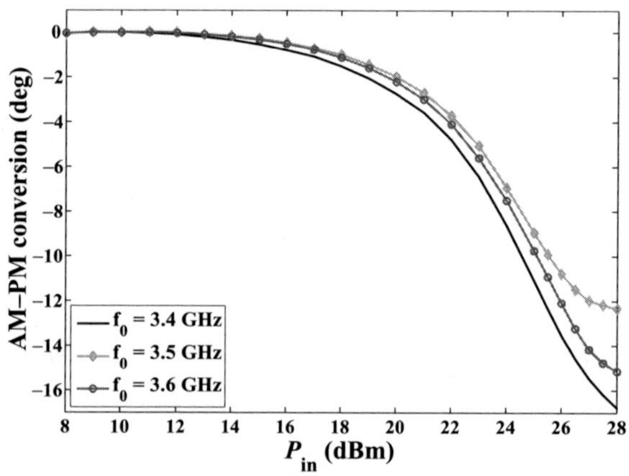

(b)

Figure 2.15 Normalised (a) AM–AM and (b) AM–PM conversion of the Freescale™ MRF7S38010H LDMOS amplifier board extracted from single-tone measurements.

third-order IMPs and the dependence on the tone spacing are clear. The power levels of the IMPs at $\Delta f = 1$ MHz have increased by up to 10 dB; the differences for the two other tone spacings are less significant. The differences between the fifth-order IMPs of the two amplifier types are also highest for the 1 MHz tone spacing.

A better illustration of the memory effects introduced by the bias-circuitry modification is given by comparing the AM–AM conversion plots for the WiMax modulated input signal. Figure 2.20 shows the upper and lower limiting curves for the two conversion plots. The widening of the AM–AM conversion due to long-term memory effects is highlighted in this way. Both conversion plots represent the same mean output power.

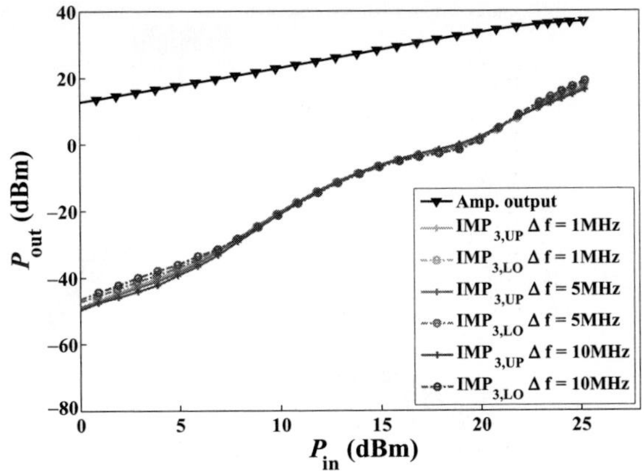

(a)

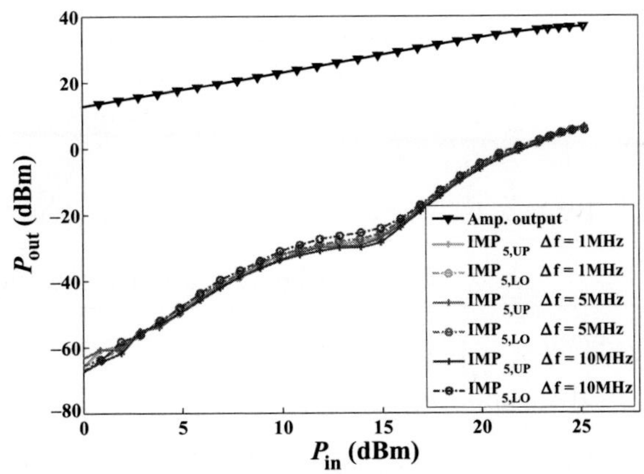

(b)

Figure 2.16 Two-tone measurement results at $f_0 = 3.5\,\text{GHz}$. The power in (a) the third-order and (b) the fifth-order IMPs are shown for three different tone spacings.

2.5.4 *Single- and two-tone amplifier characterisation*

In the previous subsection we presented typical amplifier responses to single-, two-tone and broadband signals. The subsections which now follow give an overview of typical measurement setups used for amplifier characterisation. Included is a compilation of models that have been parametrised by the use of the corresponding amplifier characterisation technique.

Single-tone measurement setups

The characterisation of the gain and phase compression of an amplifier is commonly achieved by means of a vector network analyser (VNA). The VNA performs reflection and transmission measurements by comparing the incident and reflected waves at the input and output of the DUT.

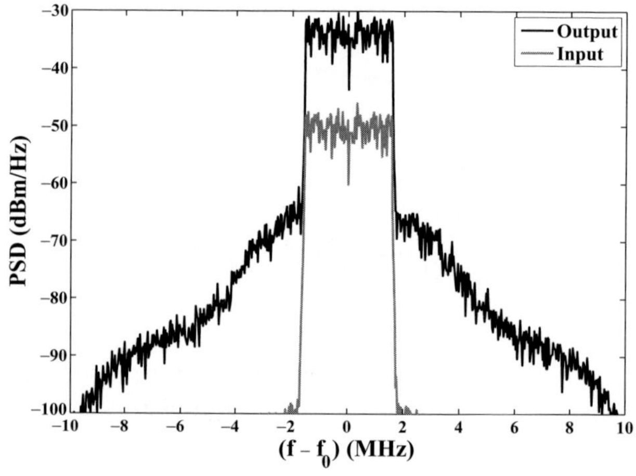

Figure 2.17 Power spectral density of the amplifier's response to a WiMax input at $f_0 = 3.5$ GHz. This input signal was designed to cover a 3.5 MHz bandwidth and shows a PAPR of 9.8 dB. The mean power of the input and output signals is 15 dBm and 31.5 dBm respectively.

Table 2.1 The effect of memory on amplifier characteristics

	Original amplifier	Modified amplifier
Gain	15.5 dB	15.2 dB
1 dB comp. point	41 dB	39.5 dB
IP_3	51 dBm	46–50 dBm
IP_5	43 dBm	43.1 dBm

A simplified block diagram of a VNA is depicted in Figure 2.21 [7]. The synthesised single-tone source (the VNA source) is connected via the RF switch to the input of the DUT. The bias tees for the DC supply at the input and output of a typical transistor amplifier are represented by the squares labelled 'T'. The incident and reflected waves at Ports 1 and 2 are all connected to the inputs of the VNA receiver by directional couplers adjacent to the two ports. These four signals are synchronously downconverted to an intermediate frequency, filtered and quantified. By using synchronous mixing the amplitude and phase relationships between the input signals are conserved. Hence, a linear relationship between the magnitude and phase of the waves at Ports 1 and 2 and the four signals measured at the VNA receiver can be established. The coefficients of this linear relationship are extracted from calibration measurements performed before the DUT is connected. In a post-processing step the gain and the return loss of the DUT can be calculated from the deduced ratios of the incident and reflected waves at the two ports. An additional power calibration of the DUT input power must be performed in order to allocate

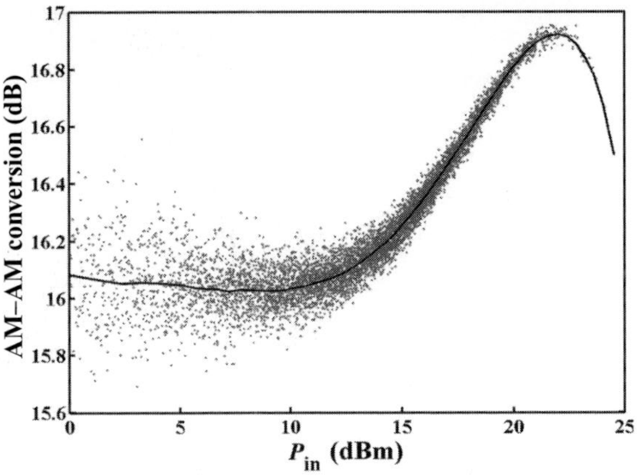

(a)

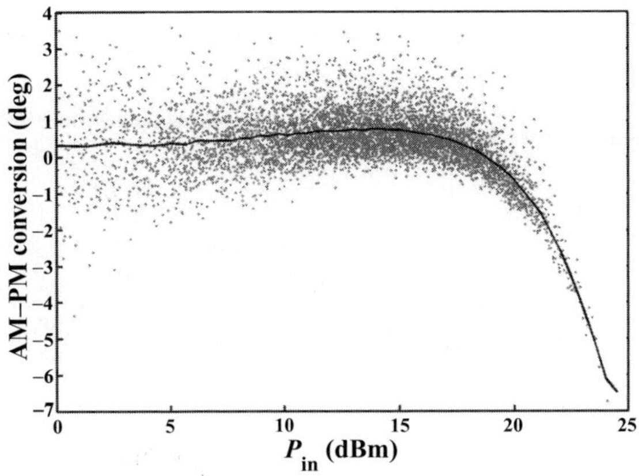

(b)

Figure 2.18 The (a) AM–AM and (b) AM–PM conversion plots for the LDMOS amplifier board, extracted from the WiMax measurements. The grey dots visualise the instantaneous AM–AM and AM–PM conversions while the solid black line represents the mean amplifier behaviour.

the correct input power level to the actual measured amplifier gain.

A more detailed overview of VNA setups, calibration and uncertainty issues can be found in, for example [35].

The concept of the (classical) VNA is limited to measurements at a single frequency set by the VNA source. Hence, the relationship between the fundamental output signal and the harmonic distortion components generated by the amplifier cannot be captured. One of the first measurement setups that overcomes this limitation was published by Lott [37]. The key element of this setup is a phase-locked signal generator with internal multiplication, as shown in Figure 2.22. The fundamental output signal of the generator is low-pass filtered to suppress the harmonics of the source. After selecting the desired power level by use of a step attenuator, this

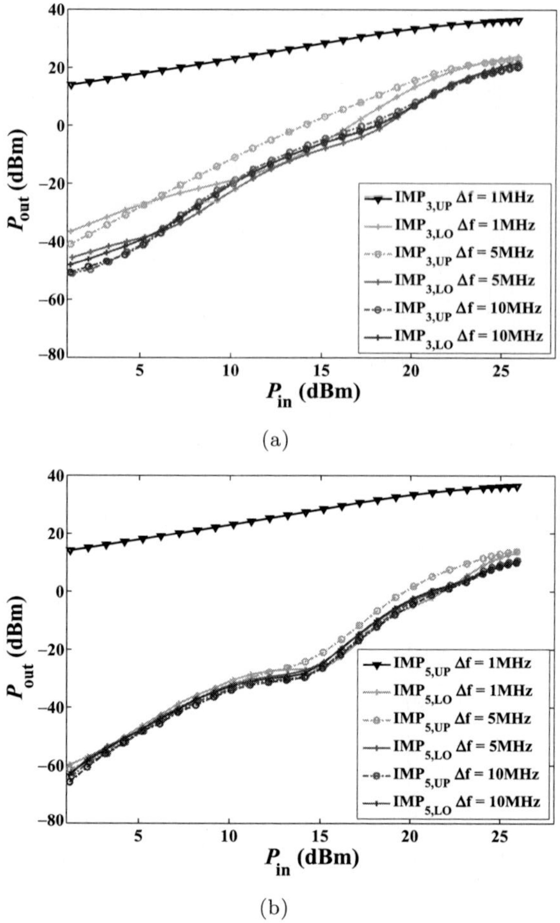

Figure 2.19 Two-tone measurement results for the nonlinear dynamic amplifier. These measurements were performed under the same conditions as those mentioned in connection with Figure 2.16.

signal is passed to the input of the DUT. A directional coupler is used to sample the input signal and feed it into a power meter. The output signal of the DUT, composed of the amplified fundamental signal and harmonic distortions, is combined with the ‘reference signal’ from the VNA Port 1 by the directional coupler. For correct operation of this measurement setup it is important that this directional coupler does not pass this reference signal back to the amplifier output. Therefore, it is essential that the coupler has a high directivity at the harmonic frequencies nf_1.

After the reference signal and the amplifier output signal have been combined, the level of the fundamental signal component is reduced by a high-pass filter. This filtering avoids saturation of the VNA Port 2 by the dominant fundamental signal component. At the VNA the magnitude and phase of the sum of the reference signal and the harmonic distortion due to the amplifier are measured and, using

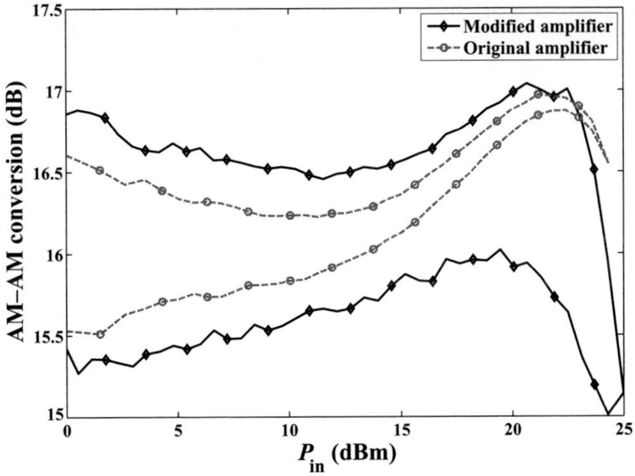

Figure 2.20 The AM–AM conversions of the original and the modified amplifier. Only the upper and the lower limiting curves of the conversion plots are shown.

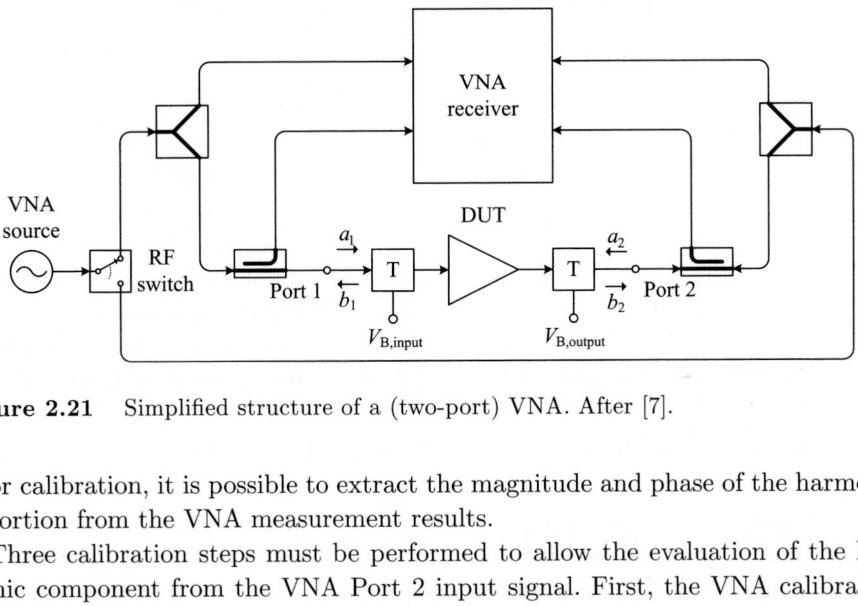

Figure 2.21 Simplified structure of a (two-port) VNA. After [7].

prior calibration, it is possible to extract the magnitude and phase of the harmonic distortion from the VNA measurement results.

Three calibration steps must be performed to allow the evaluation of the harmonic component from the VNA Port 2 input signal. First, the VNA calibration must be performed while the DUT is replaced by two matched loads. After this step the reference signal vector is defined as $1.0\angle 0°$ (magnitude 1.0 and phase 0°). The absolute power of this reference signal is measured by connecting a power meter at the output of the directional coupler. A measurement of the harmonics of a reference diode completes the calibration of the measurement setup. It is assumed that the phase relationship between the fundamental signal and the generated harmonics is known for the reference diode. The harmonic distortion components generated by the DUT are now compared with a reference signal of which the magnitude and phase are known.

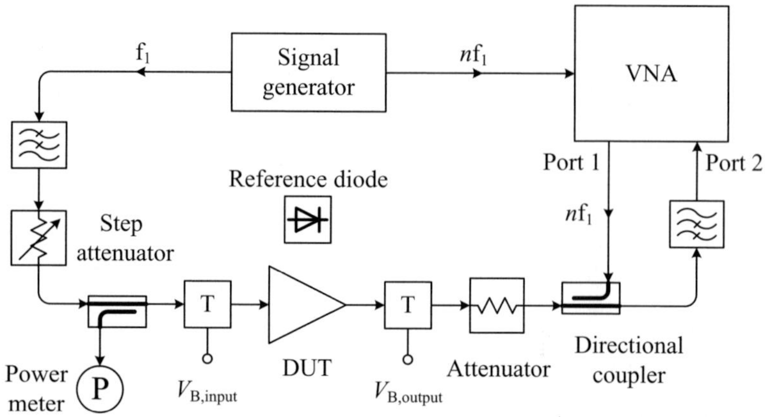

Figure 2.22 Measurement setup to characterise the fundamental signal and the harmonic distortion components in magnitude and phase as explained in [37].

Another possible way of characterising amplifiers with harmonic distortion components is the nonlinear VNA (NVNA), also called the large-signal network analyser (LSNA). Such a measurement setup is able to handle single-tone and multisine excitation signals and is discussed in subsection 2.5.6.

As stated in Section 1.3, single-tone AM–AM/AM–PM conversion measurements are commonly used for the parametrisation of nonlinear static models and models with linear memory. In Table 2.2 a compilation of models that may be fitted from single-tone measurements is presented. It is predominantly based on articles cited within this book. Only articles that explicitly specified the usage of measurement results for the model extraction were included. Others that used single-tone harmonic-balance simulations were not considered. Table 2.2 shows only low-pass equivalent models. This limitation results from the fact that circuit-level behavioural models are typically parametrised for NVNA measurements. It is of interest to note that in the nonlinear dynamic model proposed by Asbeck *et al.* [48] a set of single-tone measurements for model fitting was applied.

Two-tone measurement setups

Two-tone characterisation is, together with single-tone measurements, one of the classical approaches to evaluating amplifier characteristics. Such an excitation signal is very simple to create and shows a non-constant envelope. Important figures of merit (FOMs) for small-signal amplifiers, such as the intercept points, can be directly extracted from these measurements. Furthermore, two-tone measurements performed at different tone spacings and different input power levels are often used to detect the presence of memory effects in the amplifier behaviour [31, 47].

A typical two-tone measurement setup is shown in Figure 2.23. The two attenuators after the individual sources provide the necessary isolation between the two generators. After the two sources have been combined, the preamplifier is used to

Table 2.2 Models extracted by the use of single-tone measurements

Model type	Source(s)	Subsection or section	Comments
Power series	[38, 39]	1.3.1, 3.4	
Saleh model	[40]	1.3.1, 1.3.2, 3.5, 4.3.3	
Two-box: ARMA + Bessel series	[41, 42]	1.3.2, 3.8, 4.2	Bessel series nonlinearity extracted from AM–AM/AM–PM measurement results
Poza–Sarkozy–Berger	[43]	1.3.2, 4.3.2	
Abuelma'atti	[44]	1.3.2, 4.4.1	
Polyspectral	[28, 45]	1.3.3	Bessel series nonlinearity extracted from AM–AM/AM–PM measurement results
Augmented behavioural characterisation	[48]	1.3.4	A set of AM–AM/AM–PM measurements was used to extract the dependence of the gain on the input power
Dynamic Volterra series expansion	[47, 46]	1.3.4, 5.6.4	Static nonlinearity extracted from AM–AM/AM–PM measurement results

achieve the desired input power level. The input and output signal powers of the DUT are then measured by two power meters. Finally, the spectrum analyser evaluates the ratio of the power levels of the tones present at the amplifier output.

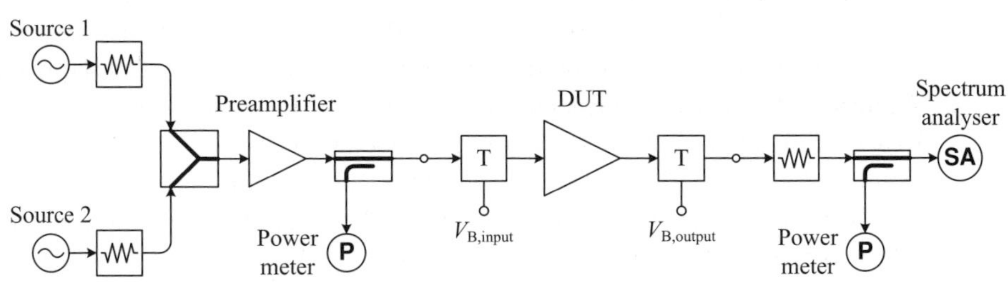

Figure 2.23 Typical two-tone measurement setup.

The big disadvantage of the classical two-tone measurement setup is its inability to characterise the AM–PM behaviour of the carriers and the distortion components. One possible way of circumventing this problem was published by Bösch and Gatti [24]. Their two-tone measurement setup used two VNAs to characterise the AM–AM and AM–PM conversion of the two carriers, as shown in Figure 2.24. The setup performs a synchronised power sweep. Hence, during a sweep a constant

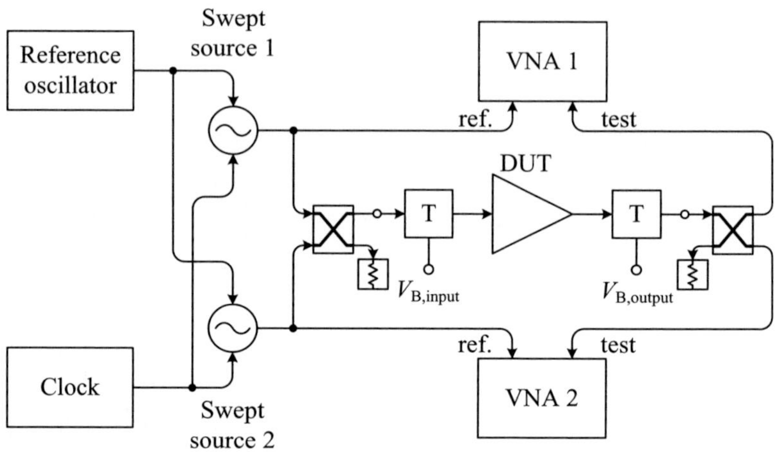

Figure 2.24 Two-tone measurement setup used by Bösch and Gatti [24].

amplitude relationship between both sources is guaranteed.

Another two-tone measurement setup characterising the amplitudes and phases of the two carriers and the distortion tones was suggested by Yang *et al.* [49] and was used to characterise the static nonlinearity of a 500 W class-AB multistage power amplifier. They fitted a Fourier series to the AM–AM curve and a rational polynomial to the AM–PM curve on the basis of their measurement results. The proposed measurement setup is presented in Figure 2.25. The upper branch of this setup is similar to the classical two-tone measurement system shown in Figure 2.23. A step attenuator is used to set the desired input power. The two power meters evaluate the input and output signal power levels of the DUT. The spectrum analyser in the upper branch extracts the power ratio between the tones present at the PA output. The key idea is to evaluate the phases of the tones by constructing a reference intermodulation distortion generator and comparing the distortion products of the PA to those generated by the reference. Yang *et al.* used a small-signal MESFET transistor to achieve a memoryless IMD generator, which operated at a very low centre frequency, 750 kHz. It was proved that the transistor showed no AM–PM conversion for the fundamental tones by harmonic balance simulation. Additionally, it presented a constant output phase for the third- and fifth-order IMD distortion from the small-signal regime up to the 1 dB compression point. The IMD generator is driven by a downconverted copy of the DUT input signal. As this signal is sampled in front of the step attenuator the IMD generator is driven at a constant power level. The output signal of the DUT is passed through a vector modulator, downconverted and added to the IMD generator output signal. The magnitude and phase of the vector modulator is now altered until the actual measured tone is cancelled. When the adjustment of the vector modulator is complete, the two-tone input signal is switched off, the RF-switch connects the vector modulator to the VNA and the corresponding cancellation phase is measured. This process of optimising and measuring the vector attenuator must be repeated for each considered tone at each power level.

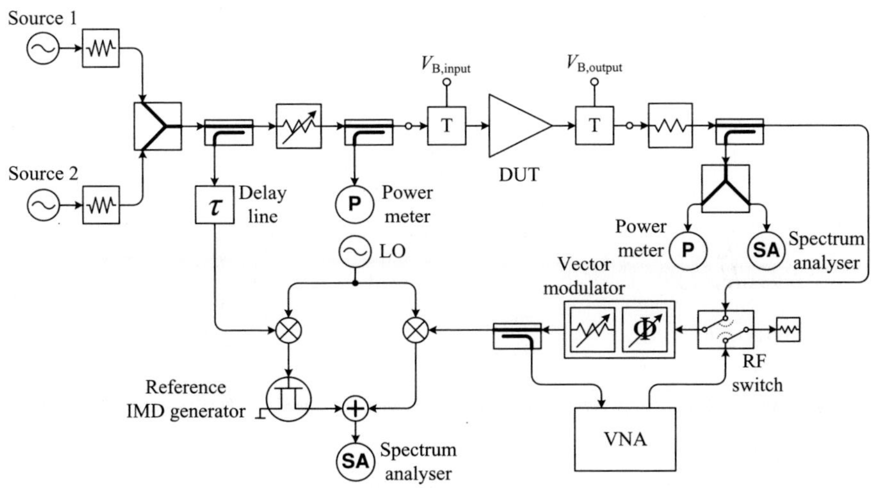

Figure 2.25 Two-tone measurement setup as presented by Yang *et al.* [49].

Table 2.3 presents a compilation of models that were fitted from two-tone measurements; the selected articles were based on the same approach as that used for Table 2.2 (see above). Models extracted from two-tone characterisation by the use of the NVNA were not added to the table. The only exception is the two-tone measurement setup developed by Le Gallou *et al.* [46, 47] to evaluate the kernels for the dynamic Volterra series expansion. In their setup an NVNA was used to extract the phase of the IMD components.

2.5.5 *Broadband-amplifier characterisation*

The expression 'broadband-amplifier characterisation setup' is used to define measurement systems that capture the low-pass equivalent DUT input and output signal in the time domain. The importance of such characterisation setups results from their structural similarity to modern communication systems. Figure 2.26 shows for comparison a simplified communication-system model (compare also the system-level co-simulation presented in Figure 7.1) and a broadband-amplifier characterisation setup (BBACS).

Table 2.3 Models extracted from two-tone measurements

Model type	Sources(s)	Section or subsection	Comments
Power series	[50]	1.3.1, 3.4	
Fourier series + rational polynomial	[49]	3.7	
Three-box model	[51]	4.3.2	Uses tone measurements in magnitude and phase to fit the complete model
Parallel-cascade Wiener model	[31, 52]	1.3.4, 5.5	
Memory polynomial	[53]	1.3.2, 5.2	
Dynamic Volterra series expansion	[46, 47]	1.3.4, 5.6.4	The dynamic Volterra kernels are extracted from two-tone measurements
Multislice model	[54]	1.4.2	

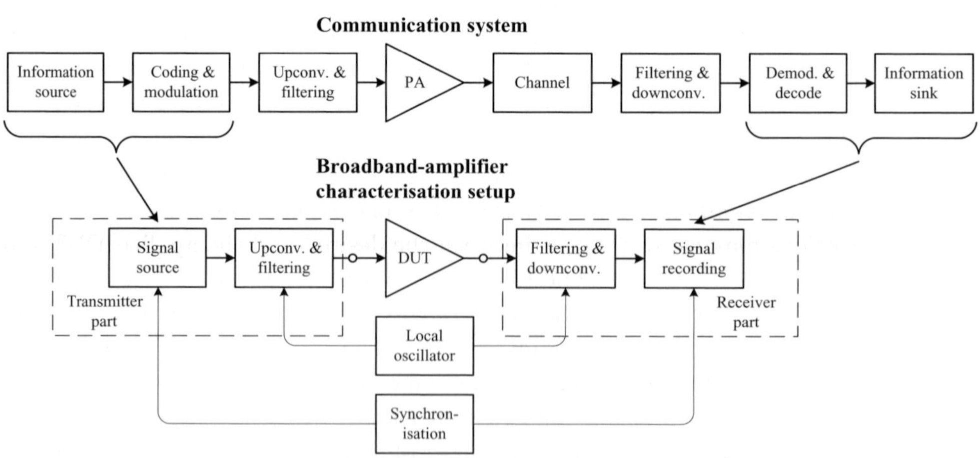

Figure 2.26 Structural comparison of a communication system to a broadband amplifier characterisation setup.

In a BBACS the coded and modulated information signal is provided by a signal source. This complex-valued baseband signal is then upconverted, filtered and amplified before it is available at the DUT input. The output signal of the DUT is attenuated to a suitable power level, filtered, downconverted and sampled. Then, instead of being demodulated and decoded, it is directly captured by the signal recording unit. The structure of the BBACS restricts the capture of harmonic distortion components from the DUT output signal as any harmonic content at the output of the amplifier will be suppressed by the following filter stage. If these harmonics are to be evaluated, an NVNA has to be used (see subsection 2.5.6).

A benefit of this measurement approach is that, compared with the previously discussed single- and two-tone measurement setups, there are no restrictions on the type of excitation signal modulation. The sampling rate at which the DUT output signal is captured is influenced only by the bandwidth of the equivalent low-pass signal, not by the carrier frequency [7]. This property distinguishes the BBACS from similar measurement setups using digital storage oscilloscopes (DSOs), which are used to record the full RF signal. Digital storage oscillators typically exhibit a significantly higher analogue bandwidth at the cost of a lower dynamic range.

An advantage of the BBACS is that slight changes in the frequency of the local oscillator will not alter the measurement results, as typically it is used synchronously in both the up- and downconversion stages.

Further important concerns about the validity of the BBACS measurement results relate to the synchronisation of the signal source and the signal recorder and the compensation of the influence of the transmitter and receiver parts of the setup. Figure 2.26 shows a block that performs the synchronisation. This unit can be either a hardware device that performs the triggering of the signal generation and the signal recording or it can be a software algorithm that takes care of the signal alignment in a post-processing stage. The synchronisation block gains even more importance if averaging techniques are used to reduce the measurement noise. For digitally modulated (noise-like) excitation signals, only an exact alignment between the measurement frames will overcome the problem that both the desired signal and the noise are reduced by the use of averaging.

Depending on the structure and the selected implementation of the BBACS, numerous imperfections may introduce linear and nonlinear distortion. These distortions may cause changes between the desired and the applied DUT input signal in the transmitter part and also a deterioration of the DUT output signal recorded in the receiver branch. The main sources of distortion are gain imbalances and DC offsets in the real and imaginary branches of the baseband signal conditioning and in the I/Q mixing stage, imperfections in the frequency response of the devices and nonlinear distortion of the active components. To avoid any degradation of the measurements, the transmitter and receiver have to be characterised and the extracted behaviour has to be used to compensate the signal generation and the signal capturing. Discussions of the measurement of frequency-translating devices and the evaluation and the compensation of the distortion in BBACS is presented in [30, 55–58].

A typical BBACS is depicted in Figure 2.27 [56, 61]. Here the input signal is loaded into the arbitrary waveform generator (AWG) by the host PC. The complex-valued baseband signal is upconverted by the vector signal generator and scaled by a variable gain amplifier. After further amplification the signal is passed to an RF switch, which either connects it to the DUT input or bypasses the DUT. The output of the second switch is downconverted and sampled in the signal analyser. The captured measurement signal is then transferred to the host PC for further processing.

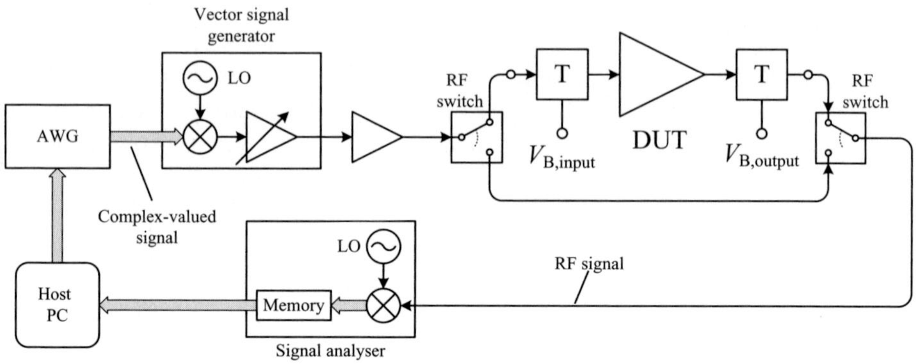

Figure 2.27 Typical BBACS, after [56]. This measurement setup was used to generate the reference data for the behavioural model comparison performed by Isaksson *et al.* [61].

A BBACS for the evaluation of the incident and reflected power waves at the input and the output of the amplifier was proposed by Macraigne *et al.* [58]. The scheme of this four-channel measurement setup is shown in Figure 2.28. An AWG produces a modulated IF signal, which is applied to the input of the vector modulator. This signal is then combined with the scaled and phase-shifted local oscillator signal to suppress the residual carrier at the vector modulator output. After LO rejection, the RF signal is boosted by a linear amplifier and is band-pass filtered to minimise the image signal. By using a step attenuator, the desired power level of the RF signal is set before the latter is applied to the input of the DUT. The load impedances at the DUT output can be changed by a tuner system. Samples of the incident and reflected waves at the input and the output of the DUT are taken by two directional couplers. These four signals are downconverted to an intermediate frequency (IF) and fed into a digital storage oscilloscope (DSO). A common trigger for the DSO and the AWG ensures synchronisation. The chosen design of the measurement system allows the replacement of the DSO by other signal-sampling devices without significant modification.

An overview of the different models that may be extracted by the use of time-domain measurement techniques is presented in Table 2.4. The second column of this compilation specifies the type of input signal that was generated by the BBACS. Again, models extracted by the use of the NVNA were not included.

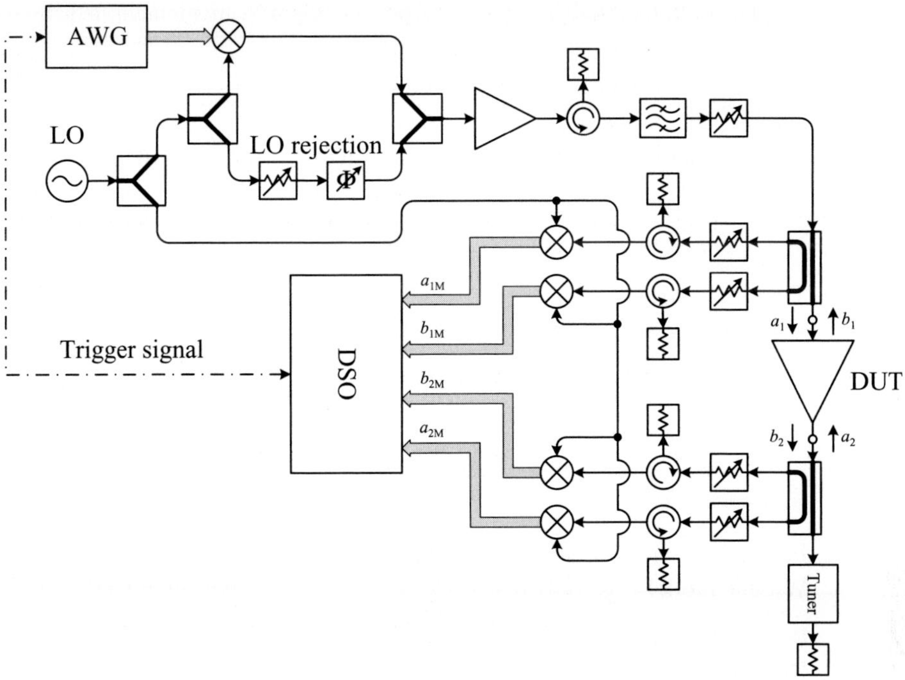

Figure 2.28 Four-channel BBACS as described in [58] for the band-pass response characterisation of microwave devices.

2.5.6 *Multisine amplifier characterisation*

Excitation design plays a significant role in the testing of power amplifier models for modern wireless communication systems. Increased complexity of the signals used in these applications involves broader bandwidths, higher carrier frequencies and more complicated baseband signal processing and modulation schemes. All this imposes stricter requirements on RF and microwave measurement equipment and makes the measurement and testing of RF or microwave PAs more challenging. However, a band-pass multisine, i.e. a periodic signal with a defined number of tones equally spaced in the frequency domain, can be easily measured and analysed with existing equipment such as the vector signal analyser (VSA) or large-signal network analyser [71]. Moreover, such multisine signals, when properly designed, can be considered as an approximation of realistic complex digitally modulated signals. However, the large number of possible multisine-parameter combinations leads to the problem of designing the optimal band-pass multisine signal [72]. Recent publications [73, 74] suggest using statistical properties of the multisine signal in the design process, rather than the range of instantaneous amplitudes of the signal to be approximated. The authors in [73] modified a method proposed in [74] to synthesise a multisine with a desired probability density function (PDF). Further modification of this procedure to shape the PDF of the multisine according to the true digitally modulated signals and their PDFs was presented in [75]. This procedure, together

Table 2.4 Models extracted from broadband time-domain measurements

Model type	Excitation signal	Source(s)	Section or subsection
Power series	IS-95 CDMA, BPSK	[38, 39, 64]	1.3.1, 3.4
Two-box: FIR + power series	bandlimited noise	[59]	1.3.2, 3.4, 4.2
Two-box: ARMA + Bessel series	OOK, BPSK	[41, 42]	1.3.2, 3.8, 4.2
Polyspectral	16-APK, BPSK, 16-QAM	[28, 30, 45]	1.3.3
Augmented behavioural characterisation	Pulsed-RF, CDMA	[62]	1.3.4
Parallel-cascade Wiener model	IS-95, bandlimited noise, CDMA	[31, 52, 59, 60]	1.3.4, 5.5
Parallel-cascade Hammerstein model	WCDMA, GSM	[61]	1.3.4
Memory polynomial	multitone, BPSK, IS-95, CDMA, QPSK	[53, 63–65]	1.3.2, 5.2
Volterra	WCDMA, GSM	[61, 66]	1.3.4, 5.6
Nonlinear impulse response	Step response, QPSK	[29]	1.3.4, 5.6.4
Neural networks	WCDMA, GSM, IS-95, CDMA2000, multitone, QPSK	[61, 67–70]	1.2.1, 5.3

with the most important factors relating to the use of a measured realistic digital signal, is explained later in this section.

Although the procedure takes into account the impact of systematic errors in the RF generator hardware on the amplitudes and phases of the desired band-pass multisine, it neglects any dynamic effects occurring between the baseband generator and the RF source output, e.g. the amplitude-dependent attenuation and phase shift of the signal envelope. A possible solution based on iterative vector predistortion of the designed multisine is described in [76]. It should also be noted that the authors in [77] later proposed that the multisine design procedure should rather focus on higher-order statistics, such as higher-order autocorrelations and power spectral density functions, rather than just on the PDF of the signal. However, this topic will not be covered in the present text.

Multisine design procedure

A complex band-pass multisine signal with N tones can be expressed as

$$x(t) = \sum_{m=0}^{N-1} A_m \exp\{j[2\pi(\mathrm{f}_0 + m\Delta\mathrm{f})t + \phi_m]\}, \tag{2.42}$$

where A_m and ϕ_m denote the amplitude and phase respectively of the multisine tones. The frequency separation $\Delta\mathrm{f}$ between the tones determines the multisine envelope period. These variables, together with the number of tones N, form the set of multisine parameters. The goal of this procedure is to find the set of these parameters that best approximates the PDF of the measured digital signal.

In this procedure, to generate and measure the digital signal we are using a VSG and a VSA respectively. The initial multisine is constructed from the measured digital signal by the sampling of a number of tones from the spectrum. These tones are selected to have a constant frequency spacing and cover most of the signal's bandwidth. Both the digital signal and the initial multisine need to be transformed to the time domain prior to the estimation of their PDF. Also, the amplitude of the time-domain multisine waveform is scaled in such a way that the peak time-domain voltage of the multisine equals that of the original digital signal.

The main idea of the procedure is to modify the PDF by substitution of the multisine time-domain amplitude values for those related to the measured digital signal. After this substitution the multisine is transformed back to the frequency domain and examined to ensure that the spectrum contains only the desired tones. In most cases the replacement of the time-domain amplitude values leads to some spurious tones in the modified multisine spectrum. These are eliminated from the spectrum by the setting of their magnitudes to a very low constant value. Finally, the multisine can be transformed back to the time domain and its PDF calculated. Depending on the result of a comparison with the digital signal's PDF these actions may be repeated, starting from the amplitude replacement step. To determine the acceptable error the convergence behaviour rather than the overall error is used. The PDF error values are plotted as a function of the iterations and the first iteration at which the error value stops decreasing significantly is found. A weighted squared error measure is used that accentuates amplitude values occurring with higher probability:

$$e = \frac{1}{N} \sum_i |(pdf_i - pdf_i')\, pdf_i|^2 \tag{2.43}$$

where pdf_i and pdf_i' signify the ith bin of the desired and optimised PDFs respectively and N represents the total number of PDF bins.

Keeping all the above in mind, the algorithm can be summarised as follows.

1. Generate the desired digital signal with a VSG and, using a VSA with optimal setup (see below), save the measured signal's frequency-domain representation to file.
2. Load the frequency spectrum into mathematical processing software.
3. Convert the band-pass signal to occupy the full spectrum and transform it to the time domain using an inverse fast Fourier transform (IFFT).
4. Calculate an estimate of its PDF in histogram form.
5. Create the initial multisine by sampling a number of tones from the spectrum to cover most of the signal's bandwidth.
6. Transform the multisine to the time domain and scale the amplitude so that the peak time-domain voltage of the multisine equals that of the original digital signal.
7. Sort the time-domain amplitudes of the digital signal into ascending order and keep track of the samples' time points (i.e. the location of the samples in the original signal).
8. Sort the multisine's time-domain amplitudes as in step 7 and replace the amplitudes of the sorted multisine signal with those of the sorted digital signal.
9. Transform the new multisine to the frequency domain.
10. Take only the frequency components specified in step 5.
11. Transform the resulting signal back to the time domain and scale the amplitude (as in step 6).
12. Calculate an estimate of the multisine's PDF and compare it with that of the original signal found in step 4, using Equation (2.43).

This procedure is iterated from step 8 until the required error is attained.

Several points related to the measurement-based character of this multisine design procedure are crucial for successful results. They will be discussed using, as an example, an unframed 16.6 MHz OFDM QPSK signal containing only '0' bits at the rate of 12 MB/s, generated at a centre frequency of 4.95 GHz. Figure 2.29 shows the PDF of this signal (black line) and the corresponding initial and optimised 54 tone multisines.

Multisine parameters

The most crucial question for a successful implementation of this procedure is the selection of the number and frequency positions of the multisine. These choices are not straightforward and depend on the properties of the digital signal under consideration and a trade-off between accuracy and complexity. For the best performance, the number of multisine tones and their frequency separation should correspond to the digital signal's frequency characteristics, such as the data rate or subcarrier spacing; more closely spaced tones do not always provide a better approximation of the PDF. In particular, it is observed that a good correspondence between the tone spacing of the multisine and the digital-signal frequency characteristics is more critical for accurate PDF approximation than a large number of sines.

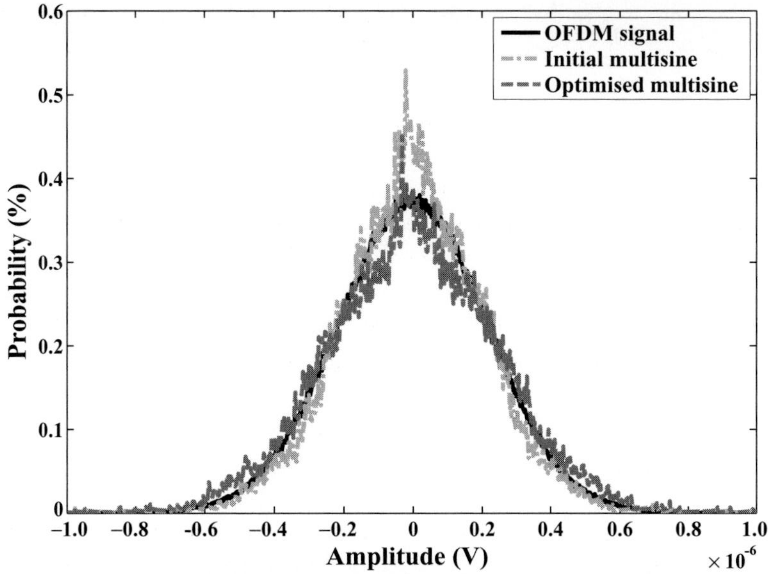

Figure 2.29 The PDF of an OFDM-modulated digital signal (black line) and 54 tone multisines before and after optimisation. (Reprinted with permission from [75], © 2006 IEEE.)

In the case considered, the lowest error was observed when the frequency separation between the multisine tones was a multiple of the digital signal's subcarrier spacing (0.3125 MHz for this reference signal). This condition, together with the bandwidth, defines a set of optimal multisine frequency characteristics. The PDF error and the frequency separation printed as a function of the number of tones are depicted in Figures 2.30 and 2.31 for the example under consideration. It can be seen that the lowest error is obtained for 54 multisines. This corresponds to a 0.3125 MHz subcarrier frequency separation (Figure 2.31).

VSA settings

As mentioned previously, the parameters of the multisine signal depend on the carrier frequency and bandwidth and the frequency characteristics of the measured digital signal. In order to facilitate fulfilling the above requirements, the measurement setup, in particular the VSA, needs to be properly adjusted. In the case considered the VSA's frequency resolution was set to a multiple of the digital signal's subcarrier spacing; it should also be high enough to provide a large number of samples across the signal's frequency bandwidth. In consequence the VSA's resolution bandwidth was set to 250 Hz and the acquired bandwidth to 32.768 MHz.

In the case of a carrier frequency higher than the VSA's specification, a tunable receiver has to be included in the measurement setup to downconvert the VSG signal to a lower IF. As a result, the spectrum of the digital signal may be measured at,

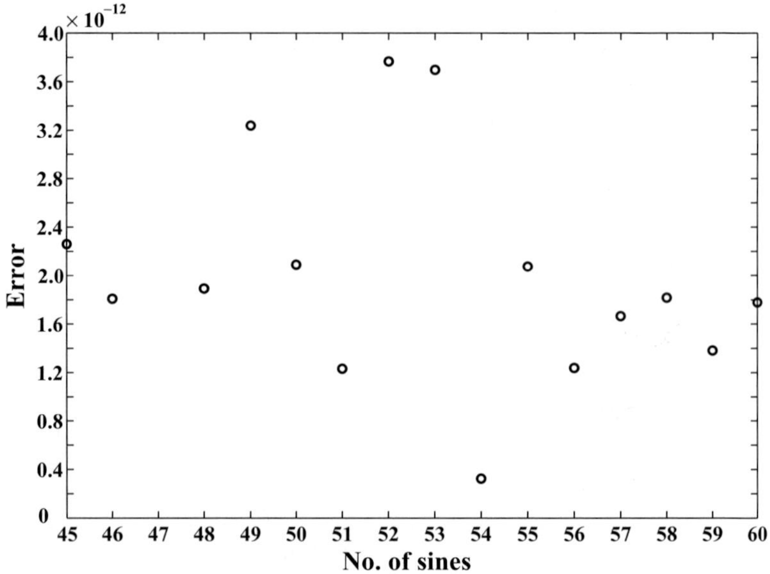

Figure 2.30 The error (2.43) as a function of the number of tones (only one iteration was made per multisine). (Reprinted with permission from [75], © 2006 IEEE.)

say, 1 GHz instead of 4.95 GHz.

Frequency-domain–time-domain transformations

In order to correctly transform the frequency-domain data to a time-domain waveform, all frequency components from DC up to the signal's maximum frequency are necessary. Additionally, a mirror spectrum at negative frequencies is required. The measured digital signal occupies only a narrow band around the carrier frequency; therefore standard IFFT implementations cannot be used directly on the spectrum data. The vector containing the frequency spectrum has to be modified to convert it from a band-pass signal to a full spectrum signal at the baseband in order to apply the inverse Fourier transform successfully and calculate the corresponding time-domain waveform.

PDF estimation

Another important question is related to the PDF estimation. In general, probability density estimation methods are divided into three classes: parametric, nonparametric and semiparametric [78]. Parametric estimation techniques make use of *a priori* knowledge of the shape of the PDF. In this case, parameters are fitted to the data. In contrast, in a nonparametric approach the shape of the PDF is derived from the data only. Semiparametric methods are a combination of the two previous techniques.

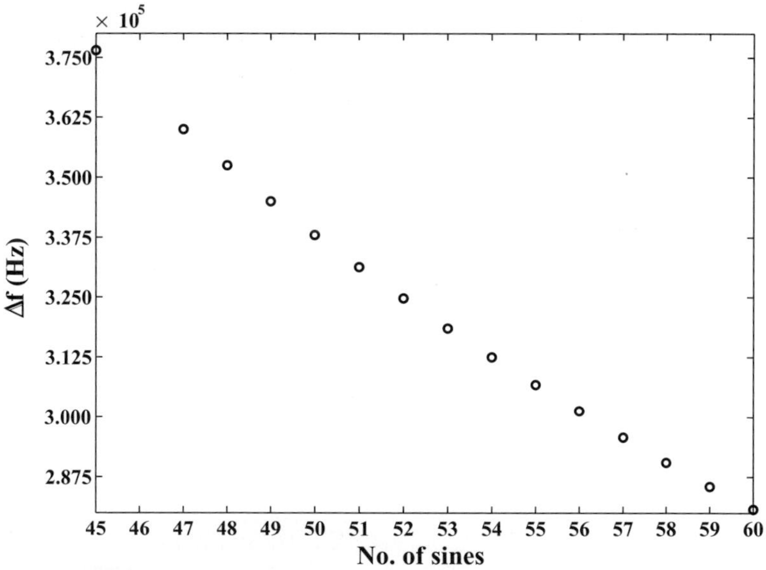

Figure 2.31 Frequency separation between multisine tones as a function of the number of tones. (Reprinted with permission from [75], © 2006 IEEE.)

In the case of an unknown PDF, nonparametric methods generally give good estimation results. One of the best known and basic nonparametric probability density estimation techniques is the histogram [79]. Especially when smoother estimators are required, kernel-based techniques are an alternative nonparametric method.

A histogram is composed of a set of bins formed by dividing the known range of a signal's sample values into uniform intervals. Subsequently all considered samples are assigned to the bins, depending on their value, and are counted. Scaling the samples counted in each bin by the total number of samples results in an estimate of the PDF.

One key question when working with histograms is the choice of the number of bins [79, 80]. Figure 2.32 shows four probability density estimates of the same digital signal. In each case a histogram with a different number of bins was constructed. It can be seen that increasing the number of bins from 100 through 1000 and 10 000 up to 100 000 leads to a proportional decrease in the probabilities and the smoothness of the histogram. However, the overall shape stays generally unchanged regardless of the number of bins. Increasing the number of bins means that each bin has a decreased width, and thus smaller differences in the samples' values can be distinguished. This is equivalent to adding bits to an A–D converter. Samples previously assigned to one big bin are now split between adjacent bins and the number per bin decreases (giving a lower probability). This increased accuracy in the histogram comes at the price of computational complexity and a histogram that is more 'noisy', i.e. less smooth, as mentioned above and as depicted in the bottom plot in Figure 2.32.

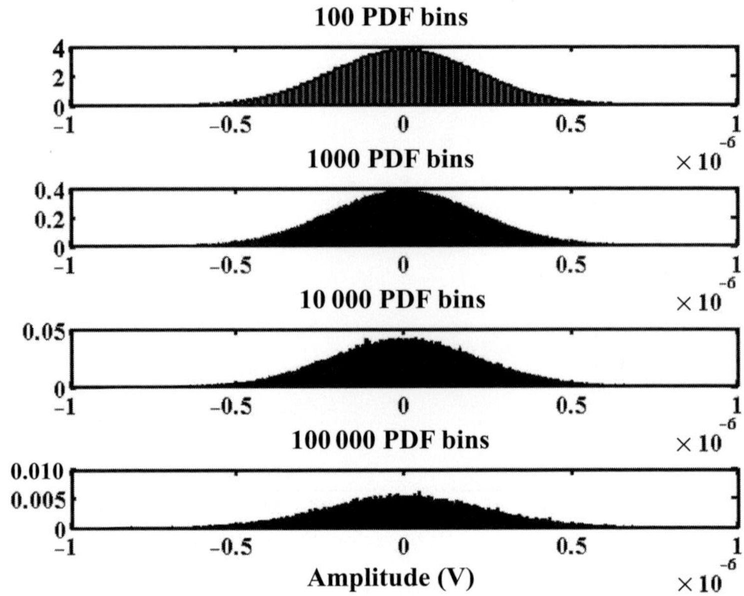

Figure 2.32 Histogram estimate of the % PDF of an OFDM digital signal calculated for different numbers of bins. (From [75] with permission, © 2006 IEEE.)

Using the above PDF estimates in the multisine generation procedure, the influence of the number of bins can be investigated. It is difficult to draw any conclusions just by examining the PDF plots of the signal and the optimised multisine. However, a logarithmic decrease in the error value (2.43) with an increase in the number of PDF bins can be observed in Figure 2.33. As seen in Figure 2.30, the absolute level of the error appears to be very small. This is mainly due to the weighting component pdf_i and the squared operator (L_2 norm) in Equation (2.43).

Usually, the decision about the optimum number of bins in a histogram is made on the basis of a compromise between a smooth PDF trace, an acceptable level of error for the application and the use of computational resources during the iterative calculations. For the present example 1000 PDF bins were chosen and so the whole amplitude range was divided into 1000 equally spaced levels. However, the ideal number of PDF bins can change with the parameters and type of digital signal.

Multisine measurements in a large-signal network analyser setup

A multisine, being a periodic modulated signal, is perfectly suited for use with an LSNA setup. An LSNA is a four-channel system that simultaneously measures the absolute values of the travelling incident and scattered voltage waves observed at the terminals of the two-port device under test. It provides accurate amplitude and phase values of all spectral components, including harmonics and intermodulation

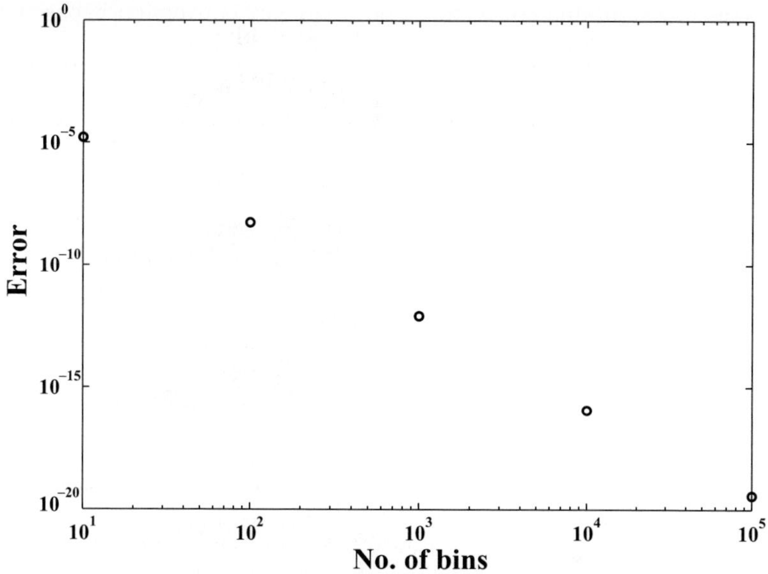

Figure 2.33 The error given in Equation (2.43) as a function of the number of PDF bins (only one iteration per histogram). (From [75] with permission, © 2006 IEEE.)

distortion products, and, for the convenience of the user, these may be expressed either as travelling waves or as current or voltage waveforms and are available in both the time and frequency domains [81]. Moreover, in order to facilitate the work with modulated signals, the corrected measured data can also be expressed in a mixed time and frequency domain (an envelope domain) as a set of time-varying complex phasors. As a result, time-dependent in-phase quadrature (IQ) traces corresponding to the narrow modulation-frequency bands around the RF fundamental frequency, and harmonic components up to 20 GHz (or 50 GHz, depending on the equipment), are available [82].

In general, depending on the applied downconversion technique, two types of LSNA measurement setup exist: sampler- and mixer-based. The major building blocks of the sampler-based LSNA system for PA characterisation are shown in Figure 2.34. Here the measurement of the four waves is performed in the time domain through a synchronous sampling with a carefully chosen sampling frequency, in a similar way to that in a DSO. Because all the frequency components in the bandwidth of interest are acquired at the same time, there is no need for a simultaneous measurement of the phase reference signal. Instead, it can be measured during the calibration procedure [82]. A modulated signal generated by an RF source passes through the LSNA RF signal path, including the attached DUT, and is terminated by a matched load. The incident waves a and reflected waves b at Ports 1 and 2 of the DUT are sensed by directional couplers placed close to the ports. The four signals, with their power levels adjusted by stepped attenuators, are simultaneously downconverted to an IF band (typically 10 MHz) by a four-channel frequency-sampling converter. The frequency conversion is based on the harmonic-sampling principle, in which different frequency components of the RF signal are mixed with different harmonics of a carefully chosen local oscillator (LO) frequency that controls the rate of a narrow-pulse generator. As a result, all frequency components of each of the four RF waves are 'squeezed' into a narrow 10 MHz IF band and, after filtering and digitisation, are sent to a PC for data correction and reconstruction. It should be noted that the procedure required to properly adjust the samplers and accurately recover the measured RF modulated signals is non-trivial, as described by Verspecht in [84].

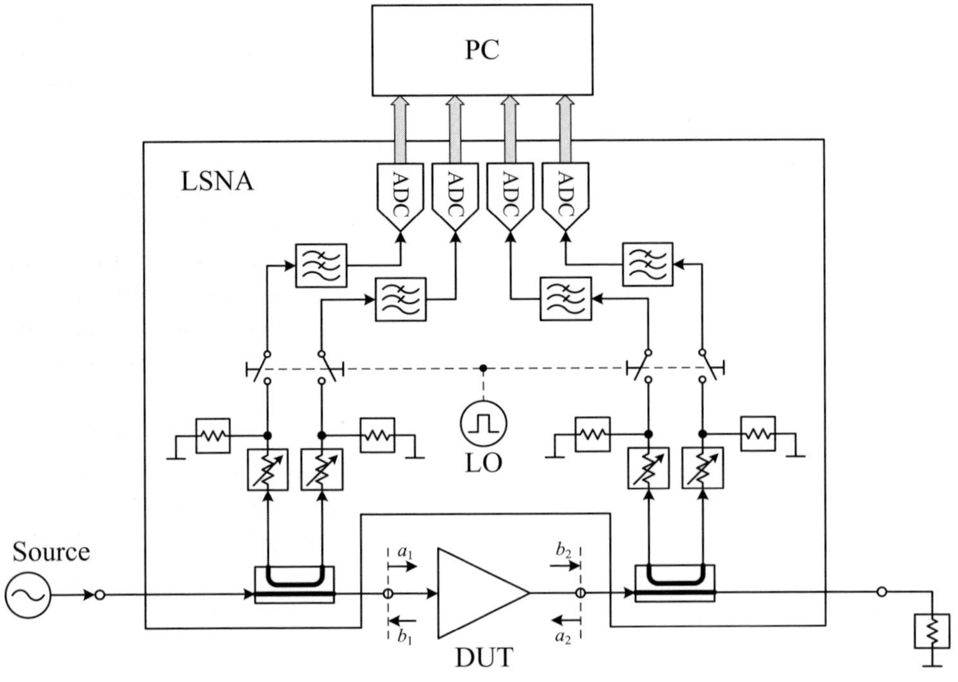

Figure 2.34 Block diagram of the sampler-based LSNA system.

The basic LSNA-based PA measurement setup from Figure 2.34 can be easily extended to perform more involved characterisation tests. For example, by inserting a load impedance tuner or by adding another RF source at the output, non-50 Ω or load-pull large-signal measurements can be performed.

The block diagram of the mixer-based LSNA is presented in Figure 2.35. Here the incident and the reflected waves are measured in the frequency domain using four separate mixers. These mixers downcovert each frequency component of the waves tone by tone to the IF band, in the same way as a spectrum analyser. At the same time a phase reference signal, provided by the comb generator, is measured in the separate channel, allowing the recovery of the relative phase relationships of the measured tones. A detailed discussion of the mixer-based LSNA is given in [83].

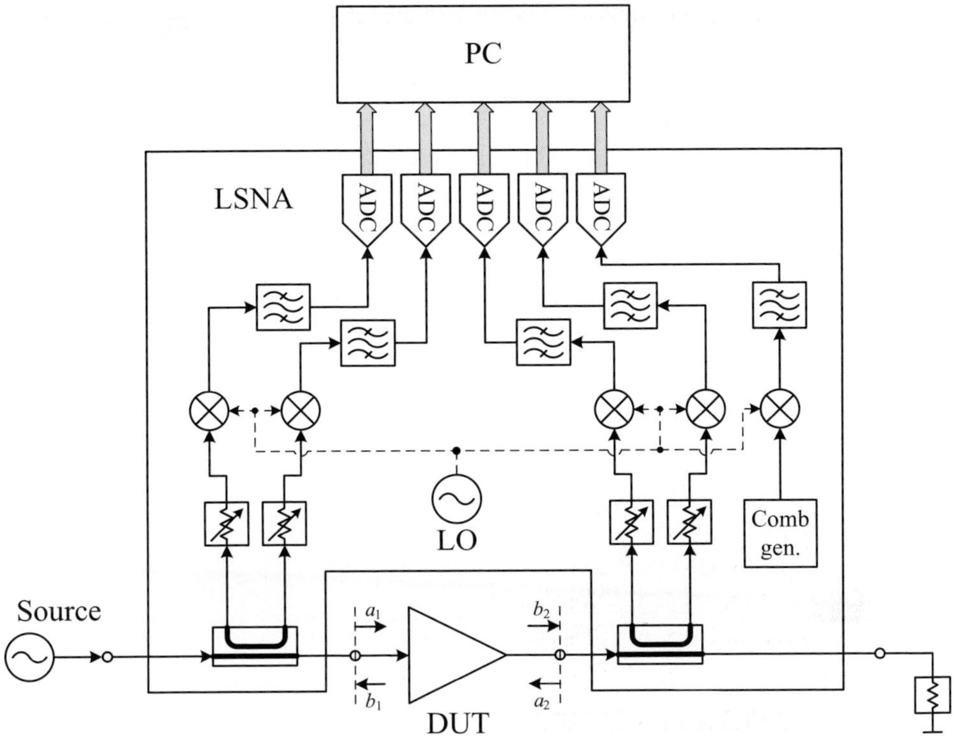

Figure 2.35 Main blocks of the mixer-based LSNA system in a basic PA measurement setup.

Table 2.5 presents an overview of the different models that may be extracted by the use of an LSNA. The second column of this compilation specifies the type of input signal.

References

[1] T. Söderström and P. Stoica, *System Identification,* Prentice Hall, 1989.

[2] H. Baher, *Analog & Digital Signal Processing,* John Wiley & Sons, 2001.

[3] J. Wood and D. Root, *Fundamentals of Nonlinear Behavioral Modeling for RF and Microwave Design,* Artech House, 2005.

[4] M. Schetzen, *The Volterra and Wiener Theories of Nonlinear Systems,* John Wiley & Sons, 1980.

[5] J. C. Pedro and N. B. Carvalho, *Intermodulation Distortion in Microwave and Wireless Circuits,* Artech House, 2003.

[6] E. A. Lee and D. G. Messerschmitt, *Digital Communication,* second edition, Kluwer/Plenum, 1994.

[7] M. C. Jeruchim, P. Balaban and K. S. Shanmugan, *Simulation of Communication Systems, Modeling, Methodology and Techniques,* second edition, Kluwer/Plenum, 2000.

Table 2.5 Models extracted from LSNA measurements.

Model type	Excitation signal	Source(s)	Section or subsection
VIOMAP	a single tone on each port	[85]	1.4.2
Describing functions	large signal + small signal tone	[86, 87]	1.4.2
Polyharmonic distortion	large signal + small signal tone	[88, 89]	1.4.2
ANN	single-tone	[90]	1.4.2, 5.3
ANN + state space	single-tone, multisine	[91, 92]	1.4.2, 5.3, 5.7
Dynamic Volterra series expansion	large signal + small signal tone	[47, 46]	1.3.4, 5.6.3

[8] A. R. Kaye, K. A. George and M. J. Eric, "Analysis and compensation of bandpass nonlinearities for communications," *IEEE Trans. Communications*, vol. 20, pp. 965–972, October 1972.
[9] S. Benedetto and E. Biglieri, *Principles of Digital Transmission, with Wireless Applications,* Kluwer/Plenum, 1999.
[10] G. Tong Zhou, Hua Qian, Lei Ding and Raviv Raich, "On the baseband representation of a bandpass nonlinearity," *IEEE Trans. Signal Processing*, vol. 53, no. 8, pp. 2953–2957, August 2005.
[11] S. Benedetto, E. Biglieri and R. Daffara, "Modeling and performance evaluation of nonlinear satellite links – a Volterra series approach," *IEEE Trans. Aerospace and Electronic Systems*, vol. 15, no. 4, pp. 2843–2849, July 1979.
[12] N. M. Blachman, "Band-pass nonlinearities," *IEEE Trans. Information Theory*, vol. 10, no. 2, pp. 162–164, April 1964.
[13] N. M. Blachman, "Detectors, bandpass nonlinearities, and their optimization: inversion of the Chebyshev transform," *IEEE Trans. Information Theory*, vol. 17, no. 4, pp. 398–404, July 1971.
[14] N. B. Carvalho and J. C. Pedro, "A comprehensive explanation of distortion sideband asymmetries," *IEEE Trans. Microw. Theory Tech.*, vol. 50, no. 9, pp. 2090–2101, September 2002.
[15] J. Brinkhoff and A. E. Parker, "Effect of baseband impedance on FET intermodulation," *IEEE Trans. Microw. Theory Tech.*, vol. 51, no. 3, pp. 1045–1051, March 2003.
[16] S. A. Maas, *Nonlinear Microwave and RF Circuits*, second edition, Artech House, 2003.
[17] J. M. Golio, *Microwave MESFETs and HEMTs*, Artech House, 1991.
[18] J. Brinkhoff and A. E. Parker, "Charge trapping and intermodulation in HEMTs," in *IEEE MTT-S Int. Microwave Symp. Dig.*, June 2004, pp. 799–802.
[19] A. E. Parker and J. G. Rathmell, "Contribution of self heating to intermodulation in FETs," in *IEEE MTT-S Int. Microwave Symp. Dig.*, June 2004, pp. 803–806.

[20] J. C. Pedro, N. B. Carvalho and P. M. Lavrador, "Modeling nonlinear behavior of band-pass memoryless and dynamic systems," in *IEEE MTT-S Int. Microwave Symp. Dig.*, June 2003, pp. 2133–2136.

[21] A. E. Parker and J. G. Rathmell, "Broad-band characterization of FET self-heating," *IEEE Trans. Microw. Theory Tech.*, vol. 53, no. 7, pp. 2424–2429, July 2005.

[22] N. L. Gallou, J. M. Nebus, E. Ngoya and H. Buret, "Analysis of low frequency memory and influence on solid state HPA intermodulation characteristics," in *IEEE MTT-S Int. Microwave Symp. Dig.*, May 2001, pp. 979–982.

[23] S. Boumaiza and F. M. Ghannouchi, "Thermal memory effects modeling and compensation in RF power amplifiers and predistortion linearizers," *IEEE Trans. Microw. Theory Tech.*, vol. 51, no. 12, pp. 2427–2433, December 2003.

[24] W. Bösch and G. Gatti, "Measurement and simulation of memory effects in predistortion linearizers," *IEEE Trans. Microw. Theory Tech.*, vol. 37, no. 12, pp. 1885–1890, December 1989.

[25] S. C. Cripps, *RF Power Amplifiers for Wireless Communications,* Artech House, 1999.

[26] P. B. Kennigton, *High-Linearity RF Amplifier Design,* Artech House, 2002.

[27] R. K. Pearson, "Selecting nonlinear model structures for computer control," *J. Process Control*, vol. 13, no. 1, pp. 1–26, February 2003.

[28] C. Silva, A. A. Moulthrop and M. Muha, "Introduction to polyspectral modeling and compensation techniques for wideband communications systems," in *ARFTG Conf. Dig.*, November 2001, pp. 1–15.

[29] A. Soury, E. Ngoya, J. M. Nébus and T. Reveyrand, "Measurement based modeling of power amplifiers for reliable design of modern communication systems," in *IEEE MTT-S Int. Microwave Symp. Dig.*, June 2003, pp. 795–798.

[30] C. Silva, "Time-domain measurement and modeling techniques for wideband communication components and systems," *Int. J. RF and Microwave CAE*, vol. 13, no. 1, pp. 5–31, January 2003.

[31] H. Ku, M. D. Mckinley and J. S. Kenney, "Quantifying memory effects in RF power amplifiers," *IEEE Trans. Microw. Theory Tech.*, vol. 50, no. 12, pp. 2843–2849, December 2002.

[32] J. Pedro, J. Madaleno and J. Garcia, "Theoretical basis for the extraction of mildly nonlinear behavioral models," *Int. J. RF and Microwave CAE*, vol. 13, pp. 40–53, January 2003.

[33] L. Ljung, *System Identification, Theory for the User,* second edition, Prentice Hall, 1999.

[34] ETSI, "Fixed radio systems; point-to-multipoint equipment; time division multiple access (TDMA); point-to-multipoint digital radio systems in frequency bands in the range 3 GHz to 11 GHz," European Telecommunication Standards Institute, EN 301 021 V1.6.1, July 2003.

[35] M. Golio, *The RF and Microwave Handbook,* CRC Press, 2001.

[36] D. E. Root, J. Wood, N. Tufillaro, D. M. Schreurs and A. Pekker, "Systematic behavioral modeling of nonlinear microwave/RF circuits in the time domain using techniques from nonlinear dynamical systems," in *Proc. IEEE Int. Workshop on Behavioral Modeling and Simulation*, October 2002, pp. 71–74.

[37] U. Lott, "Measurement of magnitude and phase of harmonics generated in nonlinear microwave two-ports," *IEEE Trans. Microw. Theory Tech.*, vol. 37, no. 10, pp. 1506–1511, October 1989.

[38] H. Gutierrez, K. Gard and M. B. Steer "Nonlinear gain compression in microwave amplifiers using generalized power-series analysis and transformation of input

3 Memoryless nonlinear models

3.1 Introduction

In this chapter we focus on memoryless nonlinear power amplifier behavioural models for use in transmitter or full communications-channel system simulations. Static envelope characteristics, i.e. static AM–AM and AM–PM characteristics, are the basis for defining a behavioural model as memoryless.

Memoryless behavioural modelling has been used for many years because of its generally easier computational implementation, its relative efficiency in system simulations and its acceptable level of accuracy in many situations. Historically, this approach has found particular favour in predicting intermodulation distortion problems in the robust multicarrier travelling-wave tube amplifiers (TWTAs) used in communication satellite transponders. Travelling-wave tube amplifiers are inherently wideband devices, with a bandwidth of e.g. 500 MHz, and have a near-constant group delay. Any memory-like effects will be those introduced by the channelising demultiplexers on the input, which mainly result in multipath and crosstalk effects, and by the zonal filters that are used on the output to remove especially harmonics but also out-of-band intermodulation products (IMPs). Generally these amplifiers operate under stable temperature conditions, even if the satellites themselves are subject to large temperature variations. The simultaneous multiple air-interface multicarrier configuration of many of these transponder amplifiers is complex by comparison with the single-carrier solid-state power amplifiers (SSPAs) encountered in mobile handsets and the like. In recent years, however, the use of air interfaces based on multicarrier modulation schemes (e.g. orthogonal frequency-division multiplex, OFDM) has grown because of their excellent spectral efficiency and reduced susceptibility to multipath fading. But the simultaneous common nonlinear power amplification of multiple air-interfaces, the parallel of the multicarrier satellite transponder, is not yet normal in SSPA applications.

Intrinsically, SSPAs have relatively narrower bandwidths and display varying degrees of short- and long-term memory effects. Some of these are linear, usually being associated with the input and output matching circuits, and others are nonlinear, occurring simply or in complex cascade within the amplifier itself. These effects are manifesting themselves more as PA technology is driven to ever higher powers, higher frequencies and wider bandwidths.

Developing general modelling approaches to cater for the presence of any or all of these memory effects has been dealt with in Chapter 1 and discussed further in Chapter 2. In later chapters we will consider models having memory effects in more detail. However, memoryless nonlinear models have the advantage of being more easily extendable for modelling strong or higher-order nonlinearities. This is in contrast with the cautionary note often expressed when using (truncated) Volterra-series-based models, as mentioned in subsection 2.4.3.

An acceptable accuracy of memoryless nonlinear models is obviously achievable if the actual PA manifests few or no memory effects. Also, in situations with memory, the dominant distortion effects are still those arising from the memoryless or equivalent memoryless PA characteristics, [1], and so system in-band and out-of-band performance estimations using memoryless models are still useful. An example of where it would, however, be problematic to use such models is in adaptive linearisation algorithms and schemes applied to amplifiers manifesting memory effects. Here the approximation shortcomings of memoryless models could give rise to instabilities and even cause the linearisation scheme to be counterproductive. However, given their ability to track the PA's dominant memoryless nonlinearity effects, these models find good use in engineering design and analysis, especially in regard to complex multicarrier systems and in first-order efforts to reduce or control aspects of this dominant distortion.

There is little doubt about the usefulness of memoryless nonlinearity modelling when care is exercised in respect of the application, with awareness of shortcomings and the significance of inaccuracies. Equally, care should be exercised in model extraction and identification for real amplifiers. This should be done through behavioural observations, ensuring the conditions of measurements used for model extraction, e.g. of the RF band and bandwidth, are those, or close to those, used in the engineering design application of targeted simulation. It can be argued that excitation signals for the extraction of memoryless behavioural models need to be selected with a view to the planned application of the model, i.e. they need to be similar to the signals in the planned application; however, as will be seen later, for PAs which are fully memoryless this does not have to be the case. Rather, single unmodulated carrier PA responses are sufficient to extract a model that can then be used effectively for the behavioural analysis of this memoryless PA subjected to quite complex input signals. However, it is good practice to examine the swept-frequency response to one or, better, two tones to establish the band limits of the memoryless property. Extracting equivalent memoryless envelope characteristics from complex signal responses is also possible whether the PA exhibits some memory or not. This is our procedure for the laterally diffused metal oxide semiconductor (LDMOS) PA used for the model comparisons in this chapter; see Section 3.3.

In high-power SSPAs memory effects are frequently observed, typically when amplifying wideband signals in which the bandwidth of the signal is comparable with the inherent bandwidth of the amplifier. The range of known memory effects may be classified as linear or nonlinear, or as short- or long-term. Behavioural modelling solutions that take account of both nonlinearities and memory effect phenomena are the themes of Chapter 4 (on linear memory) and Chapter 5 (on nonlinear memory). Where memory effects are only linear and can be transposed for modelling purposes to PA input and/or output filters, the PA model then reverts to a memoryless modelling problem.

The context for memoryless behavioural models is quite often within system-level behavioural modelling, with a view to large-system simulation; see Chapter 7. As the thrust of previous chapters indicates, the development of accurate behavioural-level models of PAs is of great assistance to the design and performance prediction of the whole RF transmitter system. This is true also for memoryless nonlinearity models although there may be an accuracy compromise in constraining the model to be memoryless. However, behavioural performance prediction using 'black-box' or response-based behavioural models of nonlinear systems will have inherently unknown aspects, simply because the physical and technological attributes of PAs are omitted in the model.

As is intuitively obvious, the closer the model is to the real device's hardware functionalities, here the PA's physical circuit, the more likely it is that the model will be robust and accurate in approximating the device's behavioural activity under all kinds of excitation and environmental conditions. This is an important goal of circuit-level models of PAs. They are conceived to simulate the typical behaviour under any input-signal conditions, with a view to the good design of biasing, input and output matching networks and so on. Seeking ever better circuit models and adjusting them to the particulars of a specific PA device or class of devices is an important ongoing research activity. Circuit-level behavioural models, however, are normally not suitable as simulation blocks for embedding within higher-level full-system simulations. Rather, they are used in a stand-alone approach to serve stated goals. Intrinsic to circuit-level models is their maintenance of frequency integrity, i.e. signals must be represented as they are in the real system. This is a major reason for not using them in system-level behavioural simulations since, for example, a 10 MHz bandwidth signal in a 5 GHz RF band will require, following the Nyquist theorem, a minimum of 10^7 samples to represent 1 ms of the signal; this corresponds to the sampling frequency f_{s1} in Figure 3.1. Apart from the inherent complexity of circuit-level PA device modelling, this high level of 'upsampling' of the envelope of the RF signal can significantly degrade the system-level simulation efficiency.

System simulations normally use non-frequency-specific band-equivalent models employing 'analytic' signal representation, and in many cases these reduce to baseband-equivalent models. This latter term does not refer to models involved in the processing of baseband signals before they are modulated. Rather, through the use of complex analytical signal representation where the positive frequency components only are present, 'complex-envelope' or 'complex-baseband-equivalent' signal representation means the positive-frequency-axis envelope of the RF signal. This is illustrated by $F\{\tilde{x}(t)\}_{@f_0}$ in Figure 3.1, where $F\{\cdot\}$ denotes the Fourier transform and $\tilde{x}(t)$ is defined in Equation (3.2) below. The RF envelope integrity is maintained, whether in the frequency or time domain, but its actual RF location f_0 is kept as a separate parameter. In this sense, the RF effects associated with the value of f_0 are removed from envelope-based system behavioural models. The value of f_0 may be set arbitrarily. In single-carrier systems it may be the carrier frequency, but for complex multicarrier systems (e.g. OFDM, FDMA) or simple narrowband or wideband noise it may be a frequency at the centre of the RF band or a notional or arbitrary frequency.

A system-level memoryless instantaneous behavioural model of an RF nonlinear PA can be described mathematically by a static nonlinear relation between the input and output signals, i.e. the instantaneous output signal $y(t)$ is related to the corresponding instantaneous input $x(t)$ by

$$y(t) = G_{\mathrm{RF}}\left(x(t)\right) \tag{3.1}$$

where G_{RF} is a complex RF memoryless nonlinear function, which is measured for a particular frequency and is regarded as static over time and unchanging over a band that is wider than the input signal of the targeted application. The nonlinear memoryless model is extracted from the measurements of G_{RF}, and the useful baseband-equivalent models are derivable from it. The RF input and output may be represented, in narrowband signal form, as centred around an arbitrary carrier or centre frequency f_0 (angular frequency ω_0):

$$x(t) = \mathrm{Re}\left[\tilde{x}(t)e^{j\omega_0 t}\right] \quad \text{and} \quad y(t) = \mathrm{Re}\left[\tilde{y}(t)e^{j\omega_0 t}\right], \tag{3.2}$$

where the signals inside the square brackets are the analytical signal representations of the input and output, and $\tilde{x}(t)$ and $\tilde{y}(t)$ are the complex-envelope input and output signals, which are variously referred to as complex baseband equivalent, complex low-pass equivalent etc. This is illustrated in Figure 3.1.

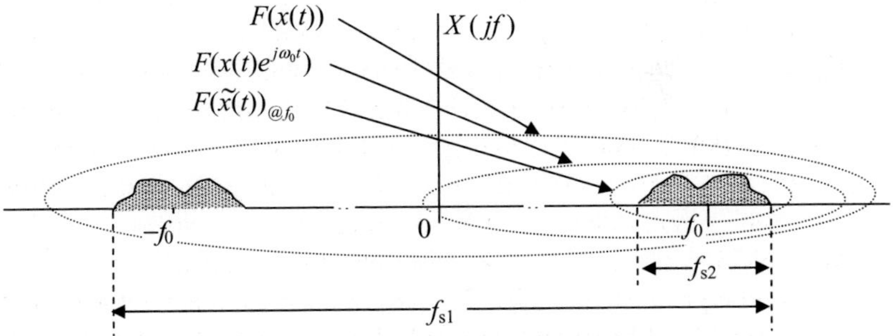

Figure 3.1 Spectra of the real signal $x(t)$, the analytical signal $\tilde{x}(t)e^{j\omega_0 t}$ and the complex-envelope signal $\tilde{x}(t)$.

To obtain a sampled version of this complex envelope (which is a single-side envelope spectrum in the frequency domain), the minimum sampling rate for the avoidance of aliasing is the bandwidth of this envelope, i.e. f_{s2} in Figure 3.1. In the example mentioned above in connection with f_{s1}, 10^4 samples will represent 1 ms of the identical signal in its baseband-equivalent form, a reduction factor of 10^3.

In regard to the memoryless model defined by Equation (3.1), the time instant on the right-hand side of the equation may be later than that on the left; the difference is equivalent to a group delay that is perfectly flat across the whole frequency band. In real systems there will be a finite group delay, inclusive of the propagation delay, and in many cases it would not be perfectly flat and thus represents a memory effect, which is large or small depending on the contributing factors. However, a perfectly flat group delay has no distortion effect in itself and so, in modelling, the delay can be set to zero, implying that the output at any instant is dependent only on the input at that same instant.

Taking the memoryless modelling approach, the baseband-envelope-equivalent memoryless models may be derived directly from Equation (3.1), e.g. along lines that will be set out in Section 3.2 and in the following sections. These models, with the memoryless assumption applied, are in harmony with the more general models introduced in Chapter 1.

The rest of this chapter is organised as follows. In Section 3.2, the most popular models for memoryless nonlinear systems will be presented and investigated:

- complex power series expansions, Section 3.4;
- Saleh models, Section 3.5;
- modified Saleh models, Section 3.6;
- Fourier series model, Section 3.7;
- Bessel–Fourier models, Section 3.8;
- the Hetrakul and Taylor model, Section 3.9;
- the Berman and Mahle model, Section 3.10.

This is followed, in Section 3.11, by a description of the approach proposed by Wiener to extract the memoryless polynomial model of a system. It is shown how this approach, once customised for input signals of a given statistical distribution, can give a better performance than a Taylor expansion. A neural network approach could also be presented at this point, but in practice such an approach has found little popularity in memoryless PA behavioural modelling. Such a model would be a special case of the more general neural network PA models, which include dynamic memory effects. These are important and have received much research attention; in this book they are considered in detail in Chapter 5.

3.2 Overview of memoryless behavioural models

Among the popular memoryless models presented in this section are the complex power series, Saleh and Bessel–Fourier series models. While generally models focus on envelope transfer characteristics, the last-mentioned model originates from a Fourier series model of the RF instantaneous transfer function. A new, modified, Saleh model which overcomes some shortcomings of the original Saleh model, particularly its limitations in approximating the AM–PM characteristics, is also presented. Use is made of a model comparison framework based on an LDMOS PA amplifying a wideband CDMA signal (WCDMA). For this, the older Hetrakul and Taylor and Berman and Mahle models are also included. There is a further discussion on model comparison at the end of the chapter.

As discussed above, if the memory duration of an RF PA is close to the period of the RF carrier and the input is a narrowband signal, or if the envelope characteristics of the amplifier are invariant over the band of interest, then the amplification process can be regarded as memoryless. Hence, the instantaneous output signal may be written as depending only on the corresponding instantaneous input, as in Equation (3.1), with no dependence on earlier or later inputs.

A general narrowband input may be written as

$$x(t) = r(t)\cos[\omega_0 t + \phi(t)] = \mathrm{Re}\left[\tilde{x}(t)e^{j\omega_0 t}\right], \tag{3.3}$$

where $r(t)$ and $\phi(t)$ are the general envelope amplitude and phase components of the input signal respectively, and where the complex baseband input $\tilde{x}(t)$ is

$$\tilde{x}(t) = r(t)e^{j\phi(t)} \tag{3.4}$$

and could contain quite complex modulation signals and noise within a band which is narrow relative to the value of the RF central frequency f_0.

The PA output, as yielded by Equation (3.1), will be composed of the amplified desired signal plus harmonics and IMPs. In the case of multicarrier input signals, e.g. OFDM signals, IMPs are readily identified and understood; in the case of complex-envelope or generic non-constant-envelope modulated (NOCEM) signals, engineers sometimes speak of spectral regrowth and of spurious output rather than IMP generation. The design goal of transmitter systems is to pass, without attenuation or further distortion, to an antenna the part $y_z(t)$ of the PA output signal that is located in the same band as the PA input. This is often called the zonal band and may be extracted from the total output and represented as

$$y_z(t) = g(r(t), \mathrm{f}_0)\cos[\omega_0 t + \phi(t) + \Phi(r(t), \mathrm{f}_0)] \tag{3.5}$$

where the nonlinear envelope characteristics are $g(r(t), \mathrm{f}_0)$ and $\Phi(r(t), \mathrm{f}_0)$ as seen at or around f_0. The function g represents a nonlinear amplitude-to-amplitude (AM–AM) modulation conversion and Φ represents a nonlinear amplitude-to-phase (AM–PM) modulation conversion. For the remainder of this chapter, assuming a memoryless amplifier, the band about f_0 for which these envelope characteristics remain unchanged is assumed to be quite broad, much broader than the likely input and output signal bandwidths. Then we may drop the functional dependence on f_0. Doing this, and also for added clarity omitting explicit indication of the temporal dependence of the input envelope, i.e. writing $r(t)$ and $\phi(t)$ as r and ϕ,

Equation (3.5) may be written in various forms as follows:

$$\begin{aligned} y_z(t) &= g(r)\cos[\omega_0 t + \phi + \Phi(r)] \\ &= g(r)\,\cos\,\Phi(r)\cos(\omega_0 t + \phi) - g(r)\sin\Phi(r)\sin(\omega_0 t + \phi) \\ &= P(r)\,\cos(\omega_0 t + \phi) - Q(r)\,\sin(\omega_0 t + \phi) \\ &= p(t) + q(t) \\ &= [P(r)\,\cos\phi - Q(r)\sin\phi]\cos\omega_0 t - [\mathrm{P}(r)\sin\phi - \mathrm{Q}(r)\cos\phi]\sin\omega_0 t \end{aligned}$$

or

$$\begin{aligned} y_z(t) &= \mathrm{Re}\left\{g(r)e^{j[\omega_0 t + \phi + \Phi(r)]}\right\} = \mathrm{Re}\left\{\left[g(r)e^{j\Phi(r)}\right]e^{j(\omega_0 t + \phi)}\right\} \\ &= \mathrm{Re}\left\{G(r)e^{j(\omega_0 t + \phi)}\right\} = \mathrm{Re}\left\{\frac{G(r)}{r}\tilde{x}(t)e^{j\omega_0 t}\right\} \\ &= \mathrm{Re}\left\{\tilde{y}(t)e^{j\omega_0 t}\right\}, \end{aligned} \tag{3.6}$$

where the in-phase, $P(r)$, and quadrature, $Q(r)$, nonlinearity forms are

$$\begin{aligned} P(r) &= \mathrm{Re}\{\mathrm{G}(r)\} = g(r)\cos\,\Phi(r), \\ Q(r) &= \mathrm{Im}\{\mathrm{G}(r)\} = g(r)\sin\,\Phi(r), \end{aligned} \tag{3.7}$$

and where

$$\tilde{y}(t) = g(r)e^{j[\phi + \Phi(r)]} = G(r)e^{j\phi} = G(r)\frac{\tilde{x}(t)}{|x(t)|} = \frac{G(r)}{r}\tilde{x}(t) \tag{3.8}$$

$$G(r) = g(r)e^{j\Phi(r)} = P(r) + jQ(r). \tag{3.9}$$

The relationships between the zonal-band RF and the envelope output signals as functions of the input signal envelope and of the memoryless AM–AM and AM–PM characteristics g and Φ of the nonlinear PA, expressed in polar form as $g(r)e^{j\Phi(r)}$ or in quadrature form as $P(r) + jQ(r)$, is thus established. Hence, given $\tilde{y}(t)$ and $\tilde{x}(t)$ measurements, $G(r)$, $g(r)$, $\Phi(r)$, $P(r)$ and $Q(r)$ are immediately found from Equation (3.8) and (3.9). Algorithmic implementation simply follows the logic suggested by the equations themselves. For example, a functional schematic implementation of the quadrature form is shown in Figure 3.2.

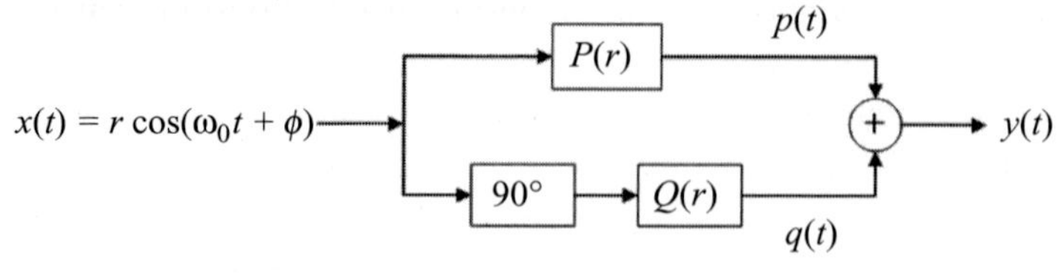

Figure 3.2 Quadrature model of a power amplifier.

The difference between the polar and quadrature descriptions is largely an algorithmic implementation difference. However, in practice the PA models and approximations used are proposed with a particular implementation in mind. Thus, for instance, in the Saleh model, Section 3.5, separate approximations are proposed for the polar and quadrature models; the Bessel–Fourier model, Section 3.8, has largely been used in a polar model form, though it may equally well be used in the quadrature form; the Hetrakul and Taylor approximations, Section 3.9, are proposed for the quadrature form; and the modified Saleh approximations proposed here, in Section 3.6, are for the polar form although they may be readily extended to the quadrature form.

Consider a general multicarrier input consisting of N narrowband modulated carriers:

$$x(t) = \sum_{l=1}^{N} A_l(t) \cos\left[\omega_l t + \phi_l(t)\right] = \mathrm{Re}\left\{\sum_{l=1}^{N} A_l(t) e^{j[\omega_l t + \phi_l(t)]}\right\} = \mathrm{Re}\left[\tilde{x}(t) e^{j\omega_0 t}\right], \tag{3.10}$$

where

$$\tilde{x}(t) = r(t) e^{j\phi(t)} = \left\{\sum_{l=1}^{N} A_l(t) e^{j[(\omega_l - \omega_0)t + \phi_l(t)]}\right\} \tag{3.11}$$

and where $A_l(t)$, $\omega_l(t)$ and $\phi_l(t)$ are the amplitude, angular frequency and phase of carrier l. The frequency ω_0 may be the band centre frequency as above or another arbitrarily chosen frequency.

This general input could include narrowband noise 'signals'. It is easy to see that the output may be represented as an infinite sum of harmonics and intermodulation product components, and so may be written as follows:

$$y(t) = \mathrm{Re}\left(\sum_{n_1, n_2, \ldots, n_N = -\infty}^{\infty} M(n_1, n_2, \ldots, n_N) e^{\sum_{l=1}^{N} j n_l [\omega_l t + \phi_l(t)]}\right) \tag{3.12}$$

where $M(n_1, n_2, \ldots, n_N)$ is the complex envelope of each component (wanted, intermodulation or harmonic), determined at any moment in time by the individual envelopes of the input multicarrier components (so that it could also be written as $M(n_1, n_2, \ldots, n_N; A_1(t), A_2(t), \ldots, A_N(t)))$ and the PA model parameters, extracted from the AM–AM and AM–PM characteristics. Setting the values of $n_1, n_2, \ldots, n_N$ yields the corresponding wanted output, harmonic or IMP components, as in Table 3.1.

The usual interest is in the zonal-band components. These are obtained from the condition

$$\sum_{l=1}^{N} n_l = 1. \tag{3.13}$$

With this condition applied, Equation (3.12) becomes the multicarrier equivalent of Equation (3.6).

The IMPs within the zonal band may be limited to those of order $\leq \gamma$ by using

Table 3.1 Examples of conditions determining which multicarrier PA model output components will be generated

$\sum_{l=1}^{N} n_l$	$\sum_{l=1}^{N} \|n_l\|$	Result
1	1	Wanted outputs (input components amplified)
1	$\leq \gamma$	The zonal-band output components with IMPs of order $\leq \gamma$
λ	1	The λ-harmonic components of the input signals
λ	γ	The γ-order IMPs in the λ-harmonic band

the condition

$$\sum_{l=1}^{N} |n_l| \leq \gamma. \tag{3.14}$$

The 'component-based' form of Equation (3.12) thus obtained may be referred to as a decomposed model. Shimbo, in [2], derives a general expression for the zonal-band output, i.e. the output for which condition (3.13) is met, as

$$M(n_1, n_2, \ldots, n_N) = \iint_0^\infty \gamma \left[\prod_{l=1}^{N} J_{n_l}(\gamma A_l(t)) \right] J_1(\gamma r) r g(r) e^{j\Phi(r)} d\gamma dr, \tag{3.15}$$

where the J_n are Bessel functions of the first kind and are defined by

$$e^{jz\cos\theta} = \sum_{n=-\infty}^{\infty} j^n J_n(z) e^{jn\theta}. \tag{3.16}$$

Actual formulations for M for different AM–AM and AM–PM modelling approaches may be found. This may be seen later for the Saleh and Bessel–Fourier models, in Sections 3.5 and 3.8 respectively.

The full characterisation of a PA, when it is memoryless as is implicit in Equations (3.6) and (3.9), reduces to the single-unmodulated-carrier AM–AM and AM–PM envelope characteristics, i.e. Equation (3.10) with $N = 1$, $A_l(t) = A$ and $\phi(t) = 0$. In other words, for truly memoryless PAs, characterisation measurements are independent of whether the signal used for parameter extraction is modulated. Unmodulated single-tone envelope-characteristic measurements are not the only way to characterise the PA, but they are quite feasible and generally acceptable in the field. Typical sets of AM–AM and AM–PM envelope characteristics may be obtained from device manufacturers or the research literature; an example is the GaN amplifier discussed in [3]. The equivalent memoryless characteristics of an LD-MOS high-power PA, which has some memory and which is used as the example for most of this chapter, are illustrated in Figure 3.3. These were extracted from measurements of an input and corresponding amplified output 3G (third generation) WCDMA signal, Figure 3.4. The input and output values are normalised and

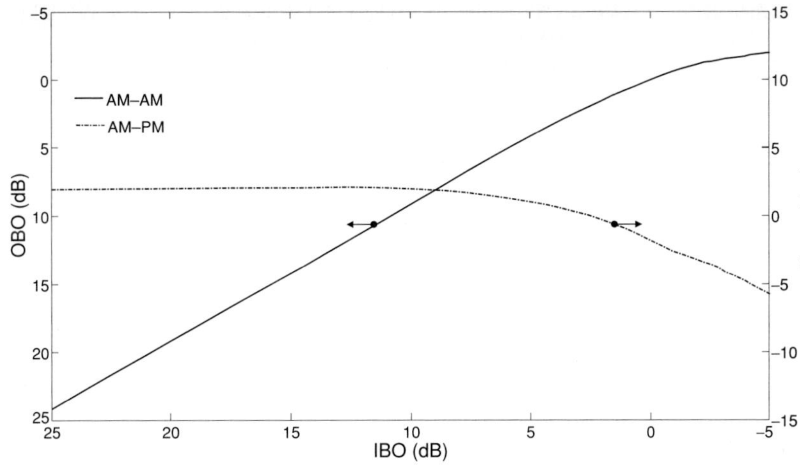

Figure 3.3 Memoryless AM–AM and AM–PM characteristics of an LDMOS PA, extracted from the WCDMA measurements shown in Figure 3.4.

graphed in dB backoff values (i.e. IBO and OBO), which are relative to the input and output 'saturation' powers, corresponding here to the powers at the 1 dB compression point. The following section provides more details on this amplifier and experimental arrangements.

3.3 A comparison of behavioural models based on PA performance prediction

A comparison of the performance of well-known memoryless behavioural models is presented here. This comparison is based on how well they predict the behaviour of an LDMOS power amplifier. The experimental platform is specified, together with a tabular summary of the comparative results, in Table 3.2. The same platform is used in the more detailed descriptions of the various models in the following sections.

The behavioural measurements* were taken on a three-stage class-AB PA with a Motorola 90 W MRF 18090A LDMOS transistor in the final stage and having the following nominal characteristics: a frequency range 1.93–1.96 GHz, maximum output power 48 dBm, 36 dB gain and a 1 dB output compression point of 53 dBm. The measurement setup allowed for an RF bandwidth of 35 MHz to be captured, within which a signal-to-noise ratio (SNR) of approximately 60 dB was achieved. This amplifier has some memory effects. The amount of memory in the device may be gauged from the AM–AM and AM–PM spread shown in Figure 3.4. This

* Grateful acknowledgement is made here to the Vienna University of Technology for this experimental work sponsored under the EU TARGET NoE.

Table 3.2 Simulation results on the LDMOS PA driven by a WCDMA signal at 5 dB IBO for different PA behavioural models; the models are polar, i.e. AM–AM and AM–PM based, unless otherwise specified

Models	NMSE[e] (dB)	ACEPR[f] (dB)	Δ ACPR[g] (dB)
Complex power series, three terms, fifth order	−30.9	−39.7	1.5
Complex power series, seven terms, 13th order	−33.4	−43.4	0.9
Original Saleh polar model[a]	−8.3	−17.1	20.2
Original Saleh polar AM–AM-only model	−27.5	−36.9	1.3
Modified Saleh polar model[b]	−32.0	−42.6	0.9
Original Saleh quadrature model: P-component model only[c]	−27.4	−36.9	1.4
Modified Saleh quadrature model: P-component model only	−29.6	−38.5	2.1
Modified Saleh quadrature model: full combined P and Q model	−31.6	−42.8	1.1
Bessel–Fourier, seven terms[d]	−33.5	−43.5	0.9
Bessel–Fourier, three terms	−31.3	−40.6	1.5
Optimised Bessel–Fourier (FOBF), three terms	−32.3	−42.5	0.9
Hetrakul and Taylor [5, 6]	−26.2	−35.5	1.9
Berman and Mahle [7] (AM–PM) and power series (AM–AM)	−32.8	−41.1	0.6

[a] See Figure 3.11; because of the 'disastrous' AM–PM modelling, the results are quite meaningless. Hence the Saleh model with AM–PM omitted is given in the next row.
[b] Modified Saleh MS-I (AM) and MS-I (PM) models from Equations (3.51) and (3.50) respectively.
[c] The extracted coefficients are complex for the quadrature (Q) component, contrary to the goal of the model, and so are omitted.
[d] In BF models, α is in the 0.6 to 0.7 range.
[e] Normalised mean-square error, Equation (3.17).
[f] Adjacent-channel error power ratio, Equation (3.20).
[g] Adjacent-channel power ratio difference, Equation (3.18).

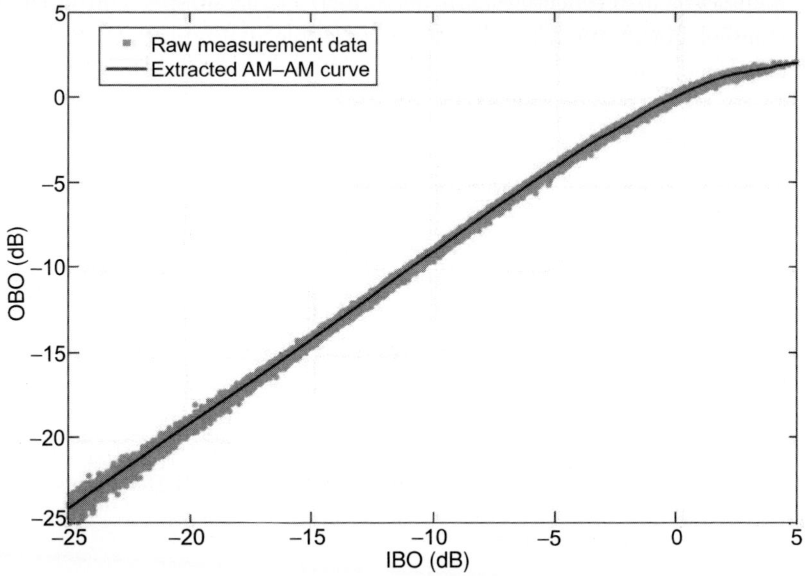

(a)

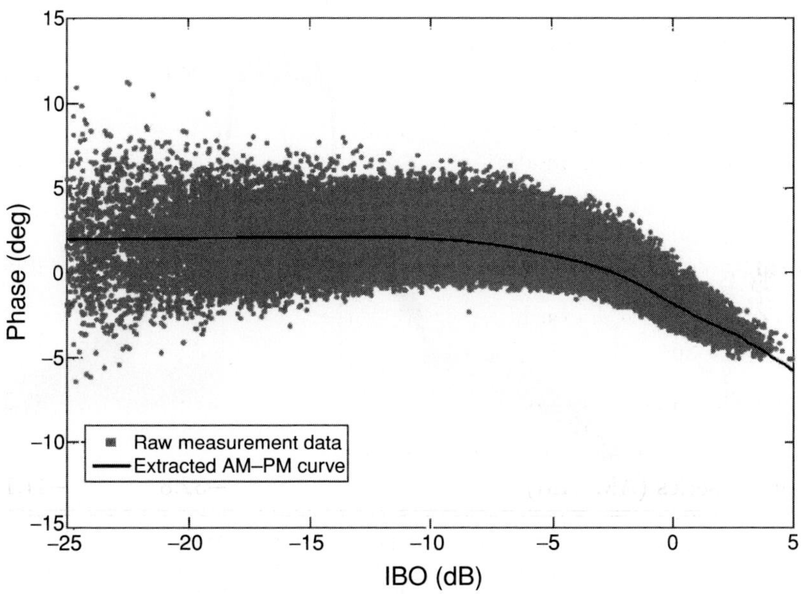

(b)

Figure 3.4 The LDMOS PA polar (a) AM–AM and (b) AM–PM measurements on the amplification of an WCDMA signal (at 5 dB IBO) and the extracted equivalent memoryless characteristics.

memory is also reflected in the asymmetry in the upper and in the lower adjacent-channel power spectral densities observable in Figure 3.5. As this translates into asymmetric upper and lower adjacent-channel power ratios (ACPRs), for the model comparisons here the poorer ACPR result is chosen.

In what follows, the medians of the AM–AM and AM–PM characteristics are used as the equivalent memoryless characteristics, labelled 'measured', from which the various behavioural models were extracted. The figures of merit results (see below) in Table 3.2 are for the target model's prediction of the output as against the measured amplifier output using a WCDMA signal different from the one used in the extraction process. This was a standard 3G WCDMA signal, with a bandwidth of 3.84 MHz, 5 MHz channel spacing and 10 dB peak-to-average power ratio (PAPR). The selected results here relate to a PA operated at 5 dB IBO from the 1 dB compression point, which, given the high PAPR WCDMA signal, can be regarded as operation well into the large-signal, nonlinear, region of the PA. Typical input and output spectra for a PA operating at 5 dB IBO are shown in Figure 3.5. In the simulations, 10^5 samples of the WCDMA signal are used, corresponding to

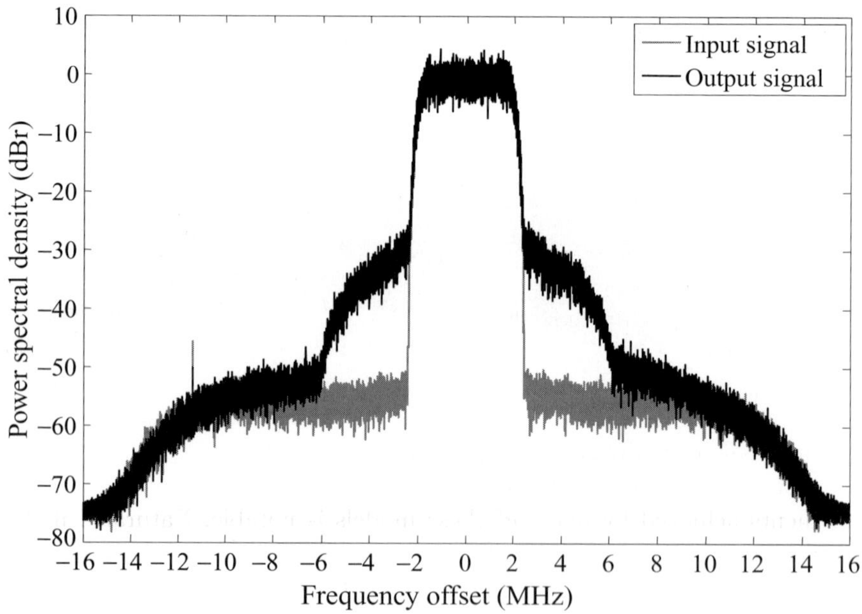

Figure 3.5 Input and output 3G WCDMA spectra in dB relative to the peak value. (for PA operation at 5 dB IBO).

a 3.125 ms duration at a sampling rate of 32 M samples per second.

The normalised mean-square error (NMSE) and the ACPR FOMs ΔACPR and ACEPR used in Table 3.2 are defined as follows. The normalised mean-square error is given by

$$\text{NMSE (dB)} = 10 \log \left\{ \frac{\sum_{i=1}^{MM} |y_{\text{measured}}(i) - y_{\text{model}}(i)|^2}{\sum_{i=1}^{MM} |y_{\text{measured}}(i)|^2} \right\} \tag{3.17}$$

where i specifies a sample and MM is the number of samples. The adjacent-channel power ratio difference is given by

$$\Delta\text{ACPR} = \text{ACPR}_{\text{measured}} - \text{ACPR}_{\text{model}}, \tag{3.18}$$

where

$$\text{ACPR} = 10 \log \left(\frac{\int_{\text{f}_{\text{adj}}} |Y(\text{f})|^2 df}{\int_{\text{f}_{\text{chan}}} |Y(\text{f})|^2 df} \right), \tag{3.19}$$

and the adjacent-channel error power ratio is given by

$$\text{ACEPR} = 10 \log \left(\frac{\int_{\text{f}_{\text{adj}}} |Y_{\text{measured}}(\text{f}) - Y_{\text{model}}(\text{f})|^2 df}{\int_{\text{f}_{\text{chan}}} |Y_{\text{measured}}(\text{f})|^2 df} \right), \tag{3.20}$$

where $Y(\text{f})$ is the Fourier transform of the corresponding signal and f_{chan} and f_{adj} are the frequency bands of the carrier channel and the standard first (upper and lower) adjacent channels. These FOMs receive a more detailed treatment in Chapter 6.

In the limited comparison of the performance of various memoryless behavioural models presented in Table 3.2, a variation in the performance of the models is apparent. The best are the seven-term Bessel–Fourier (BF) model and the 13th-order complex power series model, very closely followed by the low-order three-term Fourier-series-optimised BF model, the modified Saleh polar model and the modified Saleh quadrative model. Bearing in mind that the PA device is one having memory, the good approximations of the dominant AM–AM nonlinearity impairments achieved by many of these models is notable. Naturally, if the device were memoryless then even better results would be expected. The model being of the extracted equivalent memoryless measurements sets an upper bound to the modeling performance results achievable. These models will be discussed in detail in the following sections.

3.4 Complex power series model

A general form for an Lth-order power series memoryless model representing the instantaneous RF output as a polynomial expansion of the instantaneous RF input is expressed as

$$y(t) = \sum_{l=1}^{L} k_l x^l(t), \tag{3.21}$$

where $x(t)$ and $y(t)$ represent the input and output signals respectively and the k_l are complex coefficients.

This is an equivalent RF model and can itself be derived from the more general equivalent RF models discussed in Chapter 1, which take account of dynamic effects such as memory. For instance, it may be derived from the general time-domain Volterra series input–output relationship of a nonlinear system, as in subsection 2.4.2 and also, for example, see [8–12]:

$$y(t) = \sum_{l=1}^{\infty} \int_{-\infty}^{\infty} \int_{-\infty}^{\infty} \cdots \int_{-\infty}^{\infty} h_l(\tau_1, \tau_2, \ldots, \tau_l) d\tau_1 d\tau_2 \cdots d\tau_l \prod_{r=1}^{l} x(t - \tau_r), \tag{3.22}$$

where $x(t)$ and $y(t)$ are the system input and output respectively. Making the memoryless assumption, the kernel of the Volterra series expansion in Equation (3.22) resolves into multiple Dirac delta functions:

$$h_l(\tau_1, \tau_2, \ldots, \tau_l) = k_l \delta(t - \tau_1)\delta(t - \tau_2) \cdots \delta(t - \tau_l), \tag{3.23}$$

where the k_l are constant factors and typically are complex. Substituting Equation (3.23) into Equation (3.22) yields:

$$y(t) = \sum_{l=1}^{\infty} k_l x^l(t), \tag{3.24}$$

i.e. Equation (3.21) but with the number of coefficients not yet constrained to L. According to the approximation criteria chosen for k_n, particular polynomial expansion classes may be identified, e.g. power series or other expansions as set out below in Section 3.11.

In the following, a harmonic and IMP analysis of a nonlinear PA using a low-order power series model (three terms) and a two-tone input is shown. The same reasoning holds for higher-order power series and a larger number of input carriers or multicarrier component inputs (e.g. OFDM signals or complex signals resolved into their frequency-domain components). The two-carrier input may be expressed as

$$x(t) = r(t) \cos[\omega_0 t + \phi(t)] = A_1(t) \cos[\omega_1 t + \phi_1(t)] + A_2(t) \cos[\omega_2 t + \phi_2(t)], \tag{3.25}$$

where $A_1(t)$, $A_2(t)$, $\phi_1(t)$, $\phi_2(t)$ are the carriers' modulated amplitudes and phases and ω_1, ω_2 their angular frequencies. The third-order power series model may be expressed as

$$y(t) = k_1 x(t) + k_2 x^2(t) + k_3 x^3(t); \tag{3.26}$$

so, substituting $x(t)$ from Equation (3.25) yields

$$y(t) = k_1 \{A_1(t) \cos[\omega_1 t + \phi_1(t)] + A_2(t) \cos[\omega_2 t + \phi_2(t)]\} \\ + k_2 \{A_1(t) \cos[\omega_1 t + \phi_1(t)] + A_2(t) \cos[\omega_2 t + \phi_2(t)]\}^2 \\ + k_3 \{A_1(t) \cos[\omega_1 t + \phi_1(t)] + A_2(t) \cos[\omega_2 t + \phi_2(t)]\}^3 . \quad (3.27)$$

By considering the carriers to be unmodulated, i.e. letting $A_i(t) = A_i$ and $\phi_i(t) = \phi_i$, the general two-tone response may be written and expanded into its decomposed form:

$$\begin{aligned} y(t) &= k_1[A_1 \cos(\omega_1 t + \phi_1) + A_2 \cos(\omega_2 t + \phi_2)] + k_2[A_1 \cos(\omega_1 t + \phi_1) \\ &\quad + A_2 \cos(\omega_2 t + \phi_2)]^2 + k_3[A_1 \cos(\omega_1 t + \phi_1) + A_2 \cos(\omega_2 t + \phi_2)]^3 \\ &= k_2 A_1 A_2 + k_2 A_1 A_2 \{\cos[(\omega_1 - \omega_2)t + (\phi_1 - \phi_2)]\} \\ &\quad + \left(k_1 A_1 + \tfrac{9}{4} k_3 A_1^3\right) \cos(\omega_1 t + \phi_1) + \left(k_1 A_2 + \tfrac{9}{4} k_3 A_2^3\right) \cos(\omega_2 t + \phi_2) \\ &\quad + \tfrac{3}{4} k_3 A_1^2 A_2 [\cos((2\omega_1 - \omega_2)t + (2\phi_1 - \phi_2))] \\ &\quad + \tfrac{3}{4} k_3 A_2^2 A_1 [\cos((2\omega_2 - \omega_1)t + (2\phi_2 - \phi_1))] \\ &\quad + k_2 A_1 A_2 \cos((\omega_1 + \omega_2)t + (\phi_1 + \phi_2)) \\ &\quad + \tfrac{1}{2} k_2 A_1^2 \cos(2\omega_1 t + 2\phi_1) + \tfrac{1}{2} k_2 A_2^2 \cos(2\omega_2 t + 2\phi_2) \\ &\quad + \tfrac{1}{4} k_3 A_1^3 \cos(3\omega_1 t + 3\phi_1) + \tfrac{1}{4} k_3 A_2^3 \cos(3\omega_2 t + 3\phi_2) . \end{aligned} \quad (3.28)$$

Equation (3.28) describes a two-tone output response consisting of spectral components at DC, the fundamental frequencies ω_1 and ω_2, the second and third harmonics $2\omega_1$, $2\omega_2$ and $3\omega_1$, $3\omega_2$, the second-order IMPs at $\omega_1 \pm \omega_2$ and the third-order IMPs at $2\omega_1 \pm \omega_2$ and $2\omega_2 \pm \omega_1$. All these frequency components are shown in Figure 3.6, where the two input tones are assumed to be of equal power ($A_1 = A_2$).

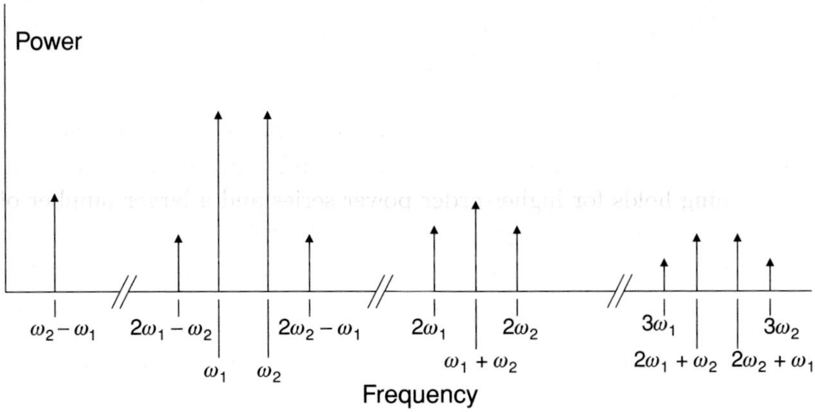

Figure 3.6 Frequency components based on a three-term memoryless power series PA model with two equipowered input tones.

The zonal components (the output components falling into the input frequency band) may be gathered into the zonal output signal $y_z(t)$:

$$y_z(t) = \left(k_1 A_1 + \tfrac{9}{4} k_3 A_1^3\right) \cos(\omega_1 t + \phi_1) + \left(k_1 A_2 + \tfrac{9}{4} k_3 A_2^3\right) \cos(\omega_2 t + \phi_2) + \tfrac{3}{4} k_3 \left[A_1^2 A_2 \cos((2\omega_1 - \omega_2) t + (2\phi_1 - \phi_2)) + A_2^2 A_1 \cos((2\omega_2 - \omega_1) t + (2\phi_2 - \phi_1))\right]. \quad (3.29)$$

This in turn may be written in the form of a general envelope transfer characteristic, following Equation (3.6):

$$y_z(t) = g(r(t)) \cos[\omega_0 t + \phi(t) + \Phi(r(t))]. \quad (3.30)$$

The symmetries of the IMPs on the frequency axis of Figure 3.6 are quite clear and are as expected. The amplitude symmetry is also clear, while it is notable from Equation (3.29) that any difference in the input carrier powers will be immediately reflected in an IMP imbalance.

For the extraction, a least-squares approximation to minimise the relative error between the experimental measurements and the values predicted by the model may be employed. For some low-noise amplifiers, small-signal amplifiers, mixers and baseband amplifiers, a low-degree complex power series model, typically with three terms, may achieve sufficient accuracy; see e.g. [13] and [14]. For large-signal PA behavioural analysis a larger number of terms should be included.

Table 3.3 Coefficients (rounded) for the third-, fifth-, seventh- and ninth-order power series terms in Figures 3.7(a), (b).

	$L=3$	$L=5$	$L=7$	$L=9$
K_1	$1.064 - j0.002$	$1.1 + j0.032$	$1.068 + j0.049$	$1.049 + j0.052$
K_3	$-0.082 - j0.014$	$-0.112 - j0.042$	$-0.06 - j0.07$	$-0.01 - j0.077$
K_5		$0.004 + j0.004$	$-0.014 + j0.014$	$-0.045 + j0.018$
K_7			$0.002 - j0.001$	$0.009 - j0.002$
K_9				$-0.0006 + j0.00007$

Figures 3.7(a), (b) show the improvements in a complex power series model of the LDMOS nonlinear PA as the number of the coefficients is increased from three to nine. The coefficients are collected in Table 3.3. The model error results for these are shown in Figures 3.8(a), (b). In Table 3.2 the results for power series with three and seven terms may be compared with those for other memoryless behavioural models.

While a lower-order complex power series has the attraction of being well known and has wide application for the approximation of well-behaved functions, its poor convergence in the presence of strong nonlinearities limits its use in PA modelling. The extraction of higher-order coefficients, as can be seen from the approximation equation, can be a little complex because their effects are difficult to separate out from those of the lower-order coefficients.

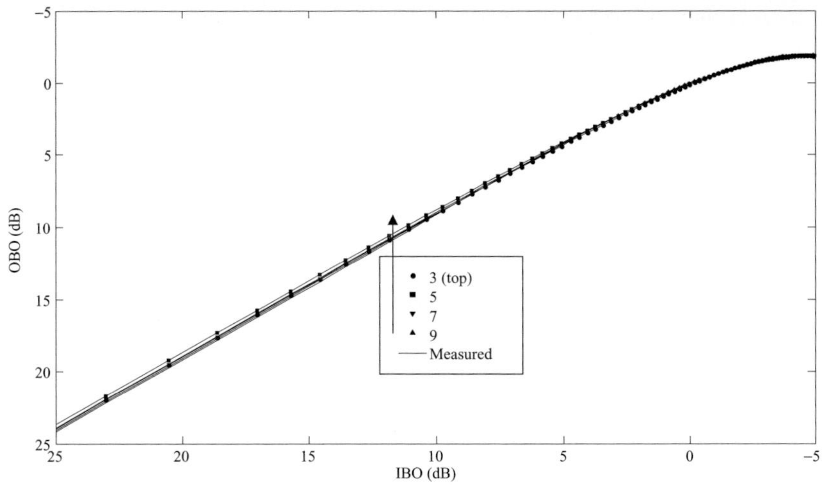

(a)

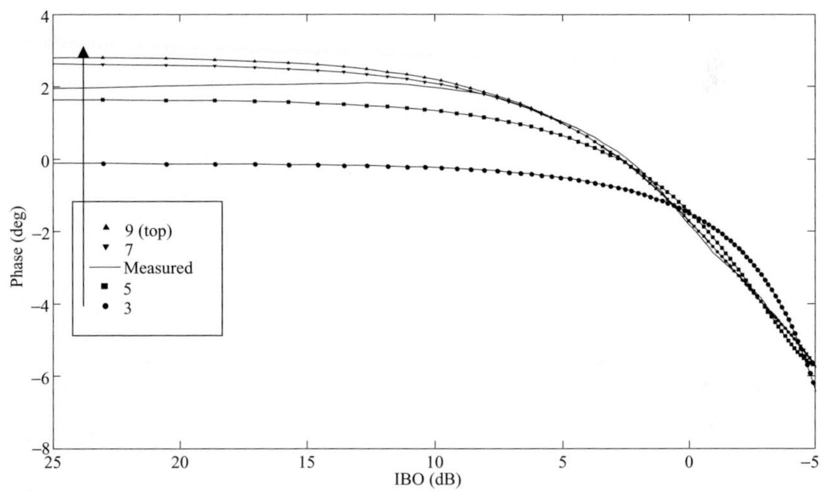

(b)

Figure 3.7 (a) The AM–AM and (b) the AM–PM measurements, and three-, five-, seven- and nine-term complex power series models. See Table 3.2.

3.5 Saleh models

Saleh, in [15], proposed a general two-parameter approximation for modelling the AM–AM and AM–PM envelope characteristics. Initially it was applied to TWTA models. It has also been used with SSPAs; however, problems have arisen, particularly in respect of the AM–PM characteristic, that have led to the proposal of a modified Saleh model, described below.

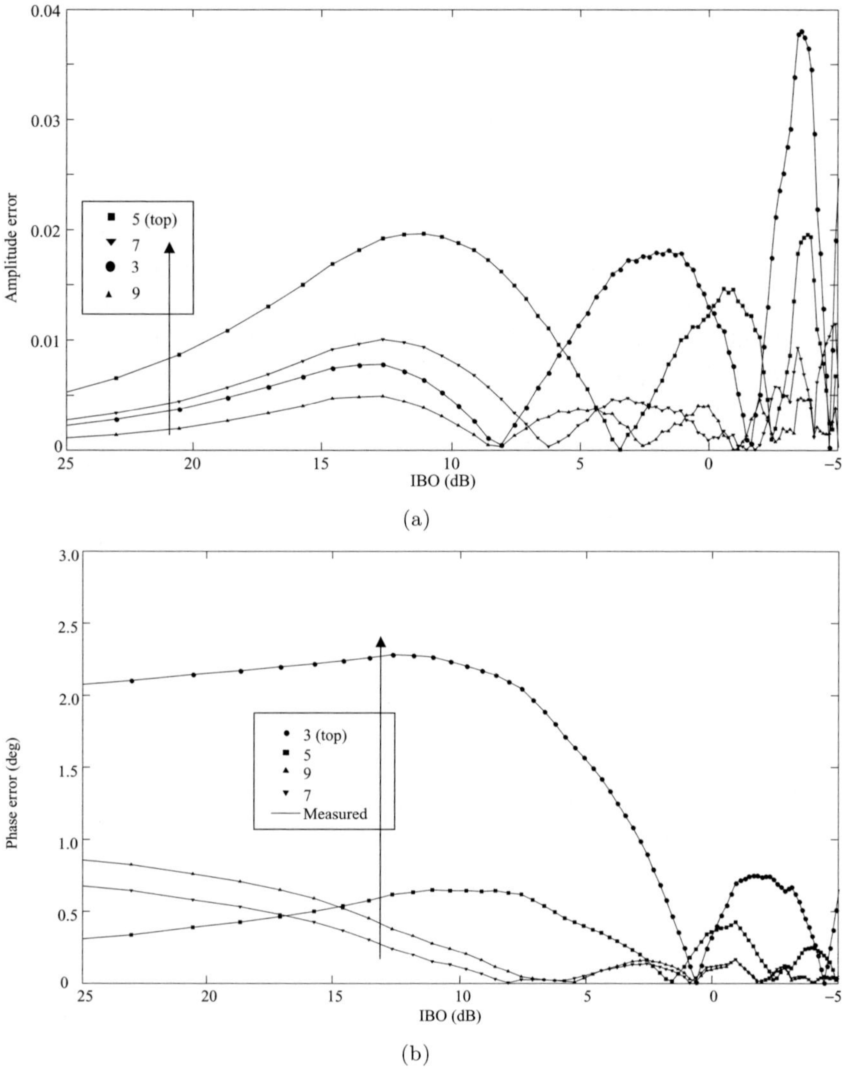

Figure 3.8 The (a) AM–AM and (b) the AM–PM measured to modelled errors for the three-, five-, seven- and nine-term complex power series models.

Saleh's general equation, from which he derived his original two-parameter models, is

$$z(r) = \frac{\alpha r^{\eta}}{(1 + \beta r^2)^{\nu}}, \tag{3.31}$$

where $n = 1, 2$ or 3, $\nu = 1$ or 2 and α and β are the model's two approximating coefficients.

Saleh polar model: The AM–AM and AM–PM two-parameter characteristic models g and Φ, which Saleh derived from Equation (3.31) and which yielded good fits to TWTA measurement data, were of the form

$$g(r) = \frac{\alpha_{\mathrm{a}} r}{1 + \beta_{\mathrm{a}} r^2}, \tag{3.32}$$

$$\Phi(r) = \frac{\alpha_{\phi} r^2}{1 + \beta_{\phi} r^2}, \tag{3.33}$$

where r represents the input envelope, Equation (3.3). The coefficients α_{a}, β_{a}, α_{ϕ} and β_{ϕ} may be extracted using a least-squares approximation to minimise the relative error between the experimental single-tone measured data and the values predicted by the model. An example of a least-squares algorithm approach to this extraction is provided in Section 3.6 below, where the modified Saleh model is considered.

Quadrature models, as in Equations (3.6) and (3.7) and Figure 3.2, with similar two-parameter forms were also proposed by Saleh [15]. These may be derived from his polar models by substituting Equations (3.32) and (3.33) into Equation (3.7) and making appropriate assumptions about the cosine and sine terms:

$$P(r) = \frac{\alpha_{\mathrm{p}} r}{1 + \beta_{\mathrm{p}} r^2}, \tag{3.34}$$

$$Q(r) = \frac{\alpha_{\mathrm{q}} r^3}{(1 + \beta_{\mathrm{q}} r^2)^2}, \tag{3.35}$$

where the quadrature-model coefficients α_{p}, β_{p}, α_{q} and β_{q} are extracted independently of the polar-model coefficients. As can be observed, the forms of $g(r)$ and $P(r)$ are identical.

A useful property, as Saleh points out, is that

$$Q(r) = \left. \frac{\partial P(r)}{\partial \beta_{\mathrm{p}}} \right|_{\alpha_{\mathrm{p}} \to \alpha_{\mathrm{q}};\ \beta_{\mathrm{p}} \to \beta_{\mathrm{q}}} . \tag{3.36}$$

Thus, if $P(r)$ is calculated for a given $r(t)$ then $Q(r)$ may be readily obtained by differentiation. This property is useful when deriving a decomposed model of a multicarrier input, as in Equation (3.37) below.

Saleh showed that, in comparison with some other quadrature models, especially that proposed by Kaye *et al.* [16], his model could provide a better fit to measurement data. Naturally, being just a two-parameter model gives it an attractive simplicity. However, for complex modulated signals, multicarrier signals etc. finding the decomposed form of the Saleh model is not trivial, as may be noted from the OFDM-like multicarrier example presented below.

Saleh reported a solution for the harmonic and intermodulation analysis of multicarrier signals using his two-parameter model. This result is significant as, for proper model implementation regardless of the complexity of the input, it is important to resolve the harmonic and IMP constituents, especially if it is desired to model highly nonlinear characteristics with good accuracy.

For a multicarrier input signal, such as an N-subcarrier OFDM input, Saleh derived a decomposed model in the form of Equation (3.12), where

$$M(n_1, n_2, \ldots, n_N) = \int_0^\infty z \left[\prod_{l=1}^{N} J_{n_l}(A_l(t)z)\right] \left\{\alpha_{\rm p}\beta_{\rm p}^{-3/2} K_1\left(z\beta_{\rm p}^{-1/2}\right)\right.$$
$$\left. + j\alpha_{\rm q}\left[\beta_{\rm q}^{-5/2} K_1\left(z\beta_{\rm q}^{-1/2}\right) - \left(\frac{z}{2\beta_{\rm q}^3}\right) K_0\left(z\beta_{\rm q}^{-1/2}\right)\right]\right\} dz. \quad (3.37)$$

Here K_0 and K_1 are modified Bessel functions of the second kind, of orders 0 and 1 respectively. Saleh's derivation is based on his quadrature model, using Shimbo's general formula [17], Equation (3.15) and a Hankel-type integral substitution.

Since it involves numerical integration, computational complexity is a disadvantage of this particular formulation and the attractiveness of the initial simplicity disappears when dealing with complex inputs. Nonetheless, the simple two-parameter form of the model, especially the polar AM–AM nonlinearity, may be employed effectively. However, care is required when the nonlinearity is strong and the out-of-band third or higher harmonic components are not negligible. The use of the model has concentrated on zonal-band outputs and has not been expanded to cater for harmonics. This downside seems implicit in Kenington [18] when he observes that the accuracy decreases as the nonlinearity of the device increases, and thus that these models are appropriate for PAs of classes A and AB but are not appropriate for modelling highly nonlinear PA classes, such as class C.

Of historical interest are Figures 3.9 and 3.10, which show the improvement in TWTA modelling achieved by the two-parameter Saleh model when compared with other suggested models (mainly extracted from Saleh's paper [15]). In Figure 3.9 the Saleh quadrature model may be compared with that proposed by Hetrakul and Taylor [5, 6] (see also Section 3.9) and with the latter's reported measurements, from which the models were extracted. In Figure 3.10 results are given for a Saleh AM–AM and AM–PM model (solid lines), the AM–AM model used by Thomas *et al.* [19] and the AM–PM model used by Berman and Mahle [7] (see also Section 3.10); the latter's reported measurement data are also shown. Again the models were extracted from this data. (The Thomas *et al.* model is omitted in our treatment as there are some ambiguities in its description in the authors' paper.)

3.6 Modified Saleh models

Applying the Saleh model directly to SSPA devices can lead to problems. This may particularly arise with the AM–PM characteristic when the shape differs from those obtained for TWTAs, which generally are positive throughout and have a shape not too different from the AM–AM characteristic. This would explain the similarity of the models for TWTAs. However, the Saleh method, using Equation (3.33), yields a poor model for the AM–PM characteristic of the LDMOS amplifier treated above. This may be seen in Figure 3.11, where the approximate shape of a typical TWTA

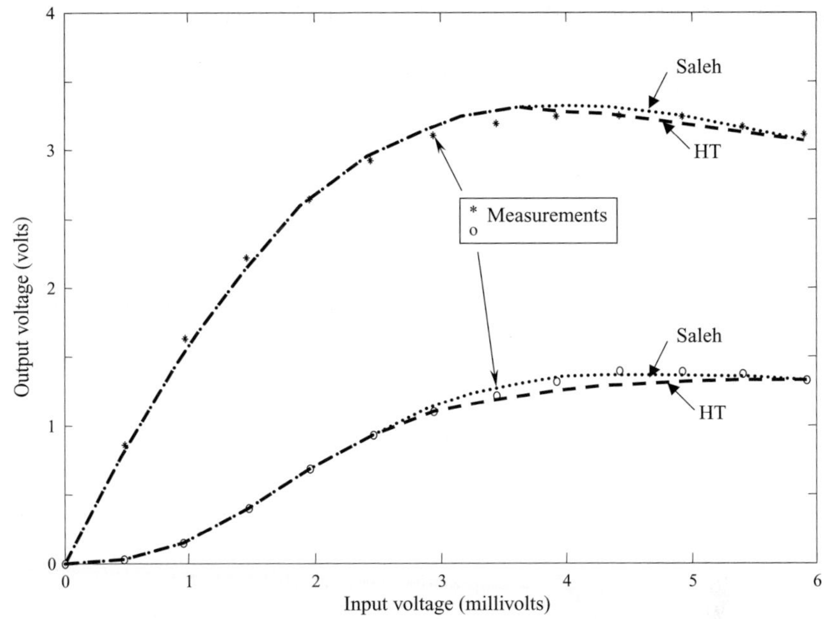

Figure 3.9 Quadrature measurements and model comparison: the Saleh model (dotted lines), the Hetrakul and Taylor model (HT) [5, 6] (broken lines) and the latter's reported measurement data (∗ and o). (From [15] with permission, © 1981 IEEE.)

AM–PM used in the original Saleh-type models is also shown.

The main difference is that the LDMOS characteristic has an inflection point and a negative-going curve that takes on zero and negative values with increasing input power. This causes optimisation problems that result in ill-defined α and β parameters. Introducing a phase shift into the measurements (so that all AM–PM values are positive), an approach that is part of the new 'modified Saleh' model presented below, does not solve the problem. The variability of the transfer characteristics of SSPAs, especially of the AM–PM characteristics, thus suggests a reconsideration of the Saleh model.

Modifying the general Saleh model, Equation (3.31), by the addition of two new parameters, an exponent γ and a phase shift ε, yields the generalised form for the 'modified Saleh' model:

$$z(r) = \frac{\alpha r^{\eta}}{(1+\beta r^{\gamma})^{\nu}} - \varepsilon. \tag{3.38}$$

For a given set of values (η, ν, γ and ε), optimum values for (α, β) can be extracted from a measurement data set $(z(r), r)$ of either the AM–AM or the AM–PM characteristics.

While this may seem to be a six-parameter model, it will be shown that it can be reduced, as in the Saleh model, to a pair of two-parameter models that model the AM–AM and the AM–PM characteristics respectively. These, especially the AM–PM model, are different from the Saleh model and so are justifiably referred

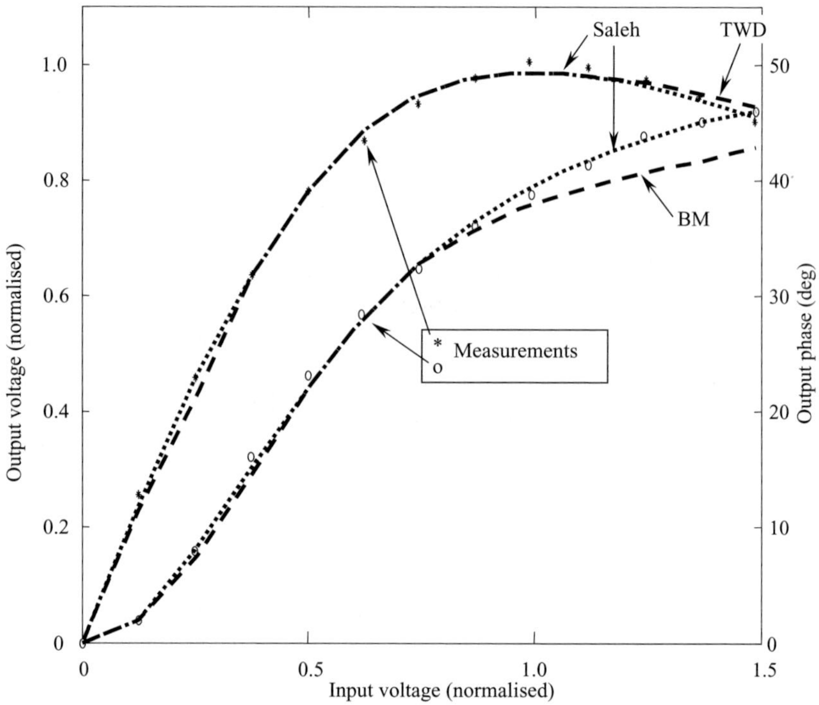

Figure 3.10 Measurements and model values for AM–AM and AM–PM characteristics: the Saleh model, the AM–AM model used by Thomas, Weidner and Durrani (TWD) [19] and the AM–PM model used by Berman and Mahle (BM) [7] and the latter's reported measurement data. (From [15] with permission, © 1981 IEEE.)

to as modified Saleh models (MS models).

Equation (3.38) may be rearranged to

$$\left(\frac{r^\eta}{z(r)+\varepsilon}\right)^{1/\nu} = \frac{\beta}{\alpha^{1/\nu}} r^\gamma + \alpha^{-1/\nu}, \tag{3.39}$$

which has the form

$$s = At + B \tag{3.40}$$

where

$$s = \left(\frac{r^\eta}{z(r)+\varepsilon}\right)^{1/\nu}, \quad t = r^\gamma, \quad A = \frac{\beta}{\alpha^{1/\nu}} \quad \text{and} \quad B = \alpha^{-1/\nu} \tag{3.41}$$

From a set of N measurements, $(z(r)_i, r_i)$, $i = 1, \ldots, N$, extracting the pair (α, β) is done by extracting values for (A, B) from the set of transformed measurements (s_i, t_i), $i = 1, \ldots, N$, obtained from Equations (3.41). This may be achieved by minimising with respect to A and B the error between the measured value s_i

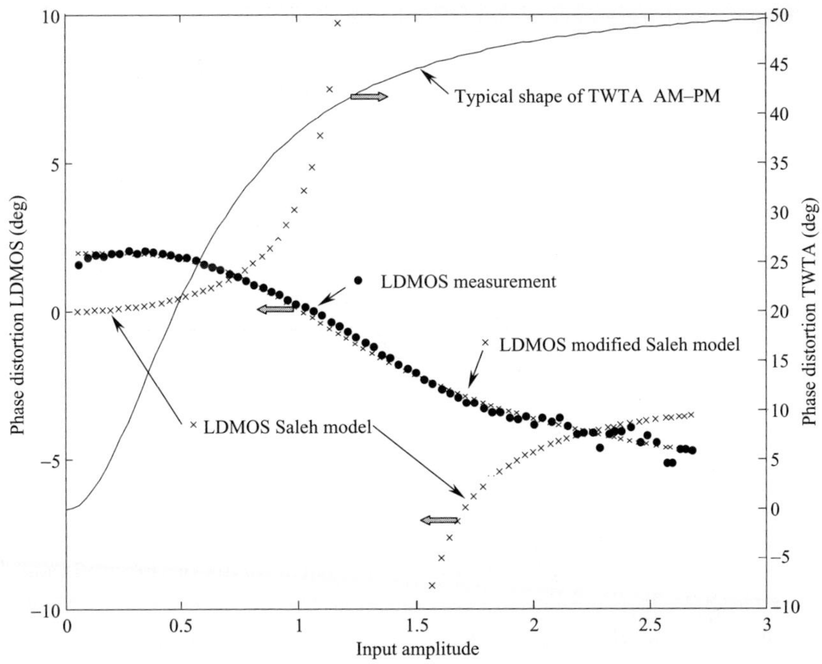

Figure 3.11 The AM–PM Saleh and modified Saleh models; a typical TWTA AM–PM characteristic is included for reference.

and the corresponding value s_i^* computed from Equation (3.40):

$$\min_{A,B}\left[\sum_{i=1}^{N}(s_i - s_i^*)^2\right] = \min_{A,B}\left[\sum_{i=1}^{N}(s_i - At_i - B)^2\right]. \tag{3.42}$$

The minimum condition may be written as the pair of equations

$$\begin{aligned} \frac{\partial\left[\sum_{i=1}^{N}(s_i - At_i - B)^2\right]}{\partial A} &= 0, \\ \frac{\partial\left[\sum_{i=1}^{N}(s_i - At_i - B)^2\right]}{\partial B} &= 0 \end{aligned} \tag{3.43}$$

and thus as

$$\begin{aligned} A\sum_{i=1}^{N} t_i^2 + B\sum_{i=1}^{N} t_i &= \sum_{i=1}^{N} s_i t_i, \\ A\sum_{i=1}^{N} t_i + BN &= \sum_{i=1}^{N} s_i. \end{aligned} \tag{3.44}$$

Hence, with reference also to Equation (3.41),

$$\alpha = \frac{1}{B^{\nu}} = \left(\frac{N\sum_{i=1}^{N} t_i^2 - \left(\sum_{i=1}^{N} t_i\right)^2}{\left(\sum_{i=1}^{N} s_i\right)\left(\sum_{i=1}^{N} t_i^2\right) - \left(\sum_{i=1}^{N} t_i\right)\left(\sum_{i=1}^{N} (s_i t_i)\right)} \right)^{\nu},$$

$$\beta = \frac{A}{B} = \frac{N\sum_{i=1}^{N} s_i t_i - \left(\sum_{i=1}^{N} s_i\right)\left(\sum_{i=1}^{N} t_i\right)}{\left(\sum_{i=1}^{N} s_i\right)\left(\sum_{i=1}^{N} t_i^2\right) - \left(\sum_{i=1}^{N} t_i\right)\left(\sum_{i=1}^{N} s_i t_i\right)}. \tag{3.45}$$

In this general form of the modified Saleh model giving the extraction of α and β, the values for η, ν, γ and ε are still open. The modified Saleh model may be applied to both the polar and quadrature forms of the nonlinear PA, Equations (3.7)–(3.9). In the following two sections, the modified Saleh AM–AM and AM–PM polar models will be derived.

3.6.1 *Modified Saleh AM–PM model*

The extraction of the AM–PM characteristic is considered first, as it is the variability of this characteristic that motivated the modification of the Saleh model. In Saleh's original model, the parameters α and β can become complex numbers for certain choices of ν; this in turn results in complex values for the modelled phase distortion. This may be seen from the minimisation process used to extract α and β above, Equations (3.45) and (3.41): assuming that ε is zero and that η, ν and γ are known, a negative value for z (i.e. Φ) and values for ν other than the odd natural numbers causes s, and thus α and β, to become complex. By the introduction of ε as a phase-shift parameter with a value sufficient to assure only positive values for the phase distortion, thus keeping the values of α and β real, the problem may be resolved. In finding the modified Saleh AM–PM model, the range of the set $(\eta, \nu, \gamma, \varepsilon)$ considered was $(-2, 0, 0, 5)$ to $(4, 4, 4, 20)$ in individual increments of $(0.1, 0.1, 0.1, 0.25)$. The values for ε are in degrees and an initial value 5 was chosen as the minimum AM–PM shift required to make all values positive; see Figure 3.11. Notably, and usefully for this LDMOS AM–PM characteristic, the value $\eta = 0$ yields the best results over all the cases considered – a significant difference from the Saleh model, Equation (3.33), where $\eta = 2$. The 'optimum' result found is given in Table 3.4. This was for 75 measurement points, though results vary little for more or fewer points as long the points are visually an acceptable description of the mean AM–PM characteristic.

A second observation is that a simple phase shift ε is not adequate to remove negative AM–PM values. In the optimisation it was found that the terms comprising the denominator of α, and also that of β, do not have a relatively smooth behaviour around the zero value of the phase, owing to the infinite asymptote present in s; see Equation (3.41). An example of the dependence of α and β on ε for this PA is shown in Figure 3.12, where the behaviour of the denominator of α or β and the numerator of β in Equation (3.45) is graphed as a function of ε.

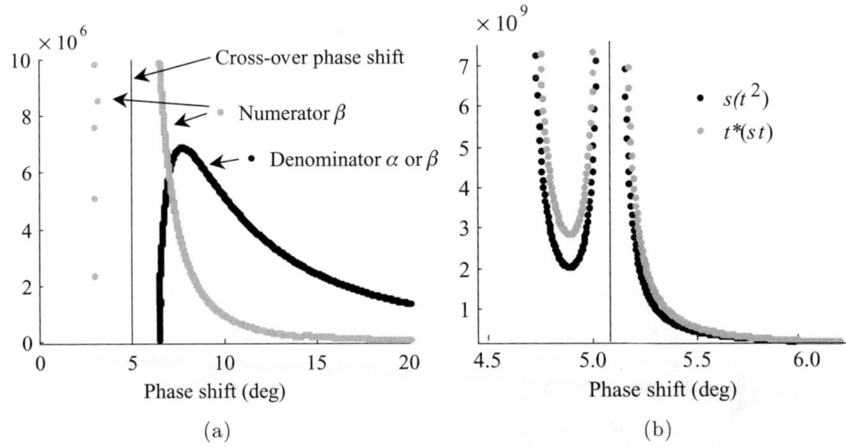

Figure 3.12 Stability analysis for the modified Saleh AM–PM model: the dependences of α and β on ε. (a) The numerator of β and the denominator of α or β (Equation (3.45)) and (b) the two terms in the denominator. The cross-over phase shift at 5.08°, which yields positive AM–PM values only, is indicated.

It is also useful to observe in Figure 3.12 that β takes on the value 1 for some values of ε. Thus β in the general modified Saleh model for this form of the AM–PM characteristic may be set to 1, removing it from the extraction process. This brings the model back to a two-parameter model involving just α and ε, which are to be optimised once values for γ and ν are chosen; the value of η is already chosen as 0 but is left unspecified here for generality. Thus the general modified Saleh AM–PM model, Equation (3.38), in this case becomes

$$\Phi(r) = z(r) = \frac{\alpha r^{\eta}}{(1 + r^{\gamma})^{\nu}} - \varepsilon. \tag{3.46}$$

For the extraction of α and ε we note that this equation has the form

$$s = \alpha t + \epsilon, \tag{3.47}$$

where

$$s = \Phi(r) = z(r), \qquad t = \frac{r^{\eta}}{(1 + r^{\gamma})^{\nu}} \qquad \text{and} \qquad \epsilon = -\varepsilon. \tag{3.48}$$

Following the same minimisation procedure as detailed above in connection with Equations (3.42)–(3.45), the solutions for α and ε in terms of s_i and t_i obtained from the measurements $(r_i, \Phi(r_i))$ via Equation (3.48) are

$$\alpha = \frac{N\sum_{i=1}^{N} s_i t_i - \left(\sum_{i=1}^{N} s_i\right)\left(\sum_{i=1}^{N} t_i\right)}{N\sum_{i=1}^{N} t_i^2 - \left(\sum_{i=1}^{N} t_i\right)^2},$$

$$\varepsilon = \frac{\left(\sum_{i=1}^{N} s_i\right)\left(\sum_{i=1}^{N} t_i^2\right) - \left(\sum_{i=1}^{N} t_i\right)\left(\sum_{i=1}^{N} s_i t_i\right)}{N\sum_{i=1}^{N} t_i^2 - \left(\sum_{i=1}^{N} t_i\right)^2}. \tag{3.49}$$

An optimisation procedure was used to find the values of the set (ν, γ) that best fit the measurements; (ν, γ) was varied over the range $(-0.5, -1)$ to $(3.5, 7)$ in increments of $(0.1, 0.1)$. The values of α and ε for three good (ν, γ) sets and the model fitting errors are presented in Table 3.4.

Table 3.4 Some optimised extraction results for the parameters α and ε, using various modified Saleh models of the LDMOS PA's AM–PM characteristic: the 'optimum' model, with α, ε, ν and γ optimised; the MS-I and MS-II models, with ν and γ set to the rational-number exponents nearest to the optimum values and with α and ε optimised; $\eta = 0$ in all cases

Modified Saleh (MS) AM–PM model,					Fit error (deg)				
Equation (3.49)	α	ε (deg)	ν	γ	Mean	Std dev.	Max.	Min.	Median
MS-I (PM)	0.161	7.1	$\frac{1}{3}$	4	0.133	0.014	0.5	0	0.1
MS-II (PM)	0.141	6.1	$\frac{1}{3}$	5	0.14	0.01	0.6	0	0.13
MS-Opt ($\beta = 1$)	0.157	7.1	0.3	4.5	0.129	0.012	0.5	0	0.11
MS-Opt (β and ε optimised), based on Equation (3.45)	$\alpha = 0.14$, $\beta = 0.37$	6	0.8	3.3	0.13	0.01	0.6	0	0.1

The optimum model, MS-Opt, has $(\nu, \gamma) = (0.3, 4.5)$. The other two near-optimum models, MS-I (PM) and MS-II (PM), use the simpler and more familiar exponent powers nearest this optimum, viz. $(\nu, \gamma) = (\frac{1}{3}, 4)$ and $(\frac{1}{3}, 5)$ respectively. These yield models almost as good.

Choosing, the MS-I (PM) model, for which $(\eta, \nu, \gamma) = (0, \frac{1}{3}, 4)$, Equation (3.38) thus becomes a model with two parameters, α and ε:

$$\Phi(r) = \frac{\alpha}{\sqrt[3]{(1 + r^4)}} - \varepsilon. \tag{3.50}$$

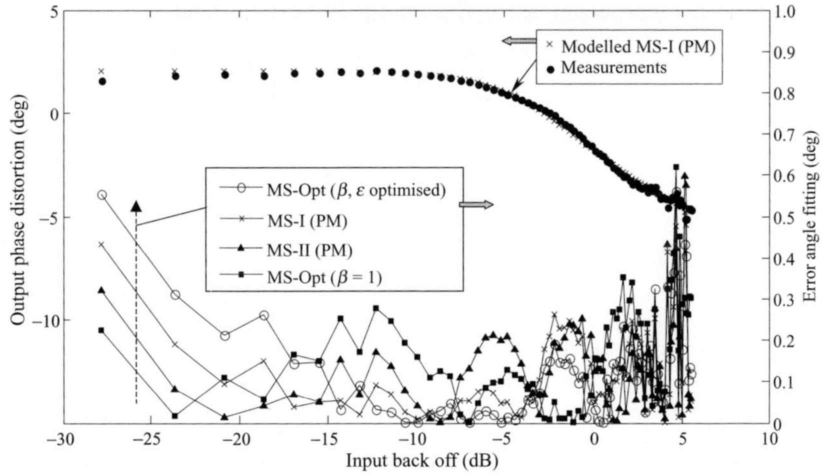

Figure 3.13 (Top) The modified Saleh model MS-I (PM) and the LDMOS PA AM–PM measurements. (Bottom) Error graphs for the MS-I (PM) model, for the two optimum modified Saleh models MS-Opt, one with $\beta = 1$ and the other with β and ε optimised, and for the MS-II (PM) model.

Figure 3.13 graphs the modified Saleh model MS-I (PM) and the LDMOS PA AM–PM measurements, along with the modelling error. For graphical comparison, also included are the error curves for the two slightly better optimum modified Saleh models MS-Opt, one with $\beta = 1$ and the other with β and ε optimised, and the error curves for the MS-II (PM) model.

3.6.2 *Modified Saleh AM–AM model*

Applying the modified Saleh model to the LDMOS PA's AM–AM characteristic yielded a different model from Saleh's. The ε parameter may be set to zero, as it is just a shifting parameter to help the optimisation when characteristics have zero and negative values, a situation that does not occur for the AM–AM characteristics. Further, and as might have been expected given the shape of the AM–AM characteristic, the best results were obtained by keeping $\eta = 1$ (i.e. as in the Saleh model). For the remaining two parameters, ν and γ, an optimum match was obtained for $\nu = 0.3$ and $\gamma = 3.7$. This optimum modified Saleh model (MS-Opt) differs, of course, with variations in the AM–AM characteristics. Using the more familiar exponent power values $\nu = \frac{1}{2}$ and $\gamma = 3$, yielded a model, MS-I, that was nearly as good as the MS-Opt model. The deviation between the two models will vary with changes in the AM–AM characteristic, but as long as it keeps the same general form, (α, β) values can be extracted that yield a good fit. Both these modified Saleh AM–AM models do better than Saleh's original model, where $\gamma = 2$.

In the classic Saleh models, $(\eta, \nu, \gamma, \varepsilon) = (1, \nu, 2, 0)$, and in his AM–AM model Saleh chose 1 for the value of ν. Marginal but monotonically improving models are yielded by increasing the value of ν; see the bottom three rows of Table 3.5 for $\nu = 1$, 2 and 3. The smallness of these marginal improvements justifies Saleh's choice of $\nu = 1$.

A summary of the model-fitting results is presented in Table 3.5, and Figure 3.14 graphs the fit of the modified Saleh model MS-I (AM) to the LDMOS PA AM–AM measurements. It also shows the error graphs (the 'misfit') for MS-I (AM), for the slightly better optimum modified Saleh model (MS-Opt) and for the poorer original Saleh AM–AM models with $\nu = 1$, 2 and 3 (Saleh1, Saleh2 and Saleh3 respectively).

Table 3.5 Saleh and modified Saleh AM–AM models; parameter selection and goodness of model fit; $\eta = 1$ and $\varepsilon = 0$ in all cases

					Fit error (%)				
AM–AM models	α	β	ν	γ	Mean	Std dev.	Max.	Min.	Median
Modified Saleh, MS-Opt (the optimum model)	1.0444	0.10195	0.3	3.7	0.5	0.1	1.5	0.02	0.4
Modified Saleh, MS-I (recommended)	1.0536	0.08594	$\frac{1}{2}$	3	0.8	0.26	2.3	0.01	0.8
Saleh1 (original, Equation (3.32))	1.09	0.0898	1	2	2.2	1.9	5.6	0.14	2.2
Saleh2	1.080	0.0393	2	2	1.8	1.3	4.8	0.0	1.8
Saleh3	1.077	0.0251	3	2	1.7	1.1	4.5	0.05	1.7

The modified Saleh AM–AM model MS-I, as derived from the general modified Saleh model, Equation (3.38), with $(\eta, \nu, \gamma, \varepsilon) = (1, \frac{1}{2}, 3, 0)$ may now be written as

$$g(r) = \frac{\alpha r}{\sqrt{(1 + \beta r^3)}}. \tag{3.51}$$

3.6.3 *Modified Saleh quadrature models*

The development of modified Saleh quadrature models follows the same reasoning as the modified Saleh polar models. In the application of the original Saleh quadrature model, Equations (3.34) and (3.35), to the LDMOS PA, difficulties similar to those occurring in the Saleh polar models were encountered, i.e. the in-phase component $P(r)$, the original Saleh quadrature P model, could be extracted without difficulty but acceptable extraction of the quadrature component $Q(r)$, i.e. the original Saleh quadrature Q model, was not possible. At best, complex values for the pair (α, β) resulted. This is not unexpected since it may readily be seen that there is a considerable similarity between the quadrature behaviour, Figure 3.15, and the polar behaviour, Figure 3.4. This is in keeping with the fact that since

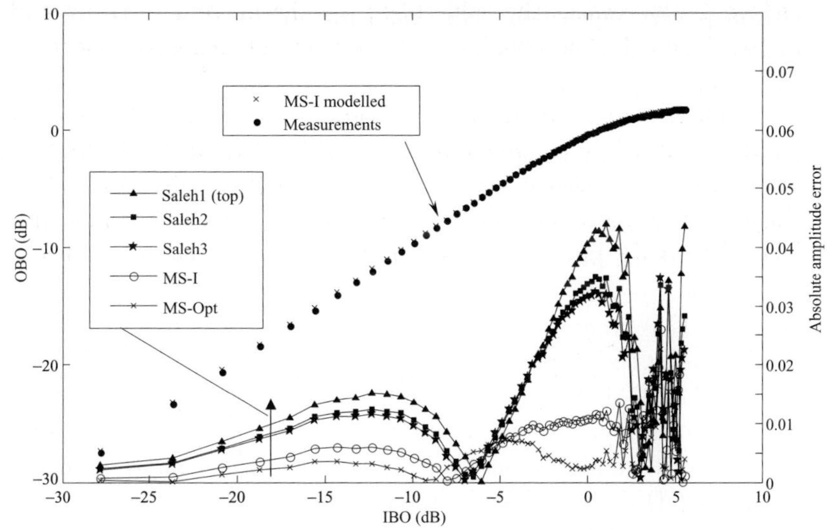

Figure 3.14 (Upper data points) Modified Saleh model MS-I (AM) and the LDMOS PA AM–AM measurements. (Lower data points, on lines) The error graphs for MS-I (AM), for the optimum modified Saleh model (MS-Opt) and for the original Saleh AM–AM models with $\eta = 1$, 2 and 3 (Saleh1, Saleh2 and Saleh3 respectively).

the phase variation is small, the quadrature P and Q characteristics will generally follow the shape of the polar AM–AM and AM–PM characteristics, Equation (3.7), respectively.

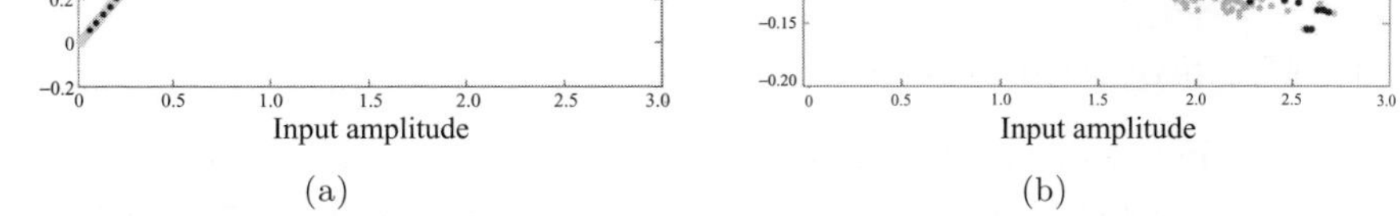

Figure 3.15 The WCDMA signal measurements (grey) and the extracted equivalent memoryless quadrature characteristics (black dots) of an LDMOS PA: (a) in-phase component P and (b) quadrature component Q (for a PA at 5 dB IBO).

As in the polar models the actual memoryless quadrature P and Q characteristics are extracted from the input and output WCDMA-derived quadrature measurements, Figure 3.15, i.e. they are derived from the polar measurements. These may be seen in Figure 3.15. It is notable also that the signal power through the

Q arm is very much less than that through the P arm. Hence its impact on the model's performance would not be expected to be significant. Modelling using only the original Saleh quadrature P model, Equation (3.34), yields quite good results; see Table 3.2. The extracted (α, β) set is $(1.09, 0.09)$. In fact the performance is almost identical to using that of the original Saleh polar AM–AM-only model.

For the modified Saleh quadrature models, the optimisation for the in-phase characteristic yields the same model, Equation (3.51), as the modified Saleh polar AM–AM model, MS-1 (AM), following the same reasoning as set out in subsection 3.6.2. This is not unexpected as the characteristics, as already mentioned, are quite similar. The extracted LDMOS PA (α, β) values were $(0.82, 0.29)$. This model alone yields a result that is better than the original Saleh quadrature P model when tested against the LDMOS amplified validation WCDMA signal and also better than the original Saleh polar AM–AM-only model, Table 3.2.

A complete modified Saleh quadrature model, formed by adding a modified Saleh quadrature Q model, yields a better result, which is comparable with the modified Saleh polar model, see Table 3.2. Applying the general modified Saleh AM–PM model, Equation (3.46), yields a choice of good modified Saleh quadrature Q models, e.g. with $(\eta, \nu, \gamma) = (0, 0.1, 4)$ and $\alpha = -0.3506$.

3.7 Fourier series model

In the memoryless Fourier series nonlinear PA behavioural model, the output signal is expressed as a complex Fourier series expansion of the instantaneous input signal. Assuming that the instantaneous voltage transfer characteristic G_{RF} in Equation (3.1) can be represented by a complex Fourier series expansion of its periodic extension, the output can be written as

$$y(t) = \sum_{k=-\infty}^{\infty} c_k e^{j\alpha k x(t)}, \tag{3.52}$$

where the c_k are the coefficients of the Fourier series,

$$c_k = \frac{1}{D} \int_a^b G(x) e^{j\alpha k x} \, dx, \tag{3.53}$$

and α is defined by the dynamic range, viz. $D = 2\pi/\alpha$, in the sense that, the period D of the periodic extension is defined by the maximum dynamic range of the input signal $x(t)$ or the maximum dynamic range for which it is desired to model the PA. In practice, of course, the right-hand side of Equation (3.52) will be constrained to a finite number of terms.

It is not unusual to consider models approximating PA nonlinearity characteristics over a normalised dynamic range up to 'XdB input overdrive with respect to saturation', i.e. $-X$dB IBO. The values of α and D are readily given in terms of X as set out in the following paragraphs.

It is usual in these situations to assume that the nonlinear characteristics are normalised with respect to the input saturation power P_s of a single unmodulated carrier. As a PA's saturation point is usually not clearly or uniquely identifiable, values are chosen arbitrarily and the corresponding P_s is thus defined.

Some authors choose the 1 dB compression point, which is suitable when one is considering a PA's small-signal performance. Another useful reference point is the 0.1 dB differential gain point, $dP_o/dP_i = 0.1$; i.e. the point where this differential gain occurs for the first time as the PA is driven into saturation, see [20, 21].

If A_s is the peak instantaneous voltage amplitude, the equivalent normalising voltage corresponding to P_s, then the relationship between A_s and P_s is given by

$$P_s = \frac{1}{T}\int_0^T (A_s \sin t)^2 \, dt = \frac{A_s^2}{2}, \tag{3.54}$$

where the measurements are assumed to be normalised to a $1\,\Omega$ input impedance.

In characterising a PA, measurements are made up to an input power in watts of P_m , corresponding to X dB input overdrive, i.e. the upper limit to the dynamic range, which equates to a peak instantaneous sinusoidal voltage amplitude A_m (normalised to a $1\,\Omega$ input impedance). Thus A_m is the actual dynamic range and A_m/A_s is the normalised dynamic range. When Equation (3.52) is used in its normalised form, i.e. with x and y normalised to their saturation input voltage A_s and output voltage, then D and α, now dimensionless, are given by

$$\alpha = \frac{2\pi}{D} = 2\pi\frac{A_s}{A_m} = 2\pi \times 10^{-X/20} \tag{3.55}$$

The usefulness of the Fourier series approximation to an unknown instantaneous transfer characteristic G is not so much in the expression itself, but in the extent to which it can be used to develop the Bessel–Fourier (BF) model, Section 3.8, and to optimise the latter when low numbers of extracted coefficients are used, subsection 3.8.3.

3.8 Bessel–Fourier models

The memoryless complex Fourier series expansion of the instantaneous voltage transfer characteristics, Equation (3.52), is the basis for deriving the memoryless complex Bessel–Fourier (BF) model. This was derived in its initial formulation by Fuenzalida *et al.* [22], Kaye *et al.* [16] being the principal source of inspiration. A simpler, more accessible, derivation was later found by O'Droma [23].

The memoryless complex BF model is eminently suitable for modelling both large- and small-signal behaviour with simple or complex multicarrier input signals. The model is highly computable and also extensible, in that more coefficients may be added for improved model accuracy. It is a decomposable model, thus enabling focused zonal- and harmonic-band behavioural analysis and the control of aliasing

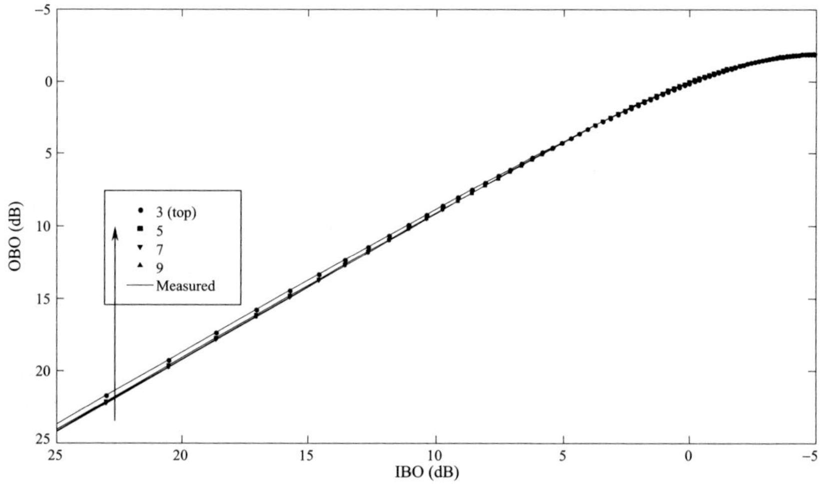

(a)

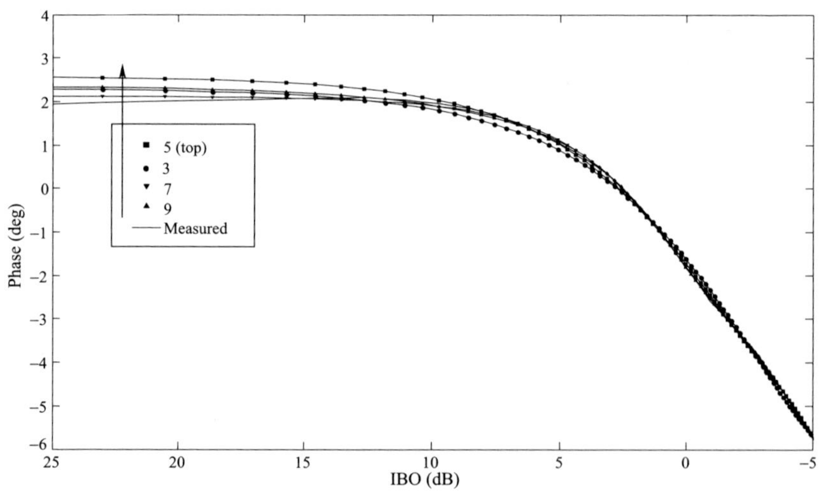

(b)

Figure 3.16 Bessel–Fourier models, using three, five, seven and nine coefficients and $\alpha = 0.7$, for the LDMOS PA: (a) AM–AM and (b) AM–PM characteristics. The measurement curves (without symbols) lie below the other curves in both (a) and (b).

coefficients from three to 20. Clearly values of $\alpha \leq 1.0$ will yield the best results. From this diagram it may also be noted that the improvement in model fit in going from ten to 20 coefficients is not significant; the average error levels out at a little under 10^{-3}. The average error here is the absolute error between the measurements and the model, averaged over the PA's dynamic range. Such graphs will also depend somewhat on the number of measurement points used in the extraction process and their distribution. The results using values of α as presented here, i.e. based on an input normalised to the 1 dB compression point, will translate fairly reliably to the

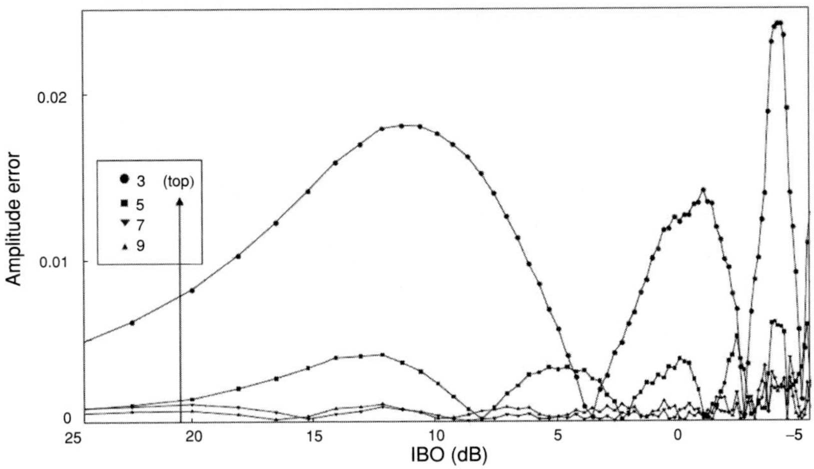

(a)

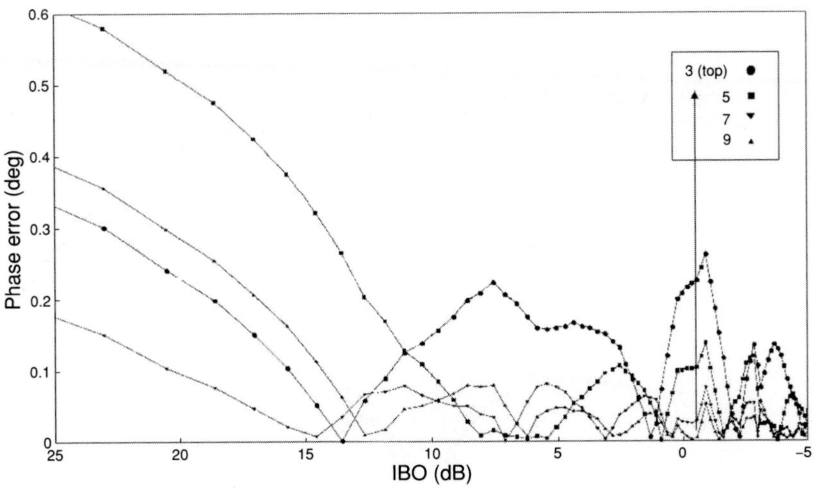

(b)

Figure 3.17 Error graphs for (a) the AM–AM and (b) the AM–PM characteristics for the Bessel–Fourier models in Figure 3.16.

modelling of any PA characteristics with a similar normalisation rule applied.

When extracting a BF model (and a Fourier series model for that matter) it is important to consider what happens at the measurement extremes, in particular at the upper measurement limit (the saturation region) of the AM–AM characteristic. Two considerations need to be addressed here: the behaviour at the saturation limit and the Gibbs effect, familiar in Fourier series approximations.

The Bessel functions given in Equation (3.16) begin to cycle as the argument increases. In the model, through the normalisation of A and the scaling effect of α, the magnitude of the argument is constrained to avoid this cycling or

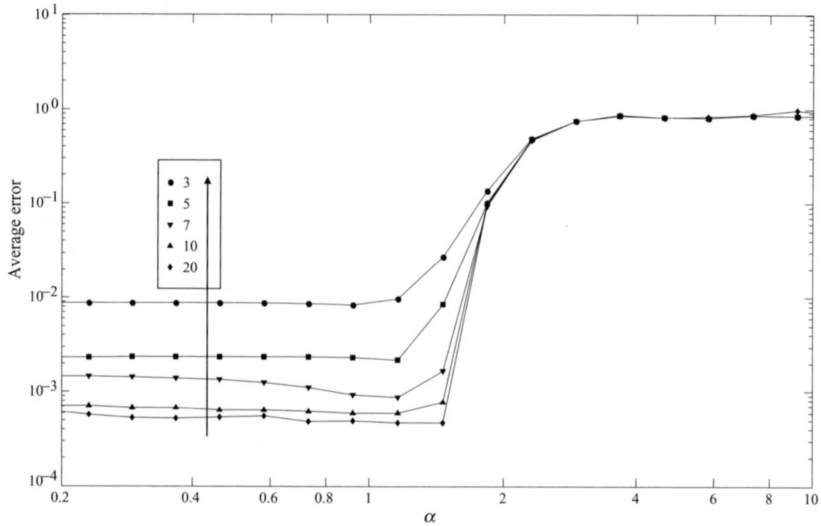

Figure 3.18 Bessel–Fourier model goodness-of-fit as a function of α, with the number of model coefficients as a parameter; the input is presumed to be normalised to the 1 dB compression point.

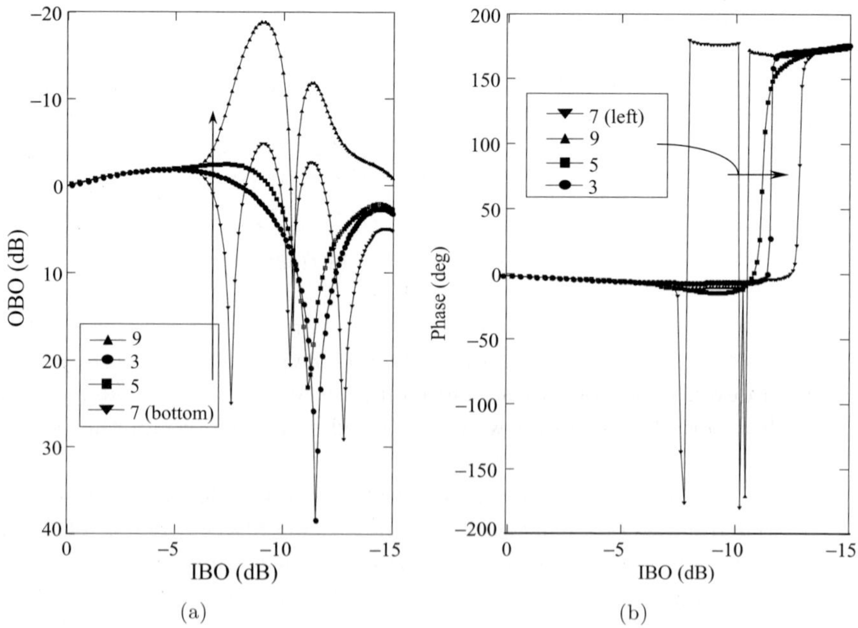

Figure 3.19 Instability beyond the modelled dynamic range for the (a) AM–AM and (b) AM–PM Bessel–Fourier models in Figure 3.16.

instability. Beyond this the instability, as shown in Figure 3.19 for the LDMOS PA model, will cause the model to yield meaningless results. To avoid such an instability impinging on the behavioural model, it is recommended that the envelope characteristics be extended well into saturation, typically by several dB beyond the input signal's expected maximum; this extension may be achieved by common-sense extrapolation if measurement is not possible. Various extrapolation options may be considered with a view to a reduction in the number of coefficients or an improvement in the resulting model accuracy, or both. As the BF-model form does not constitute a period in a periodic extension, unlike the Fourier series model, seeking to create a model from a smoothed periodic extension of the measurements has no mathematical guarantee of any improvement, and modelling tests have borne this out.

3.8.3 Fourier-series-optimised Bessel–Fourier (FOBF) model

As the BF model is derived from the Fourier series model of the instantaneous transfer characteristics, the potential to use this linkage further to improve the BF model may be considered. The main possibility lies in introducing changes to the instantaneous transfer characteristics outside the PA's dynamic range which might either reduce the number of coefficients or improve the accuracy for a given number of coefficients in the Fourier series model; this in turn would translate into fewer coefficients in the BF model, via Equation (3.59).

The algorithmic approach is to convert a good BF model of the envelope characteristics, i.e., one obtained using ten or more coefficients extracted from the envelope characteristics, into a Fourier series RF instantaneous model. Figure 3.20 shows the instantaneous Fourier series characteristics G_{RF} so found and the originating BF envelope model. An inverse Chebyshev transform could also be used for this [4]. The amplitude and phase characteristics in the Fourier series model have, respectively, odd and even symmetry about zero and about the discontinuities at the boundary (not shown here). The latter leads to the presence of significant higher-order coefficients that are visible, especially in a direct periodic extension, as the Gibbs effect. However, being outside the PA's dynamic range the discontinuity may be removed by an even reflection of the characteristics about the boundary. Figure 3.21 shows such a periodic extension with smoothed symmetry about the boundaries as well as about zero. Now, with discontinuities removed, any Gibbs effect is also removed. Applying an FFT will yield a set of complex coefficients. With this forced symmetry, even-order coefficients are driven to zero. Of the odd coefficients only the first five or so will be of any significance; this is a drop from ten or more. Using Equation (3.59) the optimised BF is obtained.

The modelling benefit of this Fourier-series-optimised BF approach is seen only when using models with low numbers of coefficients. For instance the Fourier-series-optimised BF model with three coefficients (i.e. the first, third and fifth coefficients) may be about twice as accurate as that without optimisation. Here, this is so: the accuracy error, measured as the accumulated absolute difference between model and

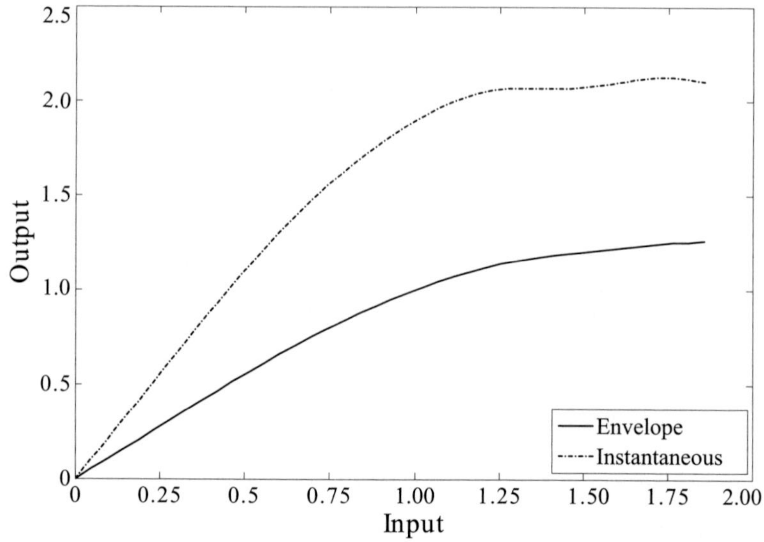

(a)

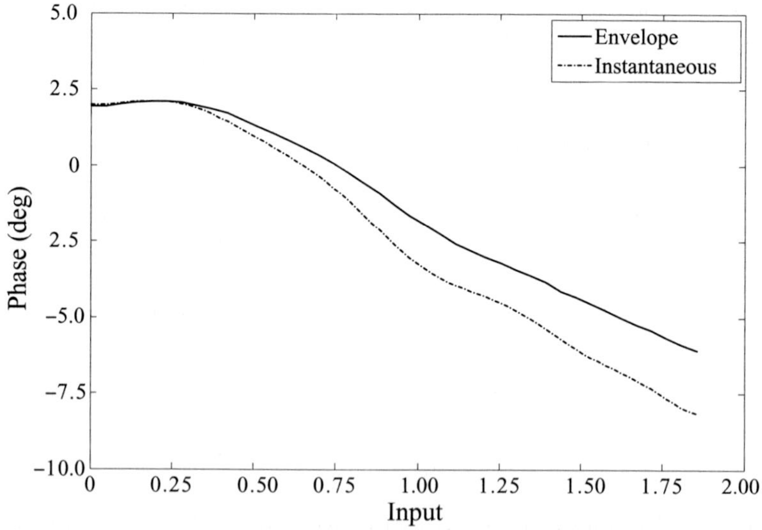

(b)

Figure 3.20 Fourier series RF instantaneous characteristic G_{RF} and the BF model of the LDMOS PA's polar envelope characteristic from which it was derived.

measurements over the dynamic range modelled, is halved to 0.0041 from 0.0085. These two models together with the LDMOS PA's extracted envelope characteristics may be seen in Figure 3.22. For higher numbers of coefficients there is little to distinguish the FOBF from the BF. This is not unexpected, as among all the techniques considered the 'un-optimised' BF model was already seen to yield the best memoryless modelling results for five or more coefficients; see Table 3.2.

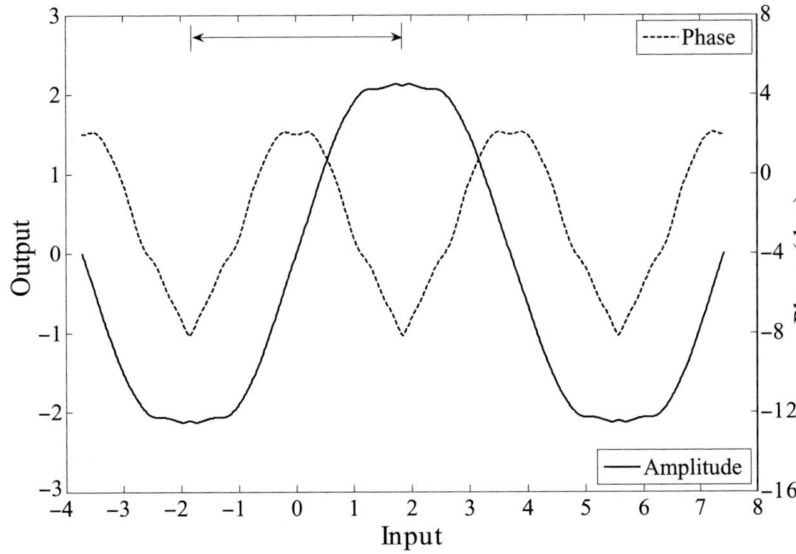

Figure 3.21 A periodic extension of the instantaneous characteristics with no boundary discontinuities, yielded by an even reflection at the boundary. The instantaneous dynamic range part representing the model is indicated by the arrows ↔.

3.9 Hetrakul and Taylor model

Hetrakul and Taylor [5, 6] developed a memoryless nonlinear behavioural quadrature model with a view to characterising satellite TWTA PAs. Such a model may be constructed from the nonlinear device's AM–AM and AM–PM envelope characterisation measurements, Equation (3.7). The in-phase and quadrature amplitude nonlinearity model equations depend on two parameters only:

$$\begin{aligned} P(r) &= \alpha_{\mathrm{p}} r e^{-\beta_{\mathrm{p}} r^2} I_0(\beta_{\mathrm{p}} r^2), \\ Q(r) &= \alpha_{\mathrm{q}} r e^{-\beta_{\mathrm{q}} r^2} I_0(\beta_{\mathrm{q}} r^2) \end{aligned} \tag{3.61}$$

where $P(r)$ and $Q(r)$ are the in-phase and quadrature components of the output and r is the input envelope signal; $I_0(\cdot)$ is the modified Bessel function of the first kind and $(\alpha_{\mathrm{p}}, \beta_{\mathrm{p}})$ and $(\alpha_{\mathrm{q}}, \beta_{\mathrm{q}})$ are the model coefficient sets. A disadvantage of this model which may be observed immediately is that coefficient extraction is a nonlinear optimisation problem in respect of the coefficients β_{p} and β_{q}, which will require appropriate optimisation algorithms, e.g. the Flecher–Powell algorithm.

The Hetrakul and Taylor model works reasonably well with TWTA-type characteristics, as was shown by Saleh when he compared it with his model in this context, Figure 3.9. But, as with the Saleh model, its modelling capacity degrades when the AM–PM characteristics deviate from the TWTA-type ones. For the LDMOS characteristics used in the example here, Figure 3.23 shows the model results versus measurements. Results for the in-phase characteristic $P(r)$ gave

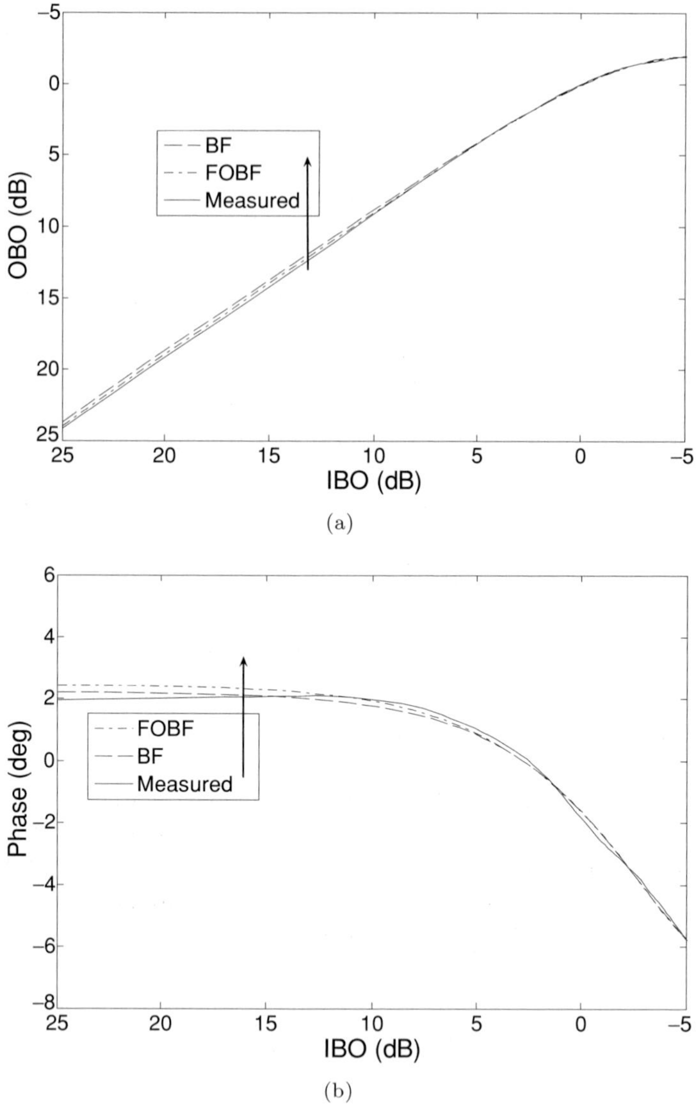

Figure 3.22 The FOBF and BF models, using three coefficients, together with the envelope measurements of the LDMOS PA: (a) AM–AM and (b) AM–PM.

reasonably good agreement, but those for the quadrature characteristic $Q(r)$ were poor; in fact, the model fails. Hence, as might be expected, the overall model results, Table 3.2, were not good. In fact they were among the poorest for the LDMOS PA models presented in this chapter. Nonetheless, they are still not completely unreasonable, for the simple reason that the signal power through the quadrature arm is small relative to that through the in-phase arm and so has little impact on the output. In fact the P model on its own would perform better.

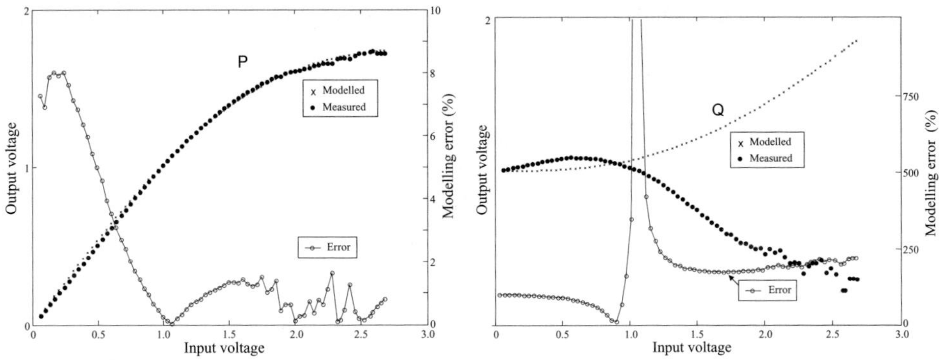

(a) (b)

Figure 3.23 Hetrakul and Taylor quadrature model result and measurements of the LDMOS PA: (a) in-phase component P and (b) quadrature component Q.

3.10 Berman and Mahle model

Berman and Mahle [7], in studying impairment effects of the nonlinear AM–PM characteristics Φ of TWTAs, proposed a three-coefficient model (α, η, γ) based on an exponential function:

$$\Phi(r) = \alpha\left(1 - e^{-\beta r}\right) + \gamma r. \tag{3.62}$$

Coefficient extraction requires a nonlinear optimisation process. An optimised result for its approximation of the LDMOS PA's AM–PM is shown in Figure 3.24.

This model is an improvement on either Saleh's model or that of Hetrakul and Taylor for this amplifier, but it lags behind what can be achieved by the modified Saleh model, by the Bessel–Fourier model or by a complex power series using higher numbers of coefficients. For a full model (i.e. including AM–AM) the Berman and Mahle (BM) model could be combined with one of the others constrained to AM–AM modelling. The data given in the last line of Table 3.2 were obtained by combining the BM model with a power series AM–AM model.

3.11 The Wiener expansion

The Taylor series expansion is a natural choice for describing a system where the output is described as a nonlinear function of the input. However, as such series approximate functions locally, the approximation error increases as one moves to the extremes of this local area, e.g. by increasing the input signal amplitude. Also, the identification of the coefficients of the Taylor series expansion is not straightforward, since all the terms of the series give simultaneous contributions. This makes it difficult to separate their effects, especially the higher-order terms as these are

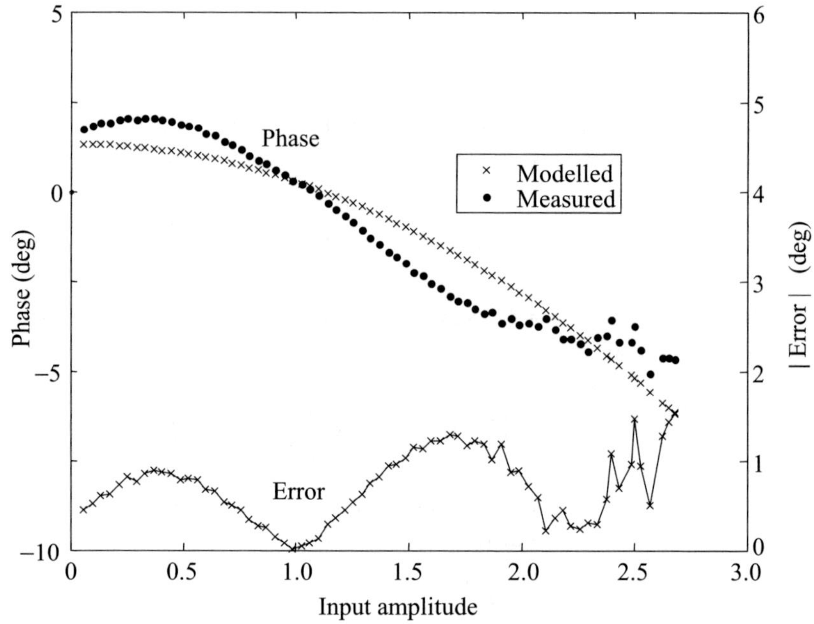

Figure 3.24 Berman and Mahle model for the AM–PM envelope characteristics of the LDMOS PA.

screened by the lower-order ones. In the 1950s Wiener [27, 12] studied an alternative expansion that could allow for easier identification; he proposed a new polynomial expansion using a polynomial orthonormal basis, together with an identification procedure able to decouple the effects of each kernel. Given a function $y(t)$, the coefficients of the Wiener expansion are found by minimising the distance, according to a chosen norm, between the truncated expansion and $y(t)$ and forcing the polynomial basis elements to be orthonormal by adopting the same norm criteria [27–30]. To formalise these concepts, a memoryless nonlinear system with input $x(t)$ and output $y(t)$ is considered; however, for greater clarity the dependence of x and y on t is largely omitted in the derivations below. The input–output relationship is approximated in terms of the *truncated* Wiener polynomial expansion,

$$y_{G,p} = \sum_{n=0}^{p} C_n \Psi_n(x), \tag{3.63}$$

where $y_{G,p}$ denotes the truncated expansion of order p, the C_n are the Wiener kernels to be extracted (there will be p kernels in the truncated approximation) and the Ψ_n are nth-degree polynomials on x defined as

$$\Psi_n(x) = \sum_{k=0}^{n} a_k x^k. \tag{3.64}$$

Through the use of a function $e(p)$ defined as

$$e(p) = y - y_{G,p}, \tag{3.65}$$

a weighted mean distance d_p between $y(x)$ and its truncated approximation $y_{G,p}(x)$ of order p may be introduced. A function $\xi^2(x)$ provides the weighting:

$$d_p = \int_{-\infty}^{\infty} \xi^2(x)e^2(p)dx = \int_{-\infty}^{\infty} \left[\xi(x)y(x) - \xi(x)\sum_{n=0}^{p} C_n\Psi_n(x)\right]^2 dx. \tag{3.66}$$

Hence, the values of C_n may be found by minimising d_p. This can be achieved as follows: for all m,

$$\frac{\partial d_p}{\partial C_m} = 0 = \int_{-\infty}^{\infty} \left[\xi(x)y(x) - \xi(x)\sum_{n=0}^{p} C_n\Psi_n(x)\right] \xi(x)\Psi_m(x)dx. \tag{3.67}$$

The polynomials Ψ_n are orthonormal; this orthonormality condition on the polynomial basis is what characterises the Wiener expansion. Thus

$$\int_{-\infty}^{\infty} \Psi_n(x)\Psi_m(x)\xi^2(x)dx = \delta_{nm}, \tag{3.68}$$

where δ_{nm} is 1 for $n = m$ and 0 otherwise: then, from (3.67),

$$C_m = \int_{-\infty}^{\infty} \Psi_m(x)y(x)\xi^2(x)dx. \tag{3.69}$$

It is here that the role of the weighting function $\xi(x)$ in realising a kernel-extraction implementation scheme becomes clear. For an input signal $x(t)$ describable also as an ergodic statistical process, with statistical distribution $\xi^2(x)$, the time average and statistical average are the same. Hence we may write

$$\overline{\Psi_m(t)y(t)} = \lim_{A\to\infty} \frac{1}{2A}\int_{-A}^{A} \Psi_m(t)y(t)dt = \int_{-\infty}^{\infty} \Psi_m(x)y(x)\xi^2(x)dx = C_m, \tag{3.70}$$

Equation (3.70) shows that the identification of each C_m can be accomplished by implementing the scheme presented in Figure 3.25. The upper box represents the nonlinear PA amplification with input signal $x(t)$ and output $y(t)$. These are the measurements for the extraction process. The box labelled $\Psi_m(t)$ applies the mth polynomial to the input-signal measurement samples. The product of these and $y(t)$ are averaged, i.e. low-pass filtered, to yield the kernel C_m, Equation (3.70).

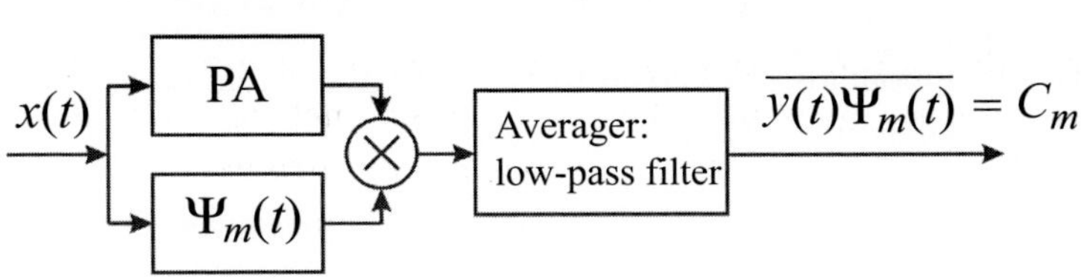

Figure 3.25 Scheme for the extraction of the Wiener kernels.

Of course, the identification process (the extraction of the C_m) could be carried out using a test input signal with statistical properties different from those of the applied signal, but in this case the model behaviour would not be optimal, in the sense that for a fixed truncation order the error would not be minimal.

It can be demonstrated [12] that the polynomial basis found above is complete and therefore that, for any system y for which

$$\int_{-\infty}^{\infty} y(x)\xi^2(x)dx \leq \infty, \tag{3.71}$$

the truncation error can be reduced to an arbitrarily small value by increasing the series order.

As examples of the Wiener-approach application we will consider two cases: the first uses signals with a uniform statistical distribution and the second uses Gaussian distributed signals, which play a major role in communication systems.

3.11.1 *Uniformly distributed signals*

In this case, the probability density distribution function is of the type:

$$\begin{aligned} \xi^2(x) &= 1 \quad \text{for } -1 \leq x \leq 1, \\ \xi^2(x) &= 0 \quad \text{otherwise.} \end{aligned} \tag{3.72}$$

The first three vector basis polynomials,

$$\begin{aligned} \Psi_0(x) &= a_0^0, \\ \Psi_1(x) &= a_1^0 + a_1^1 x, \\ \Psi_2(x) &= a_2^0 + a_2^1 x + a_2^2 x^2, \end{aligned} \tag{3.73}$$

are found by forcing their orthonormality, so that

$$\begin{aligned} &a_0^0 = \sqrt{1/2}, \\ &a_1^0 = 0, \qquad a_1^1 = \sqrt{3/2}, \\ &a_2^0 = \sqrt{5/8}, \qquad a_2^1 = 0, \quad a_2^2 = -3\sqrt{5/8}. \end{aligned} \tag{3.74}$$

Thus

$$\begin{aligned} y_3^{\text{Wiener}}(x) &= \sqrt{1/2}\,C_0 + \sqrt{3/2}\,xC_1 + \sqrt{5/8}\,(1-3x^2)C_2 \\ &= \sqrt{1/2}\,C_0 + \sqrt{5/8}\,C_2 + \sqrt{3/2}\,C_1 x - 3\sqrt{5/8}\,C_2 x^2. \end{aligned} \tag{3.75}$$

These polynomials are known to be Legendre polynomials. Considering the Taylor series expansion

$$y_3^{\text{Taylor}}(x) = a_0 + a_1 x + a_2 x^2, \tag{3.76}$$

the difference between it and the Wiener expansion for a uniformly distributed signal may be noted.

3.11.2 *Gaussian-distributed signals*

When the signals are Gaussian distributed the probability density distribution function is of the type

$$\xi^2(x) = \frac{1}{2\pi\sigma} \exp\left(-\frac{x^2}{2\sigma^2}\right), \tag{3.77}$$

where for simplicity the Gaussian mean value is assumed to be zero.

Again using forced polynomial orthonormality, and remembering that

$$\int_{-\infty}^{\infty} \xi^2(x)dx = 1, \quad \int_{-\infty}^{\infty} x\xi^2(x)dx = \int_{-\infty}^{\infty} x^3\xi^2(x)dx = 0,$$
$$\int_{-\infty}^{\infty} x^2\xi^2(x)dx = \sigma^2, \quad \int_{-\infty}^{\infty} x^4\xi^2(x)dx = 3\sigma^4, \tag{3.78}$$

we find

$$\begin{aligned} a_0^0 &= 1, \\ a_1^0 &= 0, \quad a_1^1 = \frac{1}{\sigma}, \\ a_2^0 &= -\sqrt{\frac{1}{2}}, \quad a_2^1 = 0, \quad a_2^2 = -\sqrt{\frac{1}{2\sigma^2}} \end{aligned} \tag{3.79}$$

and finally

$$y_3^{\text{Wiener}}(x) = \left(1 - \sqrt{\frac{1}{2}}\right) C_0 + \frac{1}{\sigma} C_1 x - \sqrt{\frac{1}{2\sigma^2}} C_2 x^2. \tag{3.80}$$

The polynomials obtained are known to be the Hermite polynomials.

Finally, it is important to notice that the Wiener approach can be extended to systems with memory [27, 31]. In this case, the mathematical formulation is far more complicated. However, it can be carried out in an analogous way: the memory Wiener expansion is again defined in terms of orthonormal kernels but now represented through multiple integral relationships.

3.12 Other comparative considerations

A core focus of this chapter was to present the main memoryless models as encountered today in the literature. Some modelling-accuracy comparative analysis of the models has been presented in respect of modelling an LDMOS PA amplifying a WCDMA signal, e.g. in Table 3.2. This analysis is not intended to be comprehensive, as that is not our goal.

In this chapter we first introduced fundamental ideas on memoryless behavioural modelling, setting out the terminology and the application context and its importance. The memoryless concept refers to the instantaneous nonlinear transfer characterisation of the amplifier and thus inherently refers to the AM–AM and AM–PM nonlinear envelope transfer characteristics. The inclusion and treatment of memoryless AM–PM without use of the category 'quasi-memoryless' is in accordance with the definition of behavioural modelling properties set out in subsection 2.4.2. The link between memoryless models and models with memory was addressed only briefly, since the following two chapters will deal extensively with models catering for linear and nonlinear memory.

Among the models presented, the Saleh, modified Saleh, complex power series and Bessel–Fourier models, in the forms stated in this chapter, are all quite accessible. Their parameters generally may be extracted by a linear least-squares or similar optimisation process. In the modified Saleh models there can be an initial optimisation of the η, ν, γ and ε parameters. In the Bessel–Fourier model an optimisation of the extracted parameters through a Fourier series model of the instantaneous behaviour is possible. Besides being used in PA behavioural analysis, in the forms given earlier they may be employed in direct time-domain (DTD) simulations. Much circumspection is advised in the use of this simulation approach when modelling amplifiers operating in their severely nonlinear regions (large-signal operation) and especially when processing complex input signals manifesting a significant envelope PAPR of the kind encountered in many modern RF air interfaces [26].

The models described in this chapter may be categorised:

- by their capability for behavioural analysis in the harmonic band as well as for zonal band analysis; this holds for the complex power series, Equation (3.28), the Saleh model, Equation (3.37), and especially the Bessel–Fourier model, Equation (3.57);
- by whether decomposed versions of the models are available, as in Equations (3.12) and (3.15). These are useful for the controlled calculation and analysis of IMPs and harmonics and for managing potential aliasing problems. Such versions are available for the Bessel–Fourier models, Equation (3.57) – for example it has been used in OFDM simulations [26] – and for the Saleh model, Equation (3.37). The computational complexity for the latter is relatively high;
- by their extensibility, i.e. the ease of adding new coefficients to improve accuracy, as in the series-based models, e.g. the complex power series and the Bessel–Fourier models. For the extensible models dealt with here, low numbers of coefficients yielded good accuracy well into the PA's large-signal region.

The modified Saleh model was designed to overcome some shortcomings of the popular two-parameter Saleh model, which are especially evident when modelling the AM–PM characteristics. Omitting the AM–PM part and using the dominant AM–AM nonlinearity Saleh model alone will still yield in many cases reasonable overall impairment predictions. This is all the more so where the AM–PM nonlinear variations are quite small, as frequently occurs in solid-state PAs; hence their distortion contribution is not significant. Nonetheless checking this is advisable before making a decision to discount it. Besides, there are many situations where it may be important to model the AM–PM characteristics properly, e.g. for feedback control signals in linearisation schemes, where even small effects may have a significant impact, or for timing precision in satellite positioning systems.

The above model is shown to be reliable in modelling the AM–PM characteristics and also to perform better AM–AM modelling than the original Saleh model. It is still a two-parameter model although, as for the original version, it may be argued to be more.

A drawback of many schemes, complex power series, Saleh, modified Saleh, Hetrakul and Taylor, Berman and Mahle etc., is the difficulty of deriving decomposed models for IMP analysis and for enabling the control of aliasing effects when carrying out behavioural analysis for a PA that is amplifying complex input signals. (The alternative upsampling solution for aliasing error reduction in some simulation contexts is an awkward brute-force type of solution; see Chapter 7.) Here the Bessel–Fourier behavioural model comes into its own, as in it the decomposition of IMPs and the control of aliasing are very manageable. The extraction of harmonic-zone outputs is also feasible. As a model, besides handling AM–AM and AM–PM with equal ease, it can be made as accurate as needed by increasing the number of coefficients. In Table 3.2 it can be seen to yield results among the best.

Some AM–AM and AM–PM approximating techniques are based on the statistical properties of the input signal. These properties need to be known since then a power expansion can be defined such that the coefficients can be extracted to a desired order directly from measurements on the given input signal without any interpolation or approximation. This is the basis of the so-called Wiener expansion, dealt with in Section 3.11 of this chapter.

References

[1] D. D. Silveira, M. E. Gadringer, P. L. Gilabert *et al.*, "Comparison of RF power amplifier behavioural models estimated from shared measurement data," in *Proc. TARGET Meets Industry Int. Colloq.*, Frascati, Rome, November 2006, pp. 129–132.

[2] O. Shimbo, *Transmission Analysis of Communications Systems*, vol. 2, Computer Science Press, 1988.

[3] M. O'Droma, N. Mgebrishvili, E. Bertran, A. A. Goacher, B. Bunz and Y. Lei, "Percentage linearisation: a new measure for RF power amplifier linearisation analysis," Chapter 3 in *Characterization and Modelling Approaches for Advanced Linearisation Techniques*, eds. J. A. Garca, J. M. Zamanillo and M. O'Droma, Research Signpost, pp. 63–80, 2005.

[4] N. M. Blachman, "Detectors, bandpass nonlinearities, and their optimisation: inversion of the Chebyshev transform," *IEEE Trans. Communications*, vol. 20, pp. 965–972, October 1972.

[5] P. Hetrakul and D. P. Taylor, "Nonlinear quadrature model for travelling wave tube type amplifier," *Electronics Lett.*, vol. 11, p. 50, January 1975.

[6] P. Hetrakul and D. P. Taylor, "Compensators for bandpass nonlinearities in satellite communications," *IEEE Trans. Communications*, vol. 24, pp. 546–553, May 1976.

[7] A. L. Berman and C. H. Mahle, "Nonlinear phase shift in travelling-wave tubes as applied to multiple access communication satellites," *IEEE Trans. Communications*, vol. 18, pp. 37–48, February 1970.

[8] E. Biglieri, S. Benedetto and R. Daffara, "Modeling and performance evaluation of non linear satellite links: a Volterra series approach," *IEEE Trans. Aerospace and Electronic Systems*, vol. 15, no. 4, July 1979, pp. 494–506.

[9] M. Wath, *Volterra and Integral Equations of Vector Functions*, Marcel Dekker, 2000.

[10] J. Barrett, "The use of functionals in the analysis of nonlinear physical systems," *J. Electronics and Control*, vol. 15, no. 6, 1957, pp. 567–615.

[11] W. Rugh, *Nonlinear System Theory. The Volterra/Wiener Approach*, Johns Hopkins University Press, 1981.

[12] M. Schetzen, *The Volterra and Wiener Theories of Nonlinear Systems*, John Wiley & Sons, 1980.

[13] C. Chien, *Digital Radio On Chip*, Kluwer, 2001.

[14] C. Fager, J. C. Pedro, N. B. de Carvalho, H. Zirath, F. Fortes and M. J. Rosario, "A comprehensive analysis of IMD behaviour in RF CMOS power amplifiers," *IEEE J. Solid-State Circuits*, vol. 39, no. 1, pp. 24–34, January 2004.

[15] A. A. M. Saleh, "Frequency-independent and frequency-dependent nonlinear models of TWT amplifiers," *IEEE Trans. Communications*, vol. 29, no. 11, pp. 1715–1720, November 1981.

[16] A. Kaye, D. George and M. Eric, "Analysis and compensation of bandpass nonlinearities for communications," *IEEE Trans. Communications*, vol. 20, no. 5, pp. 965–972, October 1972.

[17] O. Shimbo, "Effects of intermodulation, AM–PM conversion, and additive noise in multicarrier TWT systems," *Proc. IEEE*, vol. 59, pp. 230–238, February 1971.

[18] P. Kenington, *High linearity RF amplifier design*, Artech House, 2000.

[19] C. M. Thomas, M. Y. Weidner and S. H. Durrani, "Digital amplitude-phase keying with M-ary alphabets," *IEEE Trans. Communications*, vol. 22, pp. 168–180, February 1974.

[20] M. S. O'Droma, N. Mgebrishvili and A. Goacher, "New percentage linearisation measures of the degree of linearisation of HPA nonlinearity," *IEEE Communications Lett.*, vol. 8, no. 4, pp. 214–216, April 2004.

[21] J. A. García, J. M. Zamanillo and M. S. O'Droma eds., *Characterization and Modelling Approaches for Advanced Linearisation Techniques*, Research Signpost 37/661 (2), 2005.

[22] J. C. Fuenzalida, O. Shimbo and W. L. Cook, "Time-domain analysis of intermodulation effects caused by nonlinear amplifiers," *COMSAT Technical Review*, vol. 3, pp. 89–143, Spring 1973.

[23] M. S. O'Droma, "Dynamic range and other fundamentals of the complex Bessel function series approximation model for memoryless nonlinear devices," *IEEE Trans. Communications*, vol. 37, no. 4, pp. 397–398, April 1989.

[24] X. T. Vuong and H. J. Moody, "Comments on a general theory for intelligible crosstalk between frequency-division multiplexed angle-modulated carriers," *IEEE Trans. Communications*, vol. COM-28, no. 11, pp. 1939–1943, November 1980.

[25] Reply to [24], *IEEE Trans. Communications*, vol. COM-28, no. 11, pp. 1943–1944, November 1980.

[26] M. S. O'Droma and N. Mgebrishvili, "Signal modeling classes for linearized OFDM SSPA behavioural analysis," *IEEE Communications Lett.*, vol. 9, no. 2, pp. 127–129, February 2005.

[27] N. Wiener, *Nonlinear Problems in Random Theory*, Technology Press, MIT Wiley, 1958.

[28] M. Schetzen, "Nonlinear system modeling based on the Wiener theory," *Proc. IEEE*, vol. 69, no. 12, pp. 1557–1573, December 1981.

[29] M. Schetzen, "Determination of optimum nonlinear systems for generalized error criteria based on the use of gate functions," *IEEE Trans. Information Theory*, pp. 117–125, January 1961.

[30] M. Schetzen, "Measurement of the kernels of a nonlinear system of finite order," *Int. J. Control*, vol. 1, no. 3, pp. 251–263, March 1965.

[31] Y. W. Lee, "Contributions of Norbert Wiener to linear theory and nonlinear theory in engineering," in *Selected Papers of Norbert Wiener*, Cambridge MA.

4 Nonlinear models with linear memory

4.1 Introduction

Conventional nonlinear static models, such as AM–AM and AM–PM representations, are frequency independent and can represent with reasonable accuracy the characteristics of various amplifiers driven by narrowband input signals. However, if an attempt is made to amplify 'wideband' signals, where the bandwidth of the signal is comparable with the inherent bandwidth of the amplifier, frequency-dependent behaviour will be encountered in the system. This kind of phenomenon is described as a memory effect and can be classified as linear or nonlinear (subsection 1.2.2) or as short- or long-term (subsection 2.4.1). Taking account of both nonlinearities and memory effects when modelling a PA becomes very important in wideband system designs.

In this chapter we focus on investigating nonlinear models that include memory effects (i.e. frequency-dependent behaviour) by using linear filters. The models considered are categorised, according to the selected structure, as two-box, three-box or parallel-cascade models. In the two-box-model section the Wiener and Hammerstein models are discussed, while the three-box-model section includes the Poza–Sarkozy–Berger (PSB) model and the frequency-dependent Saleh model. These models represent some first attempts at extending the nonlinear static AM–AM and AM–PM models to cover frequency-dependent effects. The Abuelma'atti and polyspectral models are major representatives of models organised in parallel branches describing linear memory and will be introduced in the parallel-cascade-model section.

4.2 Two-box models

The two-box modelling technique is also known as the modular [1] or feed-forward block-oriented [2] approach. It is obtained by combining components taken from the following two classes: static (or memoryless) nonlinearities and causal, linear, time-invariant dynamic subsystems. Parametric and nonparametric modelling methodologies can be used [2]. Flexible arrangements of such block-structured models include two possibilities: the Wiener model (linear–nonlinear) and the Hammerstein model (nonlinear–linear) [3]. The overall output is a reliable model that can predict an amplifier's behaviour and its approximate Volterra kernels.

The most frequently used configuration is a finite nonlinear autoregressive moving-average with exogenous input (NARMAX) representation, where the linear subsystem belongs to the linear FIR class and the nonlinearity is represented by a polynomial [2]. Examples of these structures are shown in Figures 4.1 and 4.2. Another possibility for constructing a Wiener system is the use of an FIR or IIR subsystem as a linear block followed by a Bessel function nonlinearity [4].

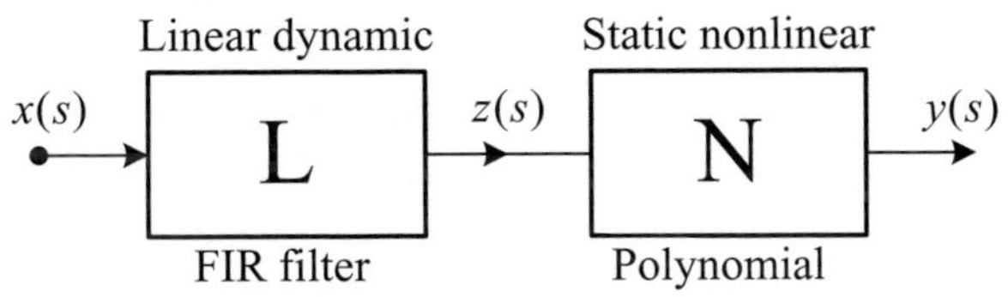

Figure 4.1 The Wiener model.

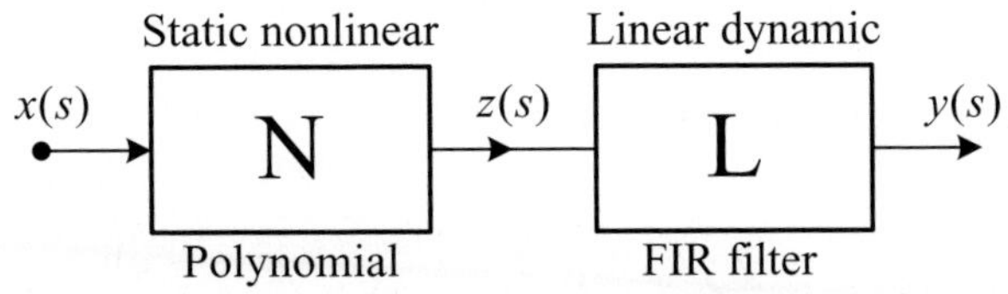

Figure 4.2 The Hammerstein model.

If the linear dynamic block is represented by an FIR filter, the output of this block for the Wiener model is

$$y_{\mathrm{W,linear}}(s) = \sum_{q=0}^{Q-1} h(q)x(s-q) \tag{4.1}$$

and for the Hammerstein model the FIR filter output is given by

$$y_{\mathrm{H,linear}}(s) = \sum_{q=0}^{Q-1} h(q)z(s-q). \tag{4.2}$$

If the static-nonlinearity block is represented by a power series, the output of this block can be formulated for the Wiener model as:

$$y_{\mathrm{W,nl}}(s) = \sum_{n=0}^{N} c^{(n)} z^n(s). \tag{4.3}$$

and for the Hammerstein model as:

$$y_{\mathrm{H,nl}}(s) = \sum_{n=0}^{N} c^{(n)} x^n(s). \tag{4.4}$$

The overall model output is then a combination of the corresponding equations for each model:

$$y_{\mathrm{W}}(s) = \sum_{n=0}^{N} c^{(n)} \left[\sum_{q=0}^{Q-1} h(q)x(s-q)\right]^n, \tag{4.5}$$

$$y_{\mathrm{H}}(s)=\sum_{q=0}^{Q-1}h(q)\left[\sum_{n=0}^{N}c^{(n)}x^{n}(s-q)\right] \tag{4.6}$$

Equations (4.5) and (4.6) are the simplest (but not an inefficient) way to represent a nonlinear amplifier with memory. It should be noted that in these equations RF input and output signals are used. The whole discussion on two-box models presented in this section can also be applied to the low-pass equivalent description. The complex-valued model coefficients may be extracted from the RF description by the transformation presented in Section 2.3. The corresponding low-pass equivalent Wiener and Hammerstein models are then given by

$$\tilde{y}_{\mathrm{W}}(s)=\sum_{m=1}^{N+1/2}\tilde{c}^{(2m-1)}\left|\sum_{q=0}^{Q-1}\tilde{h}(q)\tilde{x}(s-q)\right|^{2(m-1)}\left[\sum_{q=0}^{Q-1}\tilde{h}(q)\tilde{x}(s-q)\right] \tag{4.7}$$

and

$$\tilde{y}_{\mathrm{H}}(s)=\sum_{q=0}^{Q-1}\tilde{h}(q)\left[\sum_{m=1}^{N+1/2}\tilde{c}^{(2m-1)}\left|\tilde{x}(s-q)\right|^{2(m-1)}\tilde{x}(s-q)\right]. \tag{4.8}$$

4.2.1 *Relationship to Volterra kernels*

As Volterra kernels constitute a complete and reliable description of the system's function [1], it is of interest to find the relationship between the above models and the Volterra kernels. A finite discrete-time Volterra series model [5] is given by the following (see Equation (2.1) and (subsection 5.6.1):

$$y_{\mathrm{V}}(s)=\sum_{n=1}^{N}\Delta^{n}\sum_{q_1=0}^{Q-1}\cdots\sum_{q_n=0}^{Q-1}h_n(q_1,\ldots,q_n)x(s-q_1)\cdots x(s-q_n), \tag{4.9}$$

where h_n is the kernel of order n, s and q are discrete indexes of the sampling period, Q is the memory length and Δ is the sampling period (which will be considered as unity for simplicity). According to [1], the sampling period must be selected to achieve the maximum bandwidth of the system and of the input and output signals.

The nth-order kernel represents the nonlinear interactions among n copies of the input. For $n=0$ a constant output is produced, for $n=1$ the one-dimensional linear kernel will be produced, for $n=2$ the nonlinear interactions between two copies of the input will result in two-dimensional matrix and so on.

If Equation (4.5) is written as:

$$y_{\mathrm{W}}(s)=\sum_{n=1}^{N}c^{(n)}\left(\sum_{q_1=0}^{Q-1}\cdots\sum_{q_n=0}^{Q-1}h(q_1)\cdots h(q_n)x(s-q_1)\cdots x(s-q_n)\right), \tag{4.10}$$

the resulting relationship for the nth-order Volterra kernel [2] will be as follows:

$$h_n(q_1,\ldots,q_n)=c^{(n)}h(q_1)h(q_2)\cdots h(q_n). \tag{4.11}$$

Thus the order-n Volterra kernel is equal to the product of n copies of the impulse response function of the linear block multiplied by the nth nonlinearity coefficient.

The Hammerstein–Volterra relationship can be derived in a similar way. The equivalent Volterra kernels are given by

$$h_n(q_1, \ldots, q_n) = \begin{cases} c^{(q)} h(q) & q_1 = q_2 = \cdots = q_n, \\ 0 & \text{otherwise.} \end{cases} \tag{4.12}$$

The Volterra kernels of a Hammerstein system are only nonzero along their diagonals ($q_1 = q_2 = \cdots = q_n$) [2].

Equations (4.11) and (4.12) show how close these representations are to a general Volterra series. Furthermore, they have the advantage that they do not need the large matrices required by Volterra models when high-order kernels are used. Therefore, fifth- or even seventh-order models can be represented in a very convenient way.

4.2.2 *Model memory estimation*

The memory estimation of a block-based model is crucial since it determines the length of the FIR filter and has a direct influence on the model's performance. The system memory length can be estimated from the first-order Volterra kernel [1]. The minimum number of samples (taps) along each dimension of the Volterra kernel required for it to be represented in the discrete-time domain is given by [1]:

$$M = 2B_{\mathrm{s}}\mu \tag{4.13}$$

where B_{s} (in Hz) denotes the system bandwidth and μ is the effective kernel memory or correlation time over which the kernel has significant values. An estimation of the maximum memory length can be performed after an initial model has been extracted. Starting from an estimated point (as indicated in Figure 4.3), the first-order Volterra kernel shows no exponential decay but some oscillations around zero that can be considered as modelling noise. The value of the maximum memory length (350 samples in this case) limits the number of taps used in the model. In [6] it was determined by squaring the magnitudes of the kernels and calculating where approximately 90% of the total kernel energy is concentrated.

Such values are important information to be obtained from the data, since there is no prior knowledge of the amplifier's memory.

4.2.3 *Model estimation methods*

In the following sections some estimation methods that could be applied to either Wiener or Hammerstein models will be discussed. These methods assume the use of broadband time-domain amplifier measurements, as in subsection 2.5.5. Some remarks on the identification of two-box models from swept-tone measurements are given in [7].

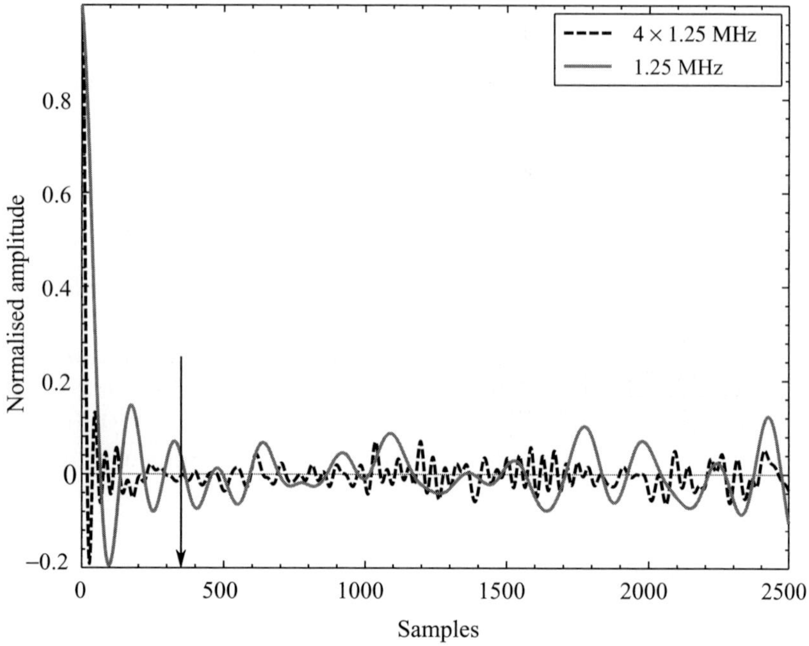

Figure 4.3 Example of a first-order kernel estimation. The arrow shows where the memory limit was obtained for the next estimation step. (Reprinted with permission from [6], © 2005 IEEE.)

At the start of such an identification process a model must be generated to estimate the amplifier memory. Assuming black-box modelling (i.e. no prior information is available) a flexible structure and a system identification method should be selected for this task. A suitable algorithm for the model estimation process can be seen in Figure 4.4. The percentage value of the variance accounted for (M_{VAF}) shown in this algorithm is a figure of merit also used for validation of the model [8]. Other figures of merit can be used as desired. Further information on figures of merit is presented in Chapter 6.

4.2.4 Linear block estimation methods

Several different methods are known for the identification of the linear dynamic block [9]. An overview of the following important representative methods will be given in this section:

- frequency-domain estimation;
- least-squares method;
- Lee–Schetzen correlation method;
- pseudo-inverse technique using the singular-value decomposition.

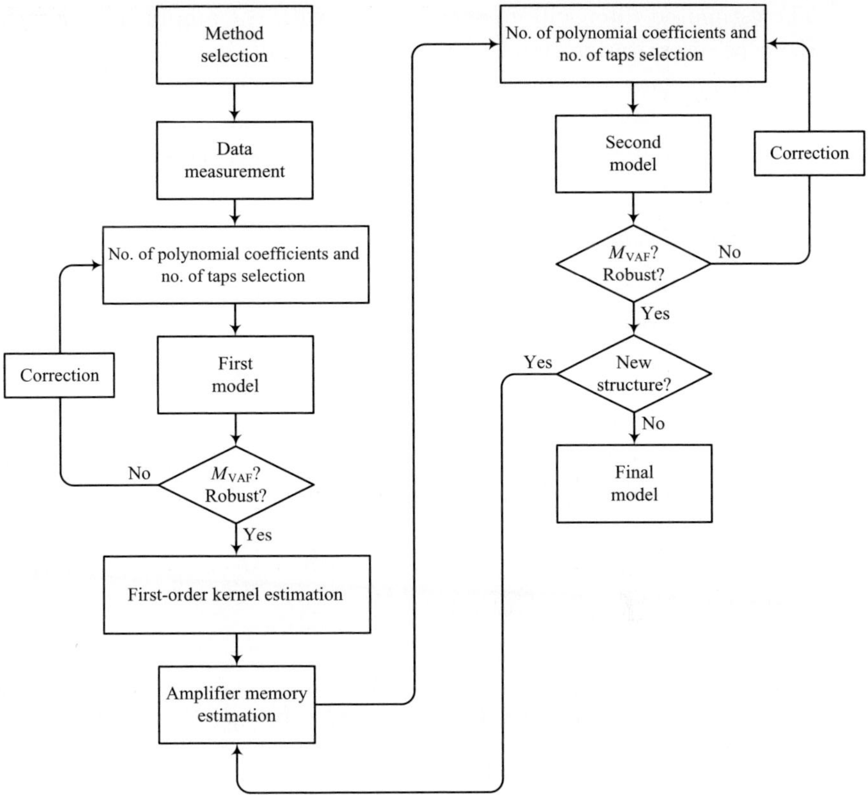

Figure 4.4 Algorithm used for model estimation. (From [6] with permission, © 2005 IEEE.)

Frequency-domain estimation

Using the input power spectrum and the input–output cross-power spectrum it is possible to estimate the frequency response as

$$\hat{H}(\mathrm{f}) = \frac{\hat{S}_{xy}(\mathrm{f})}{\hat{S}_{xx}(\mathrm{f})}, \tag{4.14}$$

where $\hat{H}(\mathrm{f})$ is the estimated frequency response, $\hat{S}_{xy}(\mathrm{f})$ is the input–output cross-power spectrum and $\hat{S}_{xx}(\mathrm{f})$ is the input power spectrum.

The cross-power spectrum $\hat{S}_{xy}(\mathrm{f})$ is defined as the Fourier transform of the cross-correlation function R_{xy}:

$$S_{xy}(\mathrm{f}) = \int_{-\infty}^{\infty} R_{xy}(\tau) \exp(-j2\pi \mathrm{f}\tau)\, d\tau \tag{4.15}$$

Other properties and definitions can be seen in [7]. The input power spectrum $\hat{S}_{xx}(\mathrm{f})$ can be derived in a similar way.

This kind of estimation was applied in [10]; the result was called an *optimal filter*.

The estimated filter will, in general, vary with the amplifier's operating point and the type of excitation signal. As stated in [10] all signal linearity between the input and output data is extracted into this filter.

The main disadvantage of this approach is that much averaging is necessary to reduce the random error to acceptable levels when the signals contain noise.

Least-squares method

The least-squares (LS) method provides another way of estimating the FIR filter. The LS equation is given by [11]:

$$\hat{\theta} = (\mathbf{X}^H\mathbf{X})^{-1}\mathbf{X}^H\mathbf{y} \tag{4.16}$$

where $\hat{\theta}$ is the parameter vector, $\mathbf{X}$ is the regression matrix and $\mathbf{y}$ is the output vector.

The regression matrix $\mathbf{X}$ for K measurements is given by

$$\mathbf{X} = \begin{bmatrix} x(Q) & x(Q-1) & \cdots & x(1) \\ x(Q+1) & x(Q) & \cdots & x(2) \\ \vdots & \vdots & \ddots & \vdots \\ x(Q+K-1) & x(Q+K-2) & \cdots & x(K) \end{bmatrix} \tag{4.17}$$

and the parameter vector $\hat{\theta}$ by

$$\hat{\theta} = \begin{bmatrix} \hat{h}(0) \\ \hat{h}(1) \\ \vdots \\ \hat{h}(Q-1) \end{bmatrix}, \tag{4.18}$$

where the vector $\hat{\mathbf{h}}$ can be interpreted as the input–output signal cross correlation multiplied by the inverse of the autocorrelation of the input [12],

$$\hat{\mathbf{h}} = \mathbf{R}_{xx}^{-1}\mathbf{R}_{xy}. \tag{4.19}$$

Here the matrix $\mathbf{R}_{xx}$ represents the input signal autocorrelation and the vector $\mathbf{R}_{xy}$ represents the input–output signal cross correlation. The main difficult is that neither $\mathbf{R}_{xx}$ nor $\mathbf{R}_{xy}$ is available during the identification process and so they need to be estimated. The expression (4.19) is also known as a *Wiener filter* and can be used for adaptive techniques (see [13]).

Lee–Schetzen correlation method

The Lee–Schetzen method described in [14] uses only cross correlations to calculate an estimate of the filter. This method can be viewed as a first approach to the initial memory estimation. It does not need the long processing time required when large matrix inversions are employed. Although this method has not been shown to generate the most compact model, it allows an important overview of the system kernels.

One example of a very large FIR estimation using the Lee–Schetzen method is shown in Figure 4.3. No previous information about the amplifier memory was available. The amplifier input test signals were white noise processes limited to 1.25 MHz and 5 MHz RF bandwidth [6]. The number of coefficients used in the FIR filter was 2500. After this maximal memory limitation, more precise and efficient models can be fitted using different methods such as the pseudo-inverse technique, which we now describe.

Pseudo-inverse technique

By applying singular-value decomposition (SVD) it is possible to find a suitable form for the least-mean-square estimator $\hat{\mathbf{h}}$ of the impulse response function $\mathbf{h}$ used to estimate the linear block. The pseudo-inverse [15, 16] is derived as follows: we have

$$\hat{\mathbf{h}} = \mathbf{R}_{xx}^{-1}\mathbf{R}_{xy} + \tilde{\mathbf{R}}_{xx}^{-1}\tilde{\mathbf{R}}_{x\mathrm{n}}, \tag{4.20}$$

$$= \mathbf{h} + \tilde{\mathbf{R}}_{xx}^{-1}\tilde{\mathbf{R}}_{x\mathrm{n}}, \tag{4.21}$$

$$= \mathbf{V}\mathbf{S}^{-1}\mathbf{V}^{H}\mathbf{V}\mathbf{S}\mathbf{V}^{H}\mathbf{h} + \mathbf{V}\mathbf{S}^{-1}\mathbf{V}^{H}\tilde{\mathbf{R}}_{x\mathrm{n}}, \tag{4.22}$$

where $\mathbf{R}_{xy} = \mathbf{R}_{xx}\mathbf{h}$, $\mathbf{R}_{xx}^{-1} = \mathbf{V}\mathbf{S}^{-1}\mathbf{V}^{H}$, $\mathbf{R}_{xx}$ is the input signal autocorrelation, $\mathbf{R}_{xy}$ is the input–output signal correlation, $\tilde{\mathbf{R}}_{x\mathrm{n}}$ is is the input–output noise correlation, $\mathbf{V}$ is a matrix composed of the singular vectors and $\mathbf{S}$ is a diagonal matrix formed by the singular values. Substituting $\mathbf{V}^{H}\mathbf{h} = \varsigma$ and $\mathbf{V}^{H}\tilde{\mathbf{R}}_{x\mathrm{n}} = \eta$ yields

$$\hat{\mathbf{h}} = \mathbf{V}\mathbf{V}^{H}\mathbf{h} + \mathbf{V}\mathbf{S}^{-1}\mathbf{V}^{H}\tilde{\mathbf{R}}_{x\mathrm{n}} = \sum_{j=1}^{T}\left(\varsigma_j + \frac{\eta_j}{s_j}\right)v_j. \tag{4.23}$$

It can be seen that only the terms for which $|\varsigma_j| \geq |\eta_j|/s_j$ give a contribution to the estimator in this summation. The minimum description length (MDL) cost function is defined as [3]:

$$\mathrm{MDL}(M) = \left(1 + M\,\frac{\log N}{N}\right)\sum\nolimits_{k=1}^{K}\left[y(k) - \hat{y}(k,M)\right]^2, \tag{4.24}$$

where M is the number of singular vectors, K is the total number of input–output realisations, $y(k)$ is the desired output and $\hat{y}(k,M)$ is the output of the M-parameter model at time k. Using Equation (4.24) it is possible to separate the necessary singular vectors for the estimator.

4.2.5 *Nonlinear block estimation methods*

The nonlinear block of a Wiener or Hammerstein cascade can be identified using the LS method for polynomials; see [12]. Equation (4.16) can still be used, but the regression matrix is defined in another way, using Equations (4.25) and (4.26) below, where $\hat{\theta}$ is the parameter vector of order N and K measurements are considered:

$$\mathbf{X} = \begin{bmatrix} 1 & x(1) & \cdots & x^N(1) \\ 1 & x(2) & \cdots & x^N(2) \\ \vdots & \vdots & \cdots & \vdots \\ 1 & x(K) & \cdots & x^N(K) \end{bmatrix}, \tag{4.25}$$

$$\hat{\theta} = \begin{bmatrix} c^{(0)} \\ c^{(1)} \\ \vdots \\ c^{(N)} \end{bmatrix}, \tag{4.26}$$

The polynomial order is chosen to minimise the MDL function (see the previous subsection). If the regressors are mutually orthogonal, they will lead to a well-conditioned Hessian. This can be achieved using orthogonal polynomials. Two kinds will be considered, Chebyshev polynomials and Hermite polynomials [17].

Chebyshev polynomials

The basis function of a Chebyshev polynomial is bounded between ± 1 for inputs between ± 1, and this leads to similar regressor variances although the regressors are not exactly orthogonal. The model input signals should be normalised to this input range. The modelling error near ± 1 is heavily weighted in this kind of polynomial [14]. This behaviour is an advantage when modelling amplifiers operating with non-Gaussian input signals. The recurrence relation for Chebyshev polynomials is

$$T_{n+1}(z) = 2zT_n(z) - T_{n-1}(z). \tag{4.27}$$

Hermite polynomials

If the input signal is a zero-mean unit-variance Gaussian distribution, the Hermite polynomials are the best choice for input signal orthogonalisation. The recurrence formula for Hermite polynomials is

$$H_{n+1}(z) = 2zH_n(z) - 2zH_{n-1}(z). \tag{4.28}$$

Another way of achieving an orthogonalised solution of the normal equations is by using modified Gram–Schmidt orthogonalisation to compute the QR factorisation of the regression matrix shown below in Equations (4.29)–(4.31):

$$\mathbf{X} = \mathbf{QR}, \tag{4.29}$$

$$\hat{\theta} = (\mathbf{R}^H \mathbf{Q}^H \mathbf{QR})^{-1} \mathbf{R}^H \mathbf{Q}^H \mathbf{y}, \tag{4.30}$$

$$\hat{\theta} = \mathbf{R}^{-1} \mathbf{Q}^H \mathbf{y}. \tag{4.31}$$

One example of the use of these techniques is shown in Figure 4.5, where an RF PA was identified at two different input power backoff levels [18].

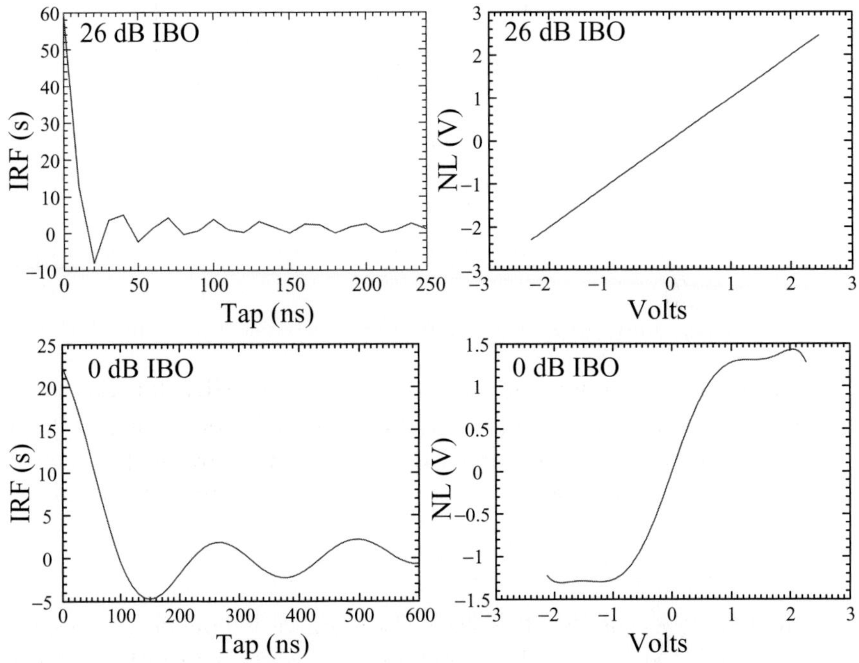

Figure 4.5 Example of a Wiener cascade identification using pseudo-inverse techniques. The input bandwidth was 1.25 MHz. The linear block impulse response function (IRF) and that for the nonlinear block (NL) are plotted. (From [18] with permission, © 2005 IEEE.)

4.3 Three-box models

The three-box model extends the two-box model by an additional filter and, therefore, by an additional degree of freedom in describing the amplifier behaviour. The three-box model structure is depicted in Figure 4.6. This structure is also known as the Wiener–Hammerstein model. Even after adding another filter to the Wiener or the Hammerstein model structure, such models can still only describe nonlinear systems with linear memory effects (see subsection 1.2.2).

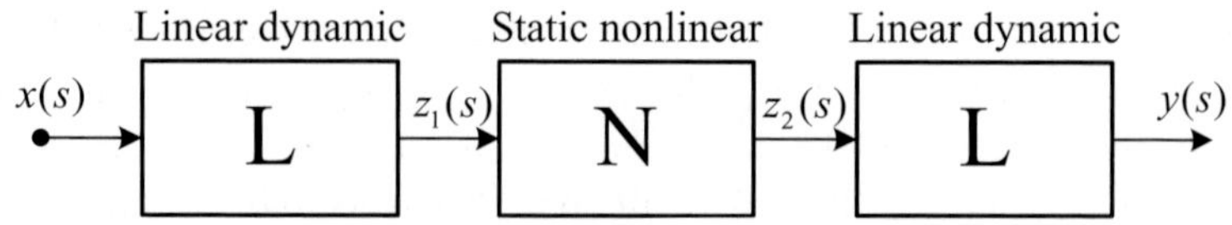

Figure 4.6 Structure of the three-box model.

After a short presentation of the general three-box model three different representatives of this model type will be investigated. Subsection 4.3.1 presents an implementation of the three-box model. Thereafter, the PSB and the frequency-dependent Saleh model are discussed. These two models were designed to allow a proper reproduction of the AM–AM/AM–PM conversion extracted by swept-tone measurements.

If the three-box model is composed of two FIR filters and a polynomial then the model output is given by:

$$y_{\mathrm{W-H}}(s) = \sum_{p=0}^{Q_{\mathrm{b}}-1} h_{\mathrm{b}}(p) \sum_{n=0}^{N} c^{(n)} \left(\sum_{q=0}^{Q_{\mathrm{a}}-1} h_{\mathrm{a}}(q) x(s-p-q) \right)^{n}, \tag{4.32}$$

where $h_{\mathrm{a}}(q)$ and $h_{\mathrm{b}}(p)$ are the impulse response functions of the input and the output filter respectively. Equation (4.32) uses the same notation as Equations (4.1)–(4.6).

To show the relationship to the discrete-time Volterra series, Equation (4.9), Equation (4.32) is rearranged in the following way:

$$\begin{aligned} y_{\mathrm{W-H}}(s) &= \sum_{n=0}^{N} \sum_{q_1=0}^{Q_{\mathrm{a}}-1} \cdots \sum_{q_n=0}^{Q_{\mathrm{a}}-1} \sum_{p=0}^{Q_{\mathrm{b}}-1} c^{(n)} h_{\mathrm{b}}(p) h_{\mathrm{a}}(q_1) \cdots h_{\mathrm{a}}(q_n) \\ &\quad \times x(s-p-q_1) \cdots x(s-p-q_n) \\ &= \sum_{n=0}^{N} \sum_{r_1=p}^{p+Q_{\mathrm{a}}-1} \cdots \sum_{r_n=p}^{p+Q_{\mathrm{a}}-1} \sum_{p=0}^{Q_{\mathrm{b}}-1} c^{(n)} h_{\mathrm{b}}(p) h_{\mathrm{a}}(r_1-p) \cdots h_{\mathrm{a}}(r_n-p) \\ &\quad \times x(s-r_1) \cdots x(s-r_n). \end{aligned} \tag{4.33}$$

The corresponding Volterra kernels are therefore

$$h_n(q_1, \ldots, q_n) = \sum_{p=0}^{Q_{\mathrm{b}}-1} c^{(n)} h_{\mathrm{b}}(p) h_{\mathrm{a}}(q_1-p) \cdots h_{\mathrm{a}}(q_n-p). \tag{4.34}$$

These kernels are composed of the convolution of n copies of the input IRF and the output IRF, weighted by the nth-order nonlinear coefficient.

For the extraction of three-box models based on broadband time-domain measurements, the same methods as presented in the last section can be used. The key question is how to apply the additional degree of freedom to improve the achievable modelling accuracy. A possibility is to evaluate the small-signal response of the amplifier by an additional single-tone measurement, as suggested by Silva *et al.*

in connection with the identification of a polyspectral model [19]. The small-signal gain characteristic is then allocated to the input filter. After filtering the time-domain input signal by the small-signal response a Hammerstein model can be extracted from the filtered input and the measured amplifier output signal. In a similar approach, the linear dependence of the measured time-domain input and output signals can be identified. This linear dependence is then reallocated to the input filter of the three-box model, and the remaining two blocks are extracted as mentioned before.

An adaptive identification of a three-box model was presented by Ibnkahla *et al.* [20]. In this case the static nonlinearity was implemented by a neural network. The proposed update algorithm was able to fit the coefficients of the two filters and the neural network simultaneously.

4.3.1 *Instantaneous quadrature model*

An interesting implementation of a three-box model was presented by Loyka [21] for the simulation of active antenna arrays. To allow a computationally efficient calculation of the model he suggested the structure presented in Figure 4.7. This implementation is called the *instantaneous quadrature technique*. In this approach,

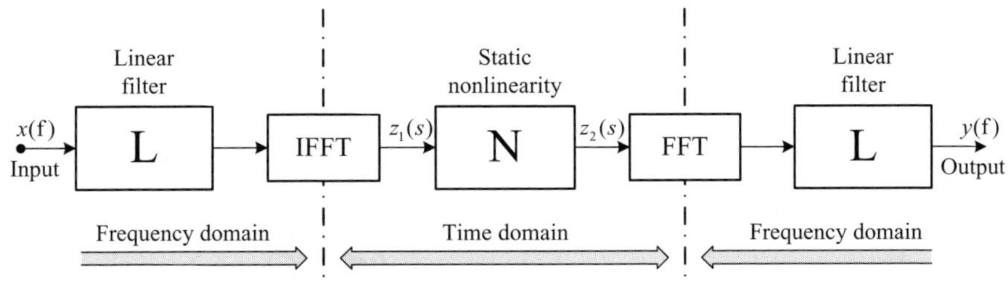

Figure 4.7 Structure of the instantaneous quadrature technique, after Loyka [21].

the band-pass input signal is passed through the first filter in a frequency-domain representation. An inverse fast Fourier transform (IFFT) is used to convert the filter output signal to the time domain and deliver it to the static nonlinear function. By means of a Hilbert transform the in-phase $z_{1,\mathrm{I}}(s)$ and quadrature $z_{1,\mathrm{Q}}(s)$ components are obtained. In Figure 4.8 the complete implementation of the nonlinear block N is shown. The time-domain output signal is given by

$$z_2(s) = z_{1,\mathrm{I}}(s) r\left(r_{z_1}(s)\right) \cos\left[\phi\left(r_{z_1}(s)\right)\right] - z_{1,\mathrm{Q}}(s) r\left(r_{z_1}(s)\right) \sin\left[\phi\left(r_{z_1}(s)\right)\right]. \quad (4.35)$$

Here, the magnitude of $z_1(s)$ is $r_{z_1}(s)$ and the functions $r(\cdot)$ and $\phi(\cdot)$ are the instantaneous AM–AM and AM–PM conversion functions. The output of the nonlinear block is then converted back into the frequency domain and passed through the second filter.

the conversion trace under consideration, two approaches are possible. In [7] the selection of a distinct point along the conversion traces was suggested. This could be, for example, the maximum output power value from the input–output power trace. However, this point is only unique if the measurement input power range also covers values above the point associated with the maximum output power (i.e. the measurement includes the falling part of the input–output power trace). For the example presented in Figure 4.9 this part of the input–output power traces was not captured as it exceeded the amplifier specification. The other possible way to evaluate the optimal alignment is to define a measure for the error between these two traces:

$$\text{Error}_{\text{AM-AM}} = \frac{\sum_{k} \left[P_{\text{out},\text{f}\neq\text{f}_{\text{ref}}}\left(P_{\text{in}}(k)\right) - P_{\text{out},\text{f}_{\text{ref}}}\left(P_{\text{in}}(k)\right)\right]^2}{\sum_{k} P^2_{\text{out},\text{f}\neq\text{f}_{\text{ref}}}\left(P_{\text{in}}(k)\right)}. \tag{4.36}$$

Here the summation over k includes those input power values where the two traces overlap. Hence, minimising this error function for all AM–AM conversion traces different from the reference trace will lead to the desired optimum filter magnitude.

In a similar way to that presented above, the magnitude of the input filter and the phase shift of the output filter can be evaluated for the AM–PM conversion of the amplifier. Figure 4.11 illustrates this approach. It is important to note that the AM–PM conversion traces used for the extraction of the filter coefficients typically include the time-delay phase shift of the amplifier. As this phase shift is usually significantly higher than that of the AM–PM conversion the resulting modelling accuracy could be insufficient. This problem can be avoided by estimating the group delay from a frequency sweep in the small-signal operation regime of the amplifier and compensating the corresponding phase shift.

Again the AM–PM conversion trace located at the centre frequency is selected as the reference. Then the desired shift along the horizontal-axis by the magnitude of the filter $|H_{\text{p}}(\text{f})|$ is made. The phase of the output filter implements the corresponding shift along the ordinate. The evaluation of the optimum alignment between the considered and the reference AM–PM conversion trace can be performed by defining an error measure similar to that in Equation (4.36). However, the use of a characteristic point for this purpose seems even more difficult than in the AM–AM case, as the AM–PM conversion traces do not show a distinct maximum.

To attain the complete PSB model the two blocks discussed above must be combined as depicted in Figure 4.12. The input filter implements the AM–PM input filter magnitude and the AM–PM output filter phase. If desired, a time-delay-based phase shift can be added to the phase response of this filter. The second filter compensates the magnitude response needed to construct the AM–PM conversion and introduces the amplitude response for the AM–AM nonlinearity. The last filter represents the AM–AM conversion output filter.

The results of fitting a PSB model from the swept-tone measurements of the Freescale™ amplifier discussed in subsection 2.5.3 are presented in Figure 4.13.

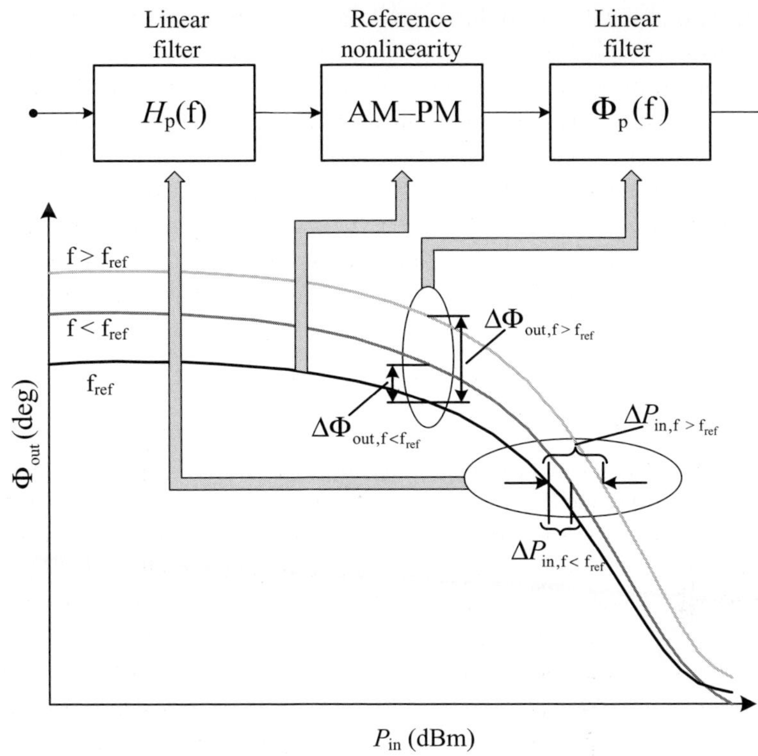

Figure 4.11 The synthesis procedure for the AM–PM portion of the PSB model, after [7, 22].

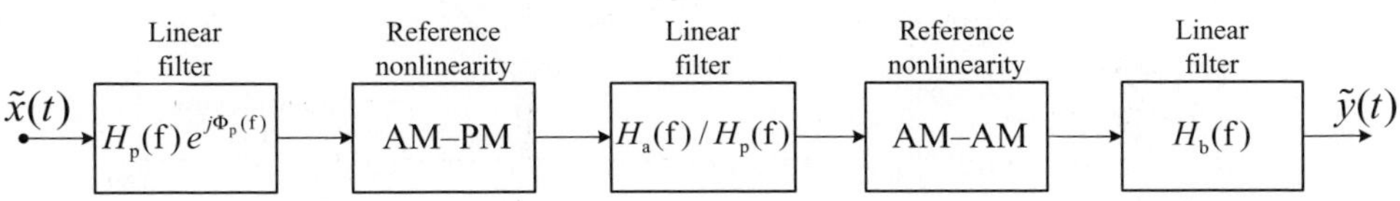

Figure 4.12 Structure of the complete PSB model as explained in [7, 22]. (© 1989 IEEE.)

The reference frequency was set to $f_{ref} = 3.5\,\text{GHz}$, which can also be recognised in the filter magnitude (and phase) response since all traces cross the 0 dB (0°) level at this frequency. By comparing the achieved $\text{Error}_{\text{Mag}}$ and the $\text{Error}_{\text{Phase}}$, shown in Figure 4.14 as functions of the frequency, the selected reference frequency is associated with a deep notch in the normalised error. Both curves highlight the fact that moving away from the reference frequency causes an increase in the corresponding error. Also of interest is that the fundamental assumption (see the start of this subsection) has much greater validity for the AM–AM than for the AM–PM conversion. At the lower end of the frequency range the relative phase error is about 40 dB higher than the magnitude error. This behaviour is not abnormal for dominantly static nonlinear SSPAs.

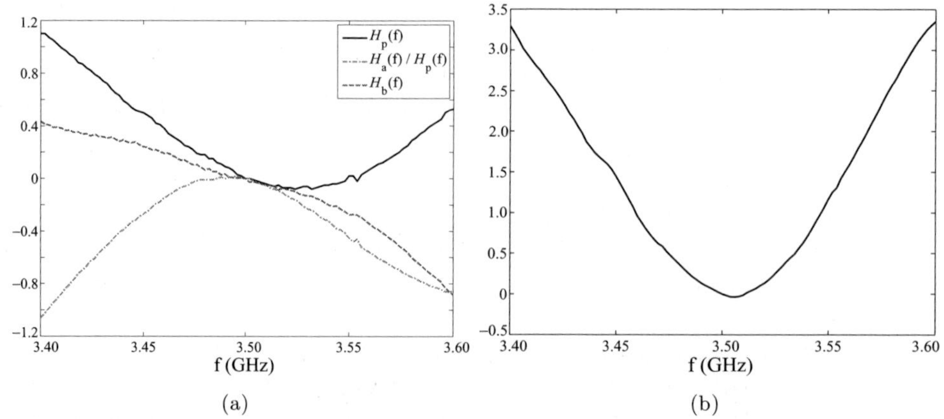

Figure 4.13 The PSB model extracted from the swept-tone measurements: (a) the filter magnitudes H(f) in dB and (b) the phase response of the input filter.

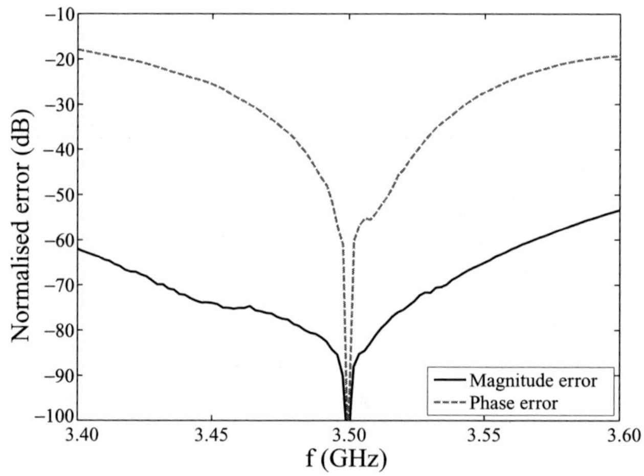

Figure 4.14 The identification errors for the magnitude and the phase responses of the Freescale™ amplifier.

An improvement in this model was presented by Clark *et al.* [23]. On the basis of the structure presented in Figure 4.6 the static nonlinear function was extracted from two-tone instead of single-tone measurements; the fundamental assumption of the PSB model still applies. These two-tone measurements were designed in the following way. The signal consisted of a large tone located at the centre frequency f_0 and a small tone at a frequency offset $f_0 \pm \Delta f$. The magnitude of the small tone was set at about 20–30 dB below the large tone. It was shown in [24] that the AM–AM/AM–PM conversion extracted by this technique, termed *dynamic carrier amplitude and phase conversion*, represents the amplifier response on two-tone and broadband input signals much better than a static nonlinearity fitted from single-tone measurements. Furthermore, the characterisation of both the AM–AM and the AM–PM conversion of the amplifier can be evaluated from power measurements of the input and output tones.

Using this characterisation technique the static nonlinearity of the three-box model is dependent on the following parameters: $g = g(r, \mathrm{f}_0, \Delta\mathrm{f})$ and $\Phi = \Phi(r, \mathrm{f}_0, \Delta\mathrm{f})$. By selecting the most appropriate Δf value for the desired excitation signals in the simulation environment, the prediction of the three-box model can be improved. In particular, modulation-signal-dependent effects, such as quasi-static thermal or bias effects, can be masked or incorporated into the model by changing Δf [7].

For the identification of the magnitude and phase of the input and output filters of the three-box model, the authors in [23] indicated several possibilities, depending on the availability of further VNA measurements or two-tone magnitude and phase measurements (performed by the use of a broadband time-domain measurement setup). These additional measurements were especially needed to allow a rigorous determination of the phase responses of the two filters.

4.3.3 *The Frequency-dependent Saleh model*

The behaviour of the memoryless Saleh model and an extension to the improved modified Saleh model were presented in Sections 3.5 and 3.6. Now, the Saleh model will be extended to represent linear memory on the basis of the solution presented in [25]. Starting from the in-phase and quadrature nonlinearities $P(r)$ and $Q(r)$, Equations (3.34) and (3.35), for any input tone of frequency f, Saleh suggested using the following two-parameter rational functions:

$$P(r, \mathrm{f}) = \frac{\alpha_{\mathrm{p}}(\mathrm{f}) r}{1 + \beta_{\mathrm{p}}(\mathrm{f}) r^2}, \tag{4.37}$$

$$Q(r, \mathrm{f}) = \frac{\alpha_{\mathrm{q}}(\mathrm{f}) r^3}{[1 + \beta_{\mathrm{q}}(\mathrm{f}) r^2]^2}. \tag{4.38}$$

This means that the in-phase and quadrature components of a swept-tone measurement are to be compared with the shapes of the two normalised frequency independent-envelope nonlinearities

$$P_0(r) = \frac{r}{1 + r^2}, \tag{4.39}$$

$$Q_0(r) = \frac{r^3}{(1 + r^2)^2}. \tag{4.40}$$

The frequency-dependent parameters α_{p}, β_{p}, α_{q} and β_{q} can then be implemented as pre- and post-filters for the two nonlinearities given in Equation (4.39) and

Equation (4.40):

$$H_{\mathrm{a,p}}(\mathrm{f}) = \sqrt{\beta_{\mathrm{p}}(\mathrm{f})}, \tag{4.41}$$

$$H_{\mathrm{b,p}}(\mathrm{f}) = \alpha_{\mathrm{p}}(\mathrm{f})/\sqrt{\beta_{\mathrm{p}}(\mathrm{f})}, \tag{4.42}$$

$$H_{\mathrm{a,q}}(\mathrm{f}) = \sqrt{\beta_{\mathrm{q}}(\mathrm{f})}, \tag{4.43}$$

$$H_{\mathrm{b,q}}(\mathrm{f}) = \alpha_{\mathrm{q}}(\mathrm{f})/\sqrt{\beta_{\mathrm{q}}^3(\mathrm{f})}. \tag{4.44}$$

Hence, the functions in Equations (4.39)–(4.44) represent three operations, filtering, memoryless nonlinearity, filtering, as shown in Figure 4.15. The $\Phi_0(\mathrm{f})$ filter at the

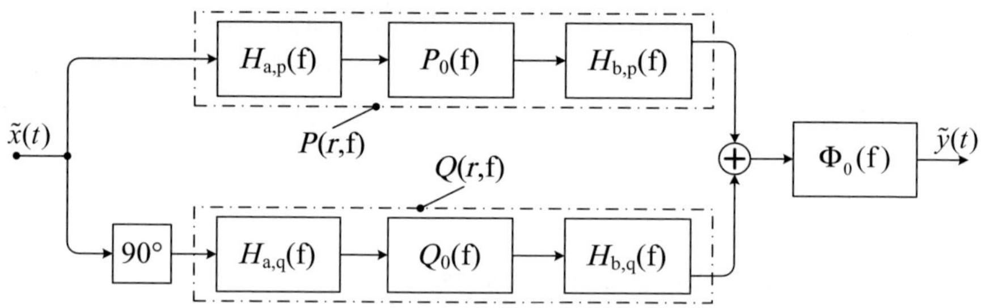

Figure 4.15 The frequency-dependent Saleh model. (Reprinted with permission from [25], © 1981 IEEE.)

output of the structure implements the so-called small-signal phase (the phase of the amplifier that is measured for $r \to 0$). The need for this filter is explained by the phase response of the frequency-independent Saleh model, Equation (3.33):

$$\lim_{r \to 0} \Phi(r) = 0 \tag{4.45}$$

As discussed in Section 3.6 the phase shift introduced by $\Phi_0(\mathrm{f})$ does not guarantee that Equations (4.39)–(4.44) will be able to represent all possible AM–PM conversion behaviours which can be produced by SSPAs.

The Saleh model is in some sense dual to the PSB model. While the PSB model constrains the AM–AM and the AM–PM curves to be of constant shape, the Saleh model constrains the in-phase and quadrature curve shapes to be independent of frequency. It is interesting to note that constraining the AM–AM and AM–PM conversion curves to maintain their shapes is completely different from maintaining the shapes of the in-phase and quadrature nonlinearities. Therefore, even if the two models represented some kind of dual structure they would behave differently. To prove this statement, it must be shown that it is impossible to represent one model in terms of the other without further limitations.

Following the approach presented in [22], the four frequency-dependent scaling factors of the PSB model and of the Saleh model will be identified by $H_{\mathrm{PSB},1}$–$H_{\mathrm{PSB},4}$ and $H_{\mathrm{Saleh},1}$–$H_{\mathrm{Saleh},4}$ respectively. The response of the PSB model to a single-tone input signal of magnitude r is given by

$$\tilde{y} = H_{\mathrm{PSB},2}\, g(H_{\mathrm{PSB},1} r) e^{j H_{\mathrm{PSB},4} + \Phi(H_{\mathrm{PSB},3} r)}, \tag{4.46}$$

where the AM–AM and AM–PM envelope nonlinearities are given by $g(\cdot)$ and $\Phi(\cdot)$.

In a similar way, the response of the Saleh model to a single-tone input is given by

$$\tilde{y} = H_{\mathrm{Saleh},2}\, P(H_{\mathrm{Saleh},1} r) + j H_{\mathrm{Saleh},4}\, Q(H_{\mathrm{Saleh},3} r). \tag{4.47}$$

The magnitude of the Saleh output can be now written as

$$|\tilde{y}| = \sqrt{\left[H_{\mathrm{Saleh},2}\, P(H_{\mathrm{Saleh},1} r)\right]^2 + \left[H_{\mathrm{Saleh},4}\, Q(H_{\mathrm{Saleh},3} r)\right]^2}. \tag{4.48}$$

The corresponding envelope of the PSB model is

$$|\tilde{y}| = H_{\mathrm{PSB},2}\, g(H_{\mathrm{PSB},1} r). \tag{4.49}$$

To achieve a Saleh model having an AM–AM conversion similar to that of the PSB model, the parameters must satisfy $H_{\mathrm{Saleh},2} = H_{\mathrm{Saleh},4}$ and $H_{\mathrm{Saleh},1} = H_{\mathrm{Saleh},3}$. With this assumption, the magnitude of the model is given by

$$|\tilde{y}| = H_{\mathrm{Saleh},2} \sqrt{P^2(H_{\mathrm{Saleh},1} r) + Q^2(H_{\mathrm{Saleh},1} r)}. \tag{4.50}$$

Additionally, the above assumption forces the AM–PM conversion traces of the Saleh model to be shifted by the same amount along the abscissa as the AM–AM traces:

$$\angle \tilde{y} = \arctan \frac{Q(H_{\mathrm{Saleh},1} r}{P(H_{\mathrm{Saleh},1} r)}. \tag{4.51}$$

This restriction on the parameters of the Saleh model is in conflict with the general PSB model, which allows independent shifts of the AM–AM and AM–PM conversion traces along the abscissa.

To fit a Saleh model from swept-tone measurements, the coefficients α_{p}, β_{p}, α_{q} and β_{q} are determined for each AM–AM/AM–PM characteristic from a least-squares best fit using Equation (3.45). According to Equations (3.39) and (3.41) the parameters η, ν, γ and ε are as follows:

$$\begin{aligned} \eta &= 1, 2, & \nu &= 1, 2, \\ \gamma &= 2, & \varepsilon &= 0. \end{aligned} \tag{4.52}$$

The summations in Equations (3.45) are over all N measurement points of the power sweep considered. On the basis of these equations a frequency-dependent Saleh model was fitted from the swept-tone amplifier measurements presented in subsection 2.5.3. The resulting magnitude responses of the four filters are presented in Figure 4.16.

The real and imaginary parts of the amplifier gain may be compared with the predictions of the corresponding models in Figure 4.17. For the in-phase part the Saleh model was incapable of reproducing the gain maxima located around $P_{\mathrm{in}} =$

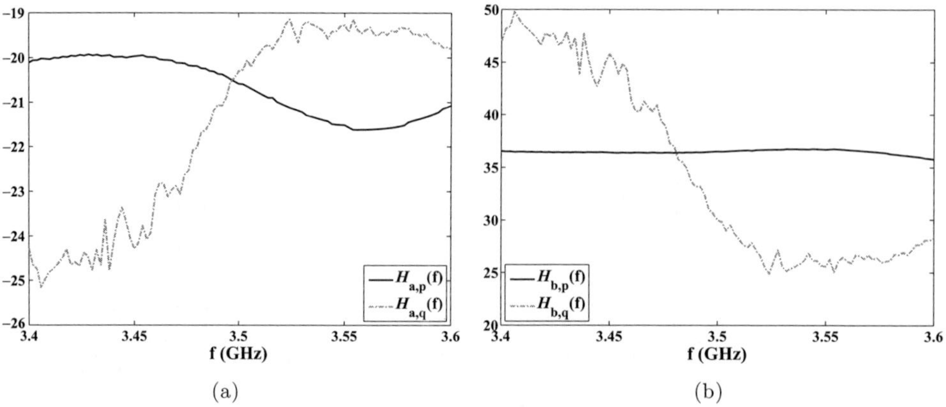

Figure 4.16 Identified (a) input filter H_a and (b) output filter H_b in dB for the P and Q branches, extracted from the swept-tone measurements of the Freescale™ amplifier discussed in subsection 2.5.3.

21 dBm. Over the input power range considered the model predicted the in-phase amplifier gain with an accuracy of ±0.6 dB.

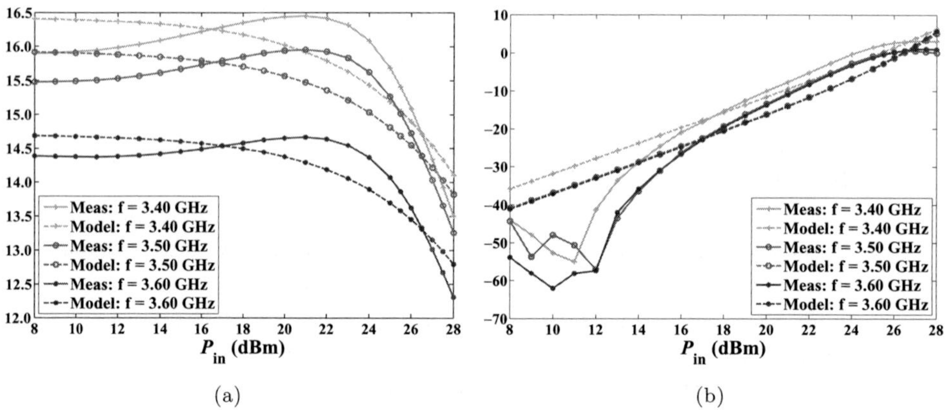

Figure 4.17 The measured and modelled (a) in-phase and (b) quadrature gain curves in dB for the nonlinearities of the Freescale™ amplifier.

The quadrature part of the nonlinear amplifier gain is significantly distorted by measurement noise for input power levels below 14 dBm. Therefore, only measurement points above this limit were considered for the parameter extraction. The gain error of the model for $P_{in} \geq 14$ dBm is +4 dB to −6 dB. Owing to the low AM–PM distortion generated by the amplifier it seems unlikely that the model can predict the quadrature nonlinearity correctly. This is reflected in the normalised modelling errors depicted in Figure 4.18. These errors were calculated from the corresponding power levels using an expression like Equation (4.36). It may be noted that at the lower end of the frequency range the quadrature nonlinearity shows the highest

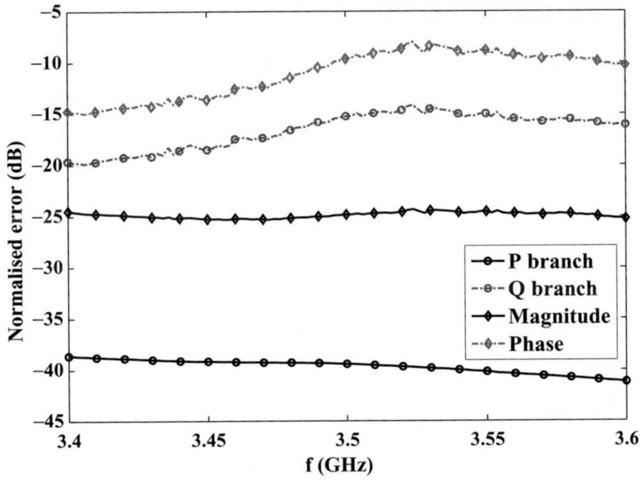

Figure 4.18 The normalised modelling error of the Saleh model for the Freescale[TM] amplifier.

modelling accuracy while the in-phase one performs worst. The resulting magnitude of the modelling error appears to be frequency independent, while the phase error follows the shape of the quadrature modelling error.

4.4 Parallel-cascade models

In this section we consider models composed of several branches connected in parallel that describe nonlinear systems with linear memory effects. The two most important representatives of this class of models are polyspectral models including linear memory and the Abuelma'atti model. An introduction to the polyspectral modelling technique was given in subsection 1.3.2. In the following subsection a discussion of the Abuelma'atti model is given.

4.4.1 *Abuelma'atti model*

The Abuelma'atti model is a frequency-independent nonlinear quadrature model that uses first-order Bessel function series to implement the in-phase and quadrature nonlinearities. The importance of envelope nonlinearities represented by Bessel function series is seen in their direct relationship to the corresponding instantaneous nonlinearities described by Fourier series. The derivation of this relationship and further properties of the Bessel–Fourier memoryless model were presented in Section 3.8. The in-phase and quadrature nonlinearities of the Abuelma'atti model may be derived in a similar way to those used in the frequency-dependent Saleh model: they are given by [26]:

$$P(r, \mathrm{f}) = \sum_{m=1}^{M} H_{\mathrm{p},m}(\mathrm{f}) J_1(\alpha m r), \tag{4.53}$$

$$Q(r, \mathrm{f}) = \sum_{m=1}^{M} H_{\mathrm{q},m}(\mathrm{f}) J_1(\alpha m r), \tag{4.54}$$

where an M-term truncated Bessel function series has been used; see Equation (3.60). The parameter α, which defines the input-envelope dynamic range of interest, was introduced in Equation (3.52). The coefficients $H_{\mathrm{p},m}$ and $H_{\mathrm{q},m}$ add a frequency dependence. The corresponding structure of the Abuelma'atti model is depicted in Figure 4.19.

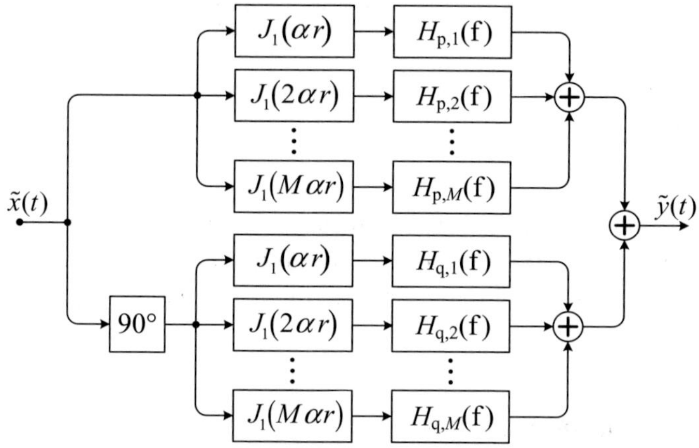

Figure 4.19 Structure of the Abuelma'atti model, after [26].

To fit this model using the swept-tone measurements of the Freescale™ amplifier, the Bessel–Fourier coefficients were evaluated by the use of a least-squares fit directly from the complex-envelope measurements. To achieve a high-accuracy Bessel–Fourier model, an even reflection of the amplifier characteristic at the boundaries was used (as presented in Figure 3.21 for the corresponding instantaneous nonlinearity).

To highlight the advantage of this approach, proposed in [27], in Figure 4.20 the real part of the envelope nonlinearity of the Freescale™ amplifier at 3.5 GHz may be compared with two Bessel–Fourier approximations. For both models the expansions were truncated after ten terms. The first BF model was parametrised between $V_{\mathrm{in}} = 0$ and $D/2$ using all coefficients and $\alpha = 2\pi/D$. In this case the Gibbs phenomenon mentioned in subsection 3.8.3 is clearly visible. In the second case the same range for V_{in} was used but the parameter α was set to π/D. Now only the odd-order coefficients are nonzero. This approximation shows a much better representation of the envelope nonlinearity and of the even reflection characteristic in the extrapolation range at $|V_{\mathrm{in}}| \geq D/2$.

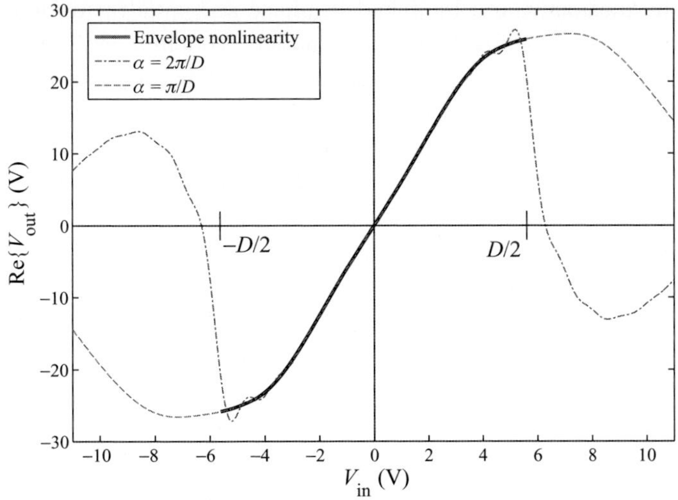

Figure 4.20 The real part of the envelope nonlinearity and two different Bessel–Fourier approximations.

Using this approach an Abuelma'atti model was parametrised from the Freescale™ amplifier measurements. To achieve comparability with the results from the Saleh model the small-signal phase was removed from the measurement results, although this compensation is not required by Abuelma'atti's model. The fitted filter magnitudes are summarised in Figure 4.21.

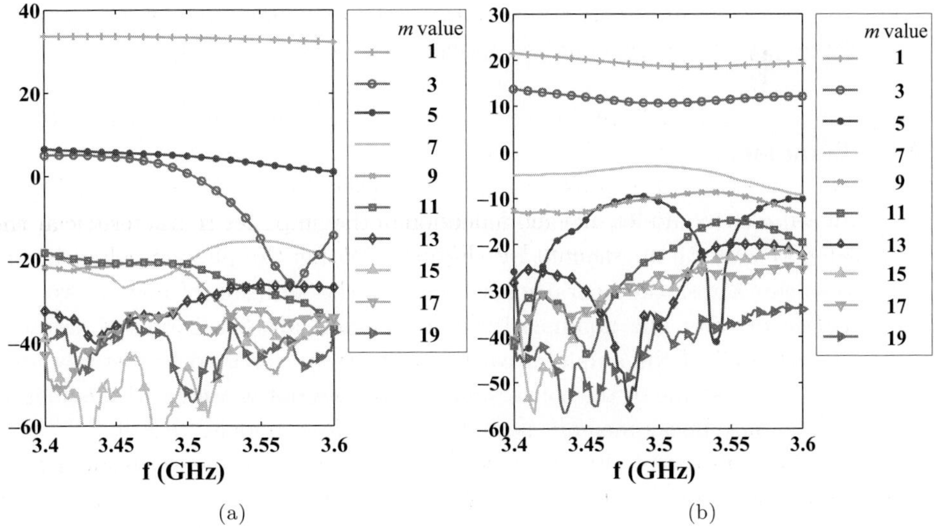

Figure 4.21 Magnitudes of (a) the in-phase filter $H_{p,m}$ (f) and (b) the quadrature filter $H_{q,m}$ (f) extracted from the Freescale™ amplifier measurements presented in subsection 2.5.3.

The measured and modelled in-phase and quadrature gain curves are depicted in Figure 4.22. Unlike the Saleh model, Abuelma'atti's model represents exactly the in-phase characteristic of the amplifier. For the identification of the quadrature nonlinearity only measurement results with $P_{\text{in}} \geq 12\,\text{dBm}$ were included. For this input power range no deviation between the measured and the modelled quadrature nonlinearity can be seen. These observations also hold for the normalised modelling

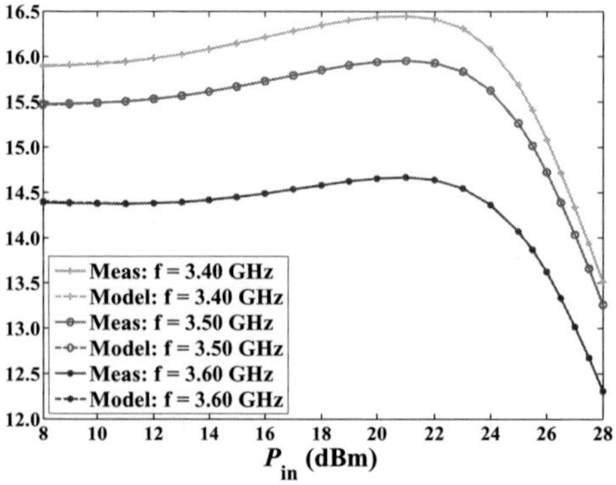

(a)

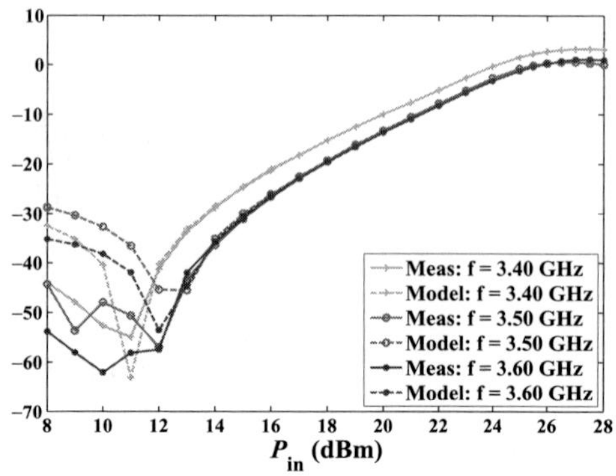

(b)

Figure 4.22 The measured and modelled (a) in-phase and (b) quadrature gain curves in dB.

error, Figure 4.23. Owing to the low AM–PM distortion of the amplifier, the quadrature error is again significantly higher than the in-phase error. The normalised modelling error of the magnitude and phase traces achieved at least $-50\,\text{dB}$. These results are about 30 dB better than those of the Saleh model, owing to the greater flexibility of the Abuelma'atti model.

4.5 Summary

In order to handle frequency-dependent behaviour, i.e. memory effects, memoryless models may be augmented by the use of either swept-tone measurements at frequencies across the operating band or by the addition of one or two linear filters before or after the nonlinearity. Such models are based on the assumption that the memory of the amplifier can be adequately captured in this way and that the qualitative shape of the conversion curves does not change with frequency. In other words, only linear memory is captured by these models.

This limitation to linear memory effects is questionable, especially when a general broadband signal is traced through these models on the assumption that such a signal can be represented as a sum of CW tones with some amplitude and phase (as is needed to process the signal through the memoryless nonlinearity). These models predict no interaction between the instantaneous tones, although it is well known that such interactions occur in a real device because of its finite memory.

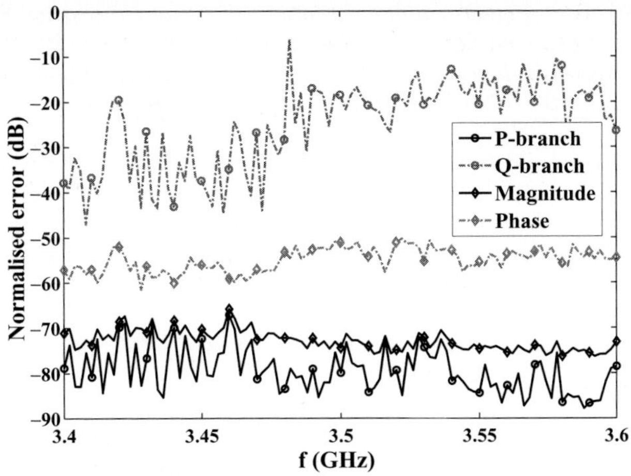

Figure 4.23 Normalised modelling error of the Abuelma'atti model.

Several other models have been proposed that lead to more accurate nonlinear modelling; in these models both linear and nonlinear memory effects are included. They are discussed in the next chapter.

References

[1] V. Z. Marmarelis, *Nonlinear Dynamic Modelling of Physiological Systems,* John Wiley & Sons, 2004.

[2] R. K. Pearson, "Selecting nonlinear model structures for computer control," *J. Process Control*, vol. 13, no. 1, pp. 1–26, February 2003.

[3] W. J. Rugh, *Nonlinear System Theory, the Volterra/Wiener Approach,* Johns Hopkins University Press, 1981.

[4] G. Chrisikos, C. J. Clark, A. A. Moulthrop, M. S. Muha and C. P. Silva, "A nonlinear ARMA model for simulating power amplifiers," in *IEEE MTT-S Int. Microwave Symp. Dig.*, June 1998, pp. 733–736.

[5] M. Schetzen, *The Volterra and Wiener Theories of Nonlinear Systems,* John Wiley & Sons, 1980.

[6] D. Silveira, M. Gadringer, H. Arthaber, M. Mayer and G. Magerl, "Modeling, analysis and classification of a PA based on identified Volterra kernels," in *Gallium Arsenide Applications Symp. Dig.*, October 2005, pp. 405–408.

[7] M. C. Jeruchim, P. Balaban and K. S. Shanmugan, *Simulation of Communication Systems, Modeling, Methodology and Techniques,* second edition, Kluwer/Plenum, 2000.

[8] A. Hagenblad and L. Ljung, "Maximum likelihood identification of Wiener models with a linear regression initialization," Technical Report from the Automatic Control Group, August 1998.

[9] T. Söderström and P. Stoica, *System Identification,* Prentice Hall, 1989.

[10] C. P. Silva, C. J. Clark, A. A. Moulthrop and M. S. Muha, "Survey of characterization techniques for nonlinear communication components and systems," in *Proc. IEEE Aerospace Conf.*, March 2005, pp. 1713–1737.

[11] S. M. Kay, *Fundamentals of Statistical Signal Processing,* Prentice Hall, 1993.
[12] O. Nelles, *Nonlinear System Identification,* Springer, 2001.
[13] S. Haykin, *Adaptive Filter Theory,* fourth edition, Prentice Hall, 2002.
[14] M. Schetzen, "Nonlinear system modeling based on the Wiener theory," *Proc. IEEE*, vol. 69, no. 12, pp. 1557–1573, December 1981.
[15] M. Lortie and R. E. Kearney, "Robust identification of time-varying system dynamics with non-white inputs and outputs noise," in *Proc. Conf. of the IEEE Eng. in Medicine and Biology Soc.*, October 1998, pp. 3036–3037.
[16] D. T. Westwick and R. E. Kearney, "Identification of physiological systems using pseudo-inverse based deconvolution," *IEEE-EBMC and CMBEC*, pp. 1405–1406, September 1997.
[17] M. Abramowitz and I. Stegun, *Handbook of Mathematical Functions,* Dover Publications, 1970.
[18] D. Silveira, M. Gadringer, H. Arthaber and G. Magerl, "RF-power amplifier characteristics determination using parallel cascade Wiener models and pseudo-inverse techniques," in *Proc. Asia Pacific Microwave Conf.*, December 2005, pp. 204–207.
[19] C. Silva, A. A. Moulthrop, and M. Muha, "Introduction to polyspectral modeling and compensation techniques for wideband communications systems," in *ARFTG Conference Dig.*, November 2001, pp. 1–15.
[20] M. Ibnkahla, N. J. Bershad, J. Sombrin and F. Castanié, "Neural network modeling and identification of nonlinear channels with memory: algorithms, applications, and analytic models," *IEEE Trans. Signal Processing*, vol. SP-46, no. 5, pp. 1208–1220, May 1998.
[21] S. L. Loyka, "The influence of electromagnetic environment on operation of active array antennas: analysis and simulation techniques," *IEEE Antennas and Propagation Magazine*, vol. 41, no. 6, pp. 23–37, December 1999.
[22] R. Blum and M. C. Jeruchim, "Modeling nonlinear amplifiers for communication simulation," in *Proc. IEEE Int. Conf. on Communications*, June 1989, pp. 1468–1472.
[23] C. J. Clark, C. P. Silva, A. A. Moulthrop and M. S. Muha, "Power-amplifier characterization using a two-tone measurement technique," in *IEEE Trans. Microw. Theory Tech.*, vol. 50, no. 6, pp. 1590–1602, June 2002.
[24] A. A. Moulthrop, C. J. Clark, C. P. Silva and M. S. Muha, "A dynamic AM/AM and AM/PM measurement technique," *IEEE MTT-S Int. Microwave Symp. Dig.*, June 1997, pp. 1455–1458.
[25] A. Saleh, "Frequency-independent and frequency-dependent nonlinear models of TWT amplifiers," *IEEE Trans. Communications*, vol. 29, no. 11, pp. 1715–1720, November 1981.
[26] M. Abuelma'atti, "Frequency-dependent nonlinear quadrature model for TWT amplifiers," *IEEE Trans. Communications*, vol. 32, no. 8, pp. 982–986, August 1984.
[27] A. R. Kaye, K. A. George and M. J. Eric, "Analysis and compensation of bandpass nonlinearities for communications," *IEEE Trans. Communications*, vol. 20, pp. 965–972, October 1972.
[28] N. M. Blachman, "Detectors, bandpass nonlinearities, and their optimization: inversion of the Chebyshev transform," *IEEE Trans. Information Theory*, vol. 17, no. 4, pp. 398–404, July 1971.

5 Nonlinear models with nonlinear memory

5.1 Introduction

This chapter contains a comprehensive overview of the approaches for modelling nonlinear PAs with nonlinear memory. The difference between linear and nonlinear memory effects was presented in subsection 1.2.2 on the basis of the model presented in Figure 1.6.

The simplest modelling approach is the memory polynomial model. It will be explained that introducing non-uniform time-delay taps yields better results. Two more elaborate approaches that are closely related to the memory polynomial model are the time-delay neural network (TDNN) model and the nonlinear autoregressive moving-average (NARMA) model.

In the case of the TDNN model, the memoryless nonlinear network is described by an artificial neural network (ANN). Since ANNs have gained importance in microwave PA behavioural modelling, the concept will be explained in a separate section. In the case of the NARMA model the output depends not only on past values of the input but also on past values of the output. Stability may be a problem with this modelling approach, but criteria are derived to check for this.

Another way to model nonlinear PAs with nonlinear memory effects is by an extension of the well-known Wiener modelling approach. By introducing parallel branches consisting of a linear time-invariant (LTI) system followed by a memoryless nonlinear system, nonlinear memory effects can be modelled adequately.

A further category of models comprises the Volterra-series-based models. It is often said that the computation of Volterra kernels is difficult when the system has complex nonlinearity. A number of extended approaches have been developed to overcome the intrinsic disadvantages of Volterra models. The parallel finite impulse response (FIR) model has a reduced computational complexity and the Laguerre–Volterra model has a reduced number of model parameters. The modified or dynamic Volterra model in its turn is primarily aimed at handling higher levels of nonlinearity. Finally, in this chapter, the link between Volterra models and TDNN models will be made.

Memory polynomial models and Volterra-series-based models are especially suitable for weakly nonlinear systems. The state-space-based behavioural models are not restricted to weakly nonlinear systems. The dynamics of the PA are determined directly from time-series data, resulting in a compact, accurate and transportable model. It will be explained how multisine excitations can make model development more efficient.

5.2 Memory polynomial model

The *baseband memory polynomial model* is widely used to describe nonlinear effects in a PA with memory effects. Two types of baseband memory polynomial PA model will be discussed and compared in this section. These are:

- a memory polynomial with unit time-delay taps;
- a memory polynomial with non-uniform time-delay taps.

Memory polynomial PA modelling approaches are based on measured data in the discrete-time domain. To convert the analogue signal to the discrete-time domain, sampling is used; this means that the value of the signal is measured at certain intervals of time (Figure 5.1). Each data point is referred to as a sample. Thus, if $V_{\text{in}}(t)$ is the input signal to the sampler, the sampler output is $V_{\text{out}}(s) = V_{\text{in}}(sT)$, where T is the sampling period, $\text{f}_{\text{sample}} = 1/T$ is the sampling frequency and $s = 0, 1, \ldots, S$, where S is the total number of samples (Figure 5.1).

Consequently, the complex envelopes of the input and output continuous-time signals $V_{\text{in}}(t)$ and $V_{\text{out}}(t)$ can be rewritten as the discrete-time signals $V_{\text{in}}(s)$ and $V_{\text{out}}(s)$ respectively.

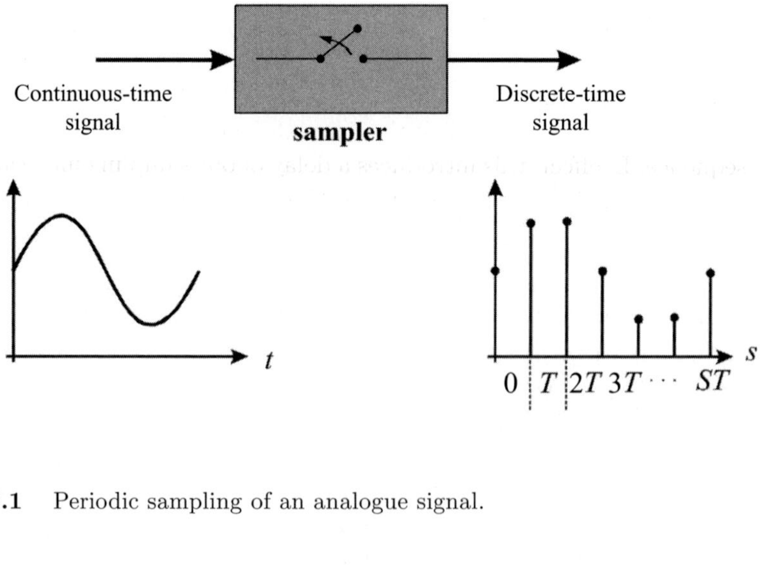

Figure 5.1 Periodic sampling of an analogue signal.

5.2.1 *Memory polynomial with unit time-delay taps*

The general form of a baseband memory polynomial PA model can be written as [1]:

$$V_{\text{out}}(s) = \sum_{q=0}^{Q}\sum_{k=1}^{K} \tilde{a}_{kq} V_{\text{in}}(s-q)\,|V_{\text{in}}(s-q)|^{2(k-1)}, \qquad (5.1)$$

where the $\tilde{a}_{kq}$ are complex memory polynomial coefficients that can be estimated by a simple least-squares method, $k = 0, 1, 2, \ldots, K$ is the polynomial order and is an integer, $V_{\text{in}}(s)$ and $V_{\text{out}}(s)$ are the measured discrete input and output complex-envelope signals of the sth sample and $q = 0, 1, \ldots, Q$ is the memory interval and is equal to the sampling interval T. The quantities Q and K are the maximum memory and polynomial orders respectively. Note that Equation (5.1) only contains odd-order terms because the signals obtained from even-order terms are far from the carrier frequency.

Equation (5.1) can also be rewritten in a compact form as

$$\begin{aligned} V_{\text{out}}(s) &= \sum_{q=0}^{Q}\sum_{k=1}^{K} \tilde{a}_{kq} F_{kq}(s-q) \\ &= \sum_{q=0}^{Q} F_q(s-q) \equiv F_0 + F_1 + \cdots + F_Q, \end{aligned} \qquad (5.2)$$

where

$$F_q \equiv F_q(s-q) = \sum_{k=1}^{K} \tilde{a}_{kq} V_{\text{in}}(s-q)\,|V_{\text{in}}(s-q)|^{2(k-1)}. \qquad (5.3)$$

Figure 5.2 shows a block diagram of the memory polynomial model with unit time-delay taps given in Equation (5.1). In this figure, the unit delay tap is denoted by the symbol Z^{-1}. When the unit delay tap Z^{-1} is applied to a member of the sequence of discrete digital values, this unit tap gives the previous value in the sequence. In effect, this introduces a delay of one sampling interval q. Thus, applying the operator Z^{-1} to an input value $V_{\text{in}}(s)$ gives the previous input, $V_{\text{in}}(s-1)$:

$$\mathrm{Z}^{-1} V_{\text{in}}(s) = V_{\text{in}}(s-q) = V_{\text{in}}(s-1). \qquad (5.4)$$

Referring to Equation (5.2), the memory polynomial can be rewritten as [2]:

$$\begin{aligned} V_{\text{out}}(s) &= \sum_{q=0}^{Q}\sum_{k=1}^{K} \tilde{a}_{kq} F_{kq}(s-q) \\ &= \sum_{k=1}^{K} \tilde{a}_{k0} F_{k0}(s) + \sum_{k=1}^{K} \tilde{a}_{k1} F_{k1}(s-1) + \cdots + \sum_{k=1}^{K} \tilde{a}_{k\,Q} F_{kQ}(s-Q) \\ &\equiv F_0 + F_1 + \cdots + F_Q \end{aligned} \qquad (5.5)$$

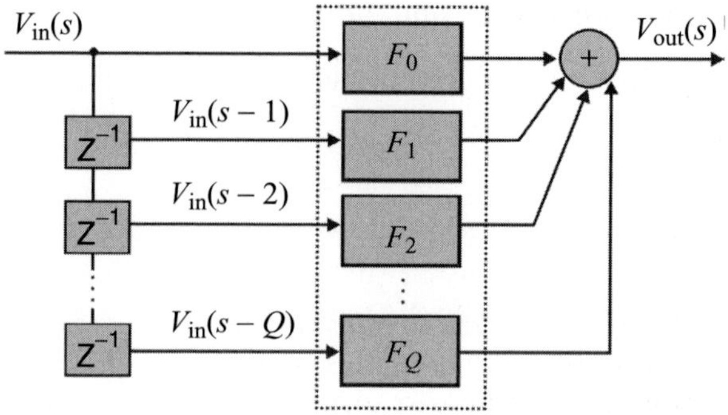

Figure 5.2 Memory polynomial PA model with unit delay taps.

so that

$$\begin{aligned} V_{\text{out}}(s) &= \sum_{q=0}^{Q} \sum_{k=1}^{K} \tilde{a}_{kq}\, F_{kq}(s-q) \\ &= \tilde{a}_{10}F_{10}(s) + \tilde{a}_{20}F_{20}(s) + \cdots + \tilde{a}_{K0}F_{K0}(s) \\ &\quad + \tilde{a}_{11}F_{11}(s-1) + \tilde{a}_{21}F_{21}(s-1) + \cdots + \tilde{a}_{K1}F_{K1}(s-1) \\ &\quad + \cdots + \tilde{a}_{1Q}F_{1Q}(s-Q) + \tilde{a}_{2Q}F_{2Q}(s-Q) + \cdots + \tilde{a}_{KQ}F_{KQ}(s-Q). \end{aligned} \tag{5.6}$$

Equation (5.6) can be written in matrix form as

$$\mathbf{V}_{\text{out}} = \mathbf{F}\ \mathbf{A}, \tag{5.7}$$

where the vectors $\mathbf{V}_{\text{out}}$ and $\mathbf{A}$ and the matrix $\mathbf{F}$ are given by

$$\mathbf{V}_{\text{out}} = [V_{\text{out}}(0)V_{\text{out}}(1)\cdots V_{\text{out}}(S-1)]^T, \tag{5.8}$$

$$\mathbf{A} = [\tilde{a}_{10}\ \tilde{a}_{20}\cdots\tilde{a}_{K0}\ \tilde{a}_{11}\tilde{a}_{21}\cdots\tilde{a}_{K1}\cdots\tilde{a}_{1Q}\tilde{a}_{2Q}\cdots\tilde{a}_{KQ}]^T, \tag{5.9}$$

$$\mathbf{F} = [\mathbf{F}_{10}\cdots\mathbf{F}_{K0}\mathbf{F}_{11}\cdots\mathbf{F}_{K1}\cdots\mathbf{F}_{1Q}\cdots\mathbf{F}_{KQ}] \tag{5.10}$$

and

$$\mathbf{F}_{kq} = [F_{kq}(-q)F_{kq}(1-q)\cdots F_{kq}(S-1-q)]^T. \tag{5.11}$$

The least-squares error method can be used to find the complex polynomial coefficients $\tilde{a}_{kq}$. The LS solution for Equation (5.7) is

$$\mathbf{A} = (\mathbf{F}^H\ \mathbf{F})^{-1}\mathbf{F}^H\mathbf{V}_{\text{out}}, \tag{5.12}$$

where $(\cdot)^H$ denotes the conjugate transpose of the matrix $\mathbf{F}$.

5.2.2 *Memory polynomial with non-uniform time-delay taps*

A memory polynomial with non-uniform time delay taps analytically is an improvement on the unit time-delay taps approach described above. The main difference between these two models is the choice of the delay tap indices for the memory polynomial. Ku and Kenney [3] reported that using delay taps obtained through an optimisation process results in better modelling of the memory effects than can be obtained using a unit delay tap. Moreover, they indicated that a memory polynomial model using the optimum delay taps could model accurately the asymmetry in the output spectrum of the PA. However, they also commented on the difficulty of deriving these optimal delay taps.

In this subsection, an approach to calculating the optimal delay taps analytically is presented [4, 5]. Different nonlinear functions were tested and it was found that a sinusoidal function gives the best modelling results. Here, the spacing of the delay taps follows a pattern similar to a sine function. Accordingly, the following simple sinusoidal function is proposed to calculate the non-uniform spacing indices:

$$p_q = \text{floor}\ (W\,|\sin\,q|) \tag{5.13}$$

where W is the maximum memory index depth. The Matlab function floor(X) returns the nearest integer less than or equal to X. A set of calculated values of such non-uniform spacing indices p_q is listed in Table 5.1.

Table 5.1 Calculated non-uniform spacing index for $W = 100$. Reprinted with permission from [5]

q	0	1	2	3	4	5	6	7
p_q	0	84	90	14	75	95	27	65

Figure 5.3 shows the improved memory polynomial with non-uniform time delay taps. In this PA model the non-uniform delay taps Z^{-p_q} are obtained using the non-uniform spacing indices p_q.

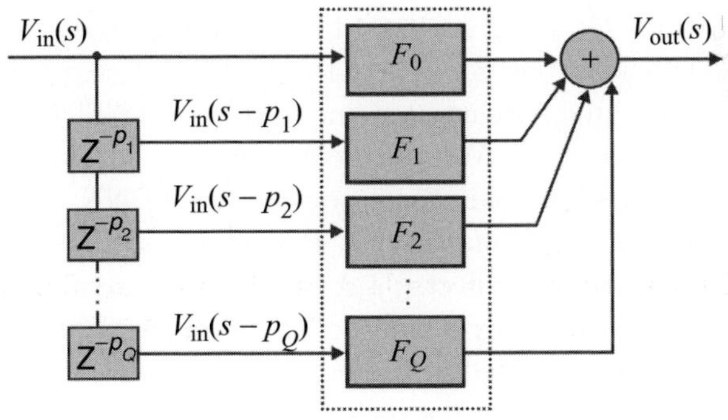

Figure 5.3 Memory polynomial PA model with non-uniform delay taps.

This approach gives superior results when modelling a PA with memory; since the memory effect is nonlinear, it is suggested that the reason could be that the non-uniform spacing index can be described by an appropriate nonlinear function rather than a linear function as in the unit time-delay tap approach. Therefore, this nonlinear function should be analysed in detail for different broadband modulated signals.

For a memory polynomial with non-uniform time-delay taps, Equations (5.1) and (5.2) now take the following form; in which q has been replaced by p_q:

$$V_{\text{out}}(s) = \sum_{q=0}^{Q} \sum_{k=1}^{K} \tilde{a}_{kq} V_{\text{in}}(s - p_q) \left| V_{\text{in}}(s - p_q) \right|^{2(k-1)} \tag{5.14}$$

and

$$\begin{aligned} V_{\text{out}}(s) &= \sum_{q=0}^{Q} \sum_{k=1}^{K} \tilde{a}_{kq} F_{kq}(s - p_q) \\ &= \sum_{q=0}^{Q} F_q(s - p_q) = F_0 + F_1 + \cdots + F_Q \end{aligned} \tag{5.15}$$

where

$$F_q \equiv F_q(s - p_q) = \sum_{k=1}^{K} \tilde{a}_{kq} V_{\text{in}}(s - p_q) \left| V_{\text{in}}(s - p_q) \right|^{2(k-1)}. \tag{5.16}$$

As described in the previous subsection, the LS error method can be used to find the complex polynomial coefficients $\tilde{a}_{kq}$.

5.3 Time-delay neural network model

5.3.1 Model description

As mentioned in Section 1.2, the *time-delay neural network model* (TDNN model) is closely related to the memory polynomial concept. The main difference is that the set of functions F_q is now replaced by a neural network, as illustrated in Figure 5.4.

The TDNN structure consists of two blocks, a linear time-invariant (LTI) system and a nonlinear memoryless system. As such, the TDNN structure can be viewed as a special case of the Wiener model (Section 4.2). For the LTI system an FIR filter (Section 1.2) is used. This provides the TDNN model with the capability of performing dynamic mappings that depend on past input values, making it suitable for modelling memory effects. The nonlinear system is represented by a neural network, more specifically a multilayer perceptron (MLP) neural network. Neural networks will be explained in the next subsection.

An alternative version, the real-valued TDNN model, will be introduced in subsection 5.3.3. This approach is more suitable for the prediction of memory effects in the case of a two-tone stimulus.

5.3.2 Artificial neural networks

Neural networks have gained recognition as a useful modelling tool in the microwave field [6]. More specifically, the modelling and simulation of nonlinear devices and circuits within a wireless communication system using behavioural models based

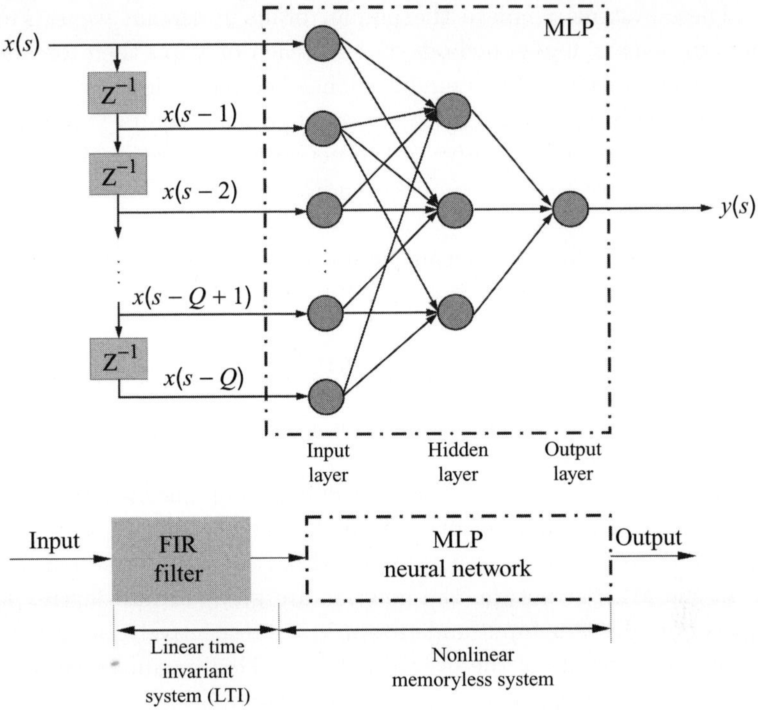

Figure 5.4 Topology of the TDNN model.

on the neural network paradigm is a field of increasing interest. The use of neural networks is not limited to TDNN models, and another PA behavioural modelling approach adopting neural networks is given in Section 5.7. Here a short theoretical overview of artificial neural networks (ANNs) in general is presented.

An ANN is a system composed of a large number of basic elements that are arranged in layers and are highly interconnected. The structure has several inputs and outputs, which may be 'trained' to react (giving y values) to the input stimuli (expressed in x values) in the desired way, as illustrated in Figure 5.5.

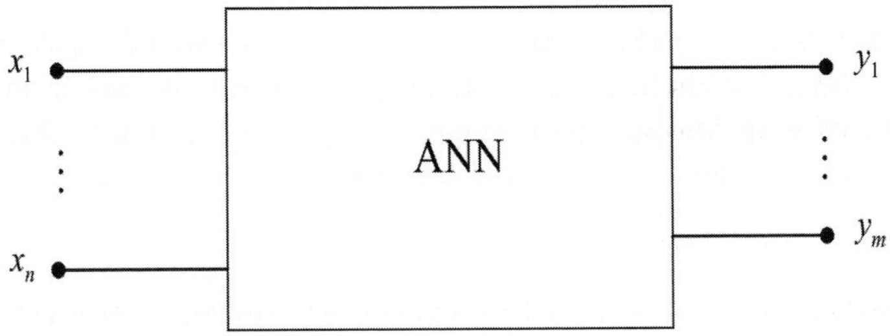

Figure 5.5 Artificial neural network schematic.

These systems emulate the human brain in certain ways. They need to learn how to behave, and somebody has to teach or train them on the basis of former knowledge of the problem environment. The basic element is the *artificial neuron.* It is basically a process unit connected to other units through synaptic connections. The processing ability of the network is stored in the inter-unit connection strengths w obtained by a process of adaptation to, or learning from, a set of training patterns.

Mathematically, the output–input relation of the ANN structure, shown in Figure 5.5, is represented by

$$y = f(x, w). \tag{5.17}$$

The definition of w and the manner in which y is computed from x and w determine the structure of the ANN.

Multilayer perception (MLP) is the most popular structure for an ANN model. It follows a general class of structures called feedforward ANNs. They have the capability to produce a general approximation of any function [7, 8].

In the MLP structure, the neurons are grouped into layers. The first and last layers are called the input and output layers respectively, because they represent the inputs and outputs of the overall network. The remaining layers are called hidden layers. This description is illustrated in Figure 5.6 [9].

Assume that the total number of layers is L. The first layer is the input layer, the Lth layer is the output layer and layers 2 to $L-1$ are hidden layers. Let the number of neurons in the lth layer be N_l, $l = 1, 2, \ldots, L$. Let w_{ij}^l represent the weight of the link between the jth neuron of the $(l-1)$th layer and the ith neuron of the lth layer, $1 \leq j \leq N_{l-1}, 1 \leq i \leq N_l$. Let x_i represent the ith external input to the MLP and z_i^l be the output of the ith neuron of the lth layer. There is an extra weight parameter for each neuron, w_{i0}^l, representing the bias of the ith neuron of the lth layer.

For an MLP with $j = 0, 1, 2, \ldots, N_{l-1}$, $i = 1, 2, \ldots, N_l$, and $l = 2, 3, \ldots, L$, the weight $\mathbf{w}$ then takes the following form:

$$\mathbf{w} = \begin{bmatrix} w_{10}^2 \; w_{11}^2 \; w_{12}^2 \; \cdots \; w_{N_L N_{L-1}}^L \end{bmatrix}^T. \tag{5.18}$$

Each neuron in the input layer receives an external input. Every neuron in the other layers, including the output layer, receives inputs from the neurons in the previous layer and processes them. The processed information is then available at the output of the neuron. Figure 5.7 illustrates this mechanism [9]. As an example, a neuron of the lth layer receives stimuli from the neurons of the $(l-1)$th layer, that is, $z_1^{l-1}, z_2^{l-1}, \ldots, z_{N_{l-1}}^{l-1}$. Each input is first multiplied by the corresponding weight parameter and the resulting products are added to produce a weighted sum γ. This weighted sum is passed through a neuron activation function $\sigma(\cdot)$ to produce the output of the neuron. Although various functions can be used as activation

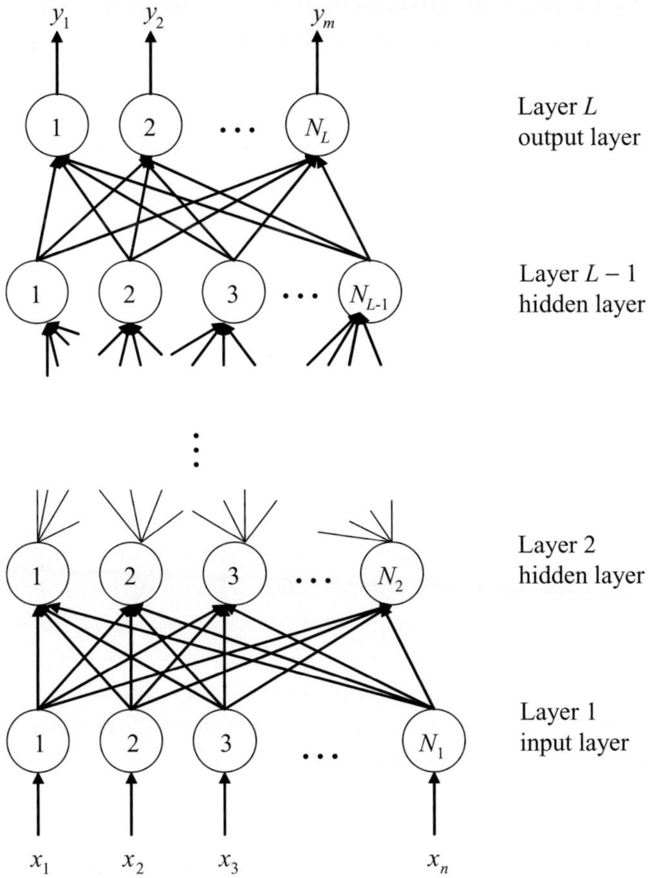

Figure 5.6 Multilayer perceptron structure. (Reproduced by permission from [9], © 2000 Artech House Inc.)

functions, the one most commonly used is the sigmoid function

$$\sigma(\gamma) = \frac{1}{1 + e^{-\gamma}}. \tag{5.19}$$

The sigmoid function is a smooth switch function having the following asymptotes:

$$\sigma(\gamma) \to \begin{cases} 1, & \gamma \to +\infty, \\ 0, & \gamma \to -\infty. \end{cases} \tag{5.20}$$

The fact that the asymptotes reach a constant value is the reason for the popularity of the sigmoid function. An ANN model may get evaluated outside its validity range during a circuit simulation, and this could escalate to divergence problems if the asymptotes of the activation function are not bounded.

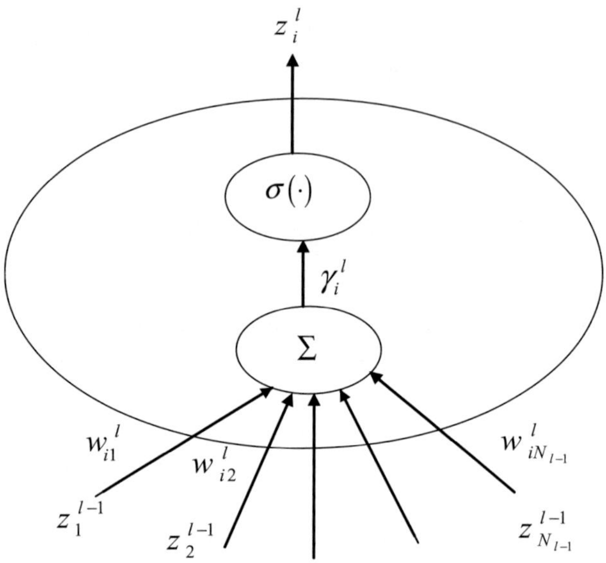

Figure 5.7 Information processing by ith neuron in lth layer. The neuron activation function is $\sigma(\cdot)$; $\sum$ indicates the summation of stimuli from different neurons. (Reproduced by permission from [9], © 2000 Artech House, Inc.)

The simplest form of an MLP-structured ANN is the three-layer topology illustrated in the nonlinear block on the right of Figure 5.4. This topology, with just one hidden layer, is usually sufficient to model PAs.

The weights $\mathbf{w}$ are determined through a learning or training phase, which requires a set of input–output data for the PA to be modelled. During the training, the weights in the network topology are iteratively modified in such a way that when an input vector is presented, the trained network predicts the output vector with minimum error. Various algorithms exist to perform this training, among which the back-propagation algorithm [10], shown in Figure 5.8, is commonly used for modelling microwave PAs. The main principle behind the back-propagation algorithm is the minimisation of the sum of the squared output errors averaged over the training set, using a gradient-descent search [11].

5.3.3 *Real-valued TDNN model*

In realistic situations involving digitally modulated excitations, the input–output data of the PA to be modelled are complex. Although the authors of [12] have proposed complex-value-based neural networks, the training algorithm becomes very complicated as the activation functions must also be complex. One approach is to

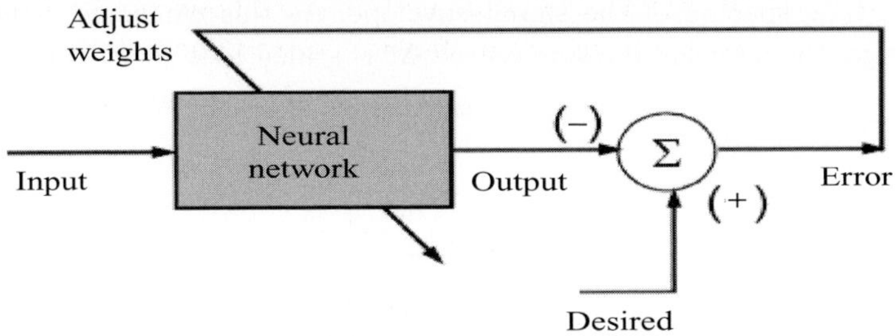

Figure 5.8 Training of an artificial neural network.

implement two separate TDNNs – this is often used for conventional ANN models – so that the complex input–output data may be processed separately, either in rectangular form (real and imaginary) or polar form (magnitude and phase).

In this subsection, the specific focus is on developing a model to predict the dynamic AM–AM and AM–PM nonlinear characteristics. The approach of using two TDNNs is illustrated in Figure 5.9. To develop the model, only the measured complex input and output data are required; there is no need to understand the internal mechanisms of the PA.

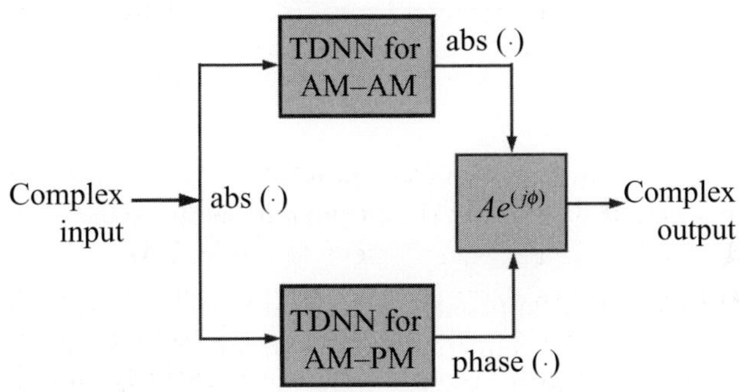

Figure 5.9 Real-valued TDNN model for modelling AM–AM and AM–PM characteristics. (From [13] with permission.)

However, this AM–AM and AM–PM structured TDNN model can be implemented only for a multitone excitation signal or a realistic telecommunication digitally modulated signal with a fixed envelope bandwidth frequency. In other words, the topology of Figure 5.9 can only model the dynamic AM–AM and AM–PM characteristics for an input with a particular baseband envelope frequency. This is too restrictive, as the goal is to have a global model valid for different envelope frequencies, which is thus able to give a good representation of the memory effects.

Therefore, an improved TDNN model based on a two-tone stimulus is presented in Figure 5.10. The topology now incorporates an additional input $\Delta\tilde{f}$ for defining the tone spacing of the signal envelope. As this can have values up to the MHz range, the additional input vector $\Delta\tilde{f}$ is scaled to a set of tone-spacing frequencies in the range {–1, 1}, because having inputs of the same order of magnitude enhances the ANN training process.

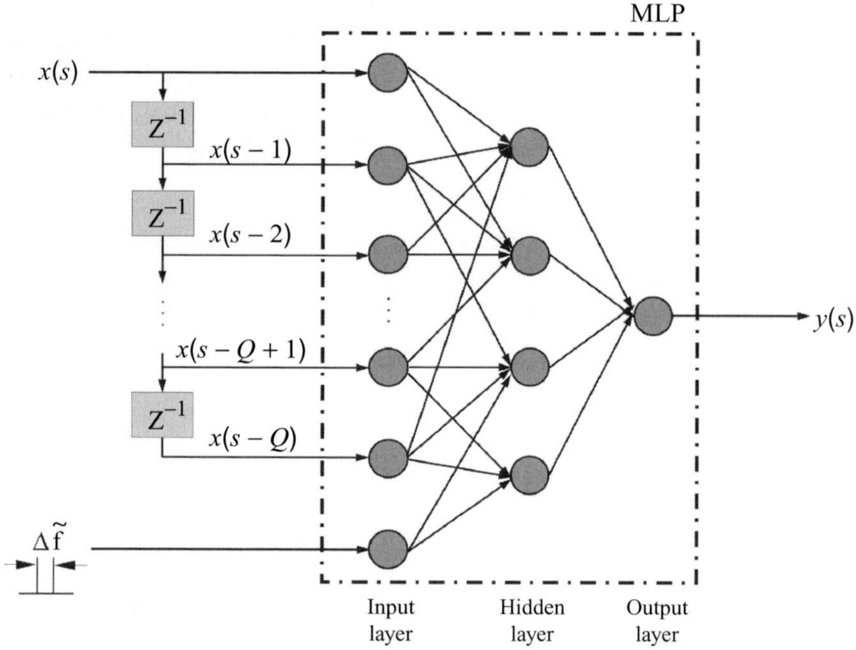

Figure 5.10 Improved TDNN topology for the precise prediction of memory effects based on a two-tone stimulus. (From [13] with permission.)

To illustrate this approach, a simulation-based example is considered using a class-AB metal-oxide-semiconductor field effect transistor (MOSFET) amplifier in the circuit simulator. To collect data, a simulation is performed in which the input power and tone spacing are swept, thus providing artificial measurements. This TDNN model was trained using the Levenberg–Marquardt back-propagation algorithm. The network topology in this example consisted of seven neurons at the input layer, 15 neurons in the hidden layer and one neuron at the output layer. The activation function used in the hidden layer was a sigmoid function. The 'measured' and modelled dynamic AM–AM and AM–PM characteristics for a quadrature phase-shift keying (QPSK) input signal are shown for comparison in Figures 5.11(a), (b).

5.4 Nonlinear autoregressive moving-average model

The *nonlinear autoregressive moving-average model* (the NARMA model) is an extension of the memory polynomial model (Section 5.2). The advantage of the

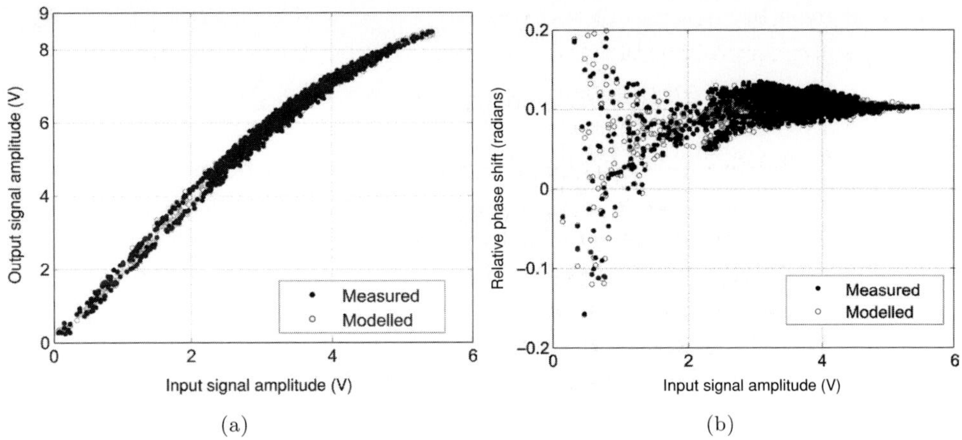

Figure 5.11 Measured and modelled dynamic (a) AM–AM and (b) AM–PM characteristics for a QPSK input signal.

NARMA model lies in the introduction of a nonlinear feedback path (involving IIR terms). This may permit the number of delayed samples required to model the PA to be reduced in comparison with a model using only FIR terms. Reducing the complexity of the PA model acquires importance when the latter is part of a linearisation scheme such as digital predistortion. However, a main weakness of the NARMA model is the need to ensure its stability, since the use of nonlinear feedback paths can result in overall system instability. Thus a suitable stability test is required; such a test based on small-gain theory is presented in subsection 5.4.2.

5.4.1 Model description

Figure 5.12 shows a block diagram of the NARMA model. The general expression

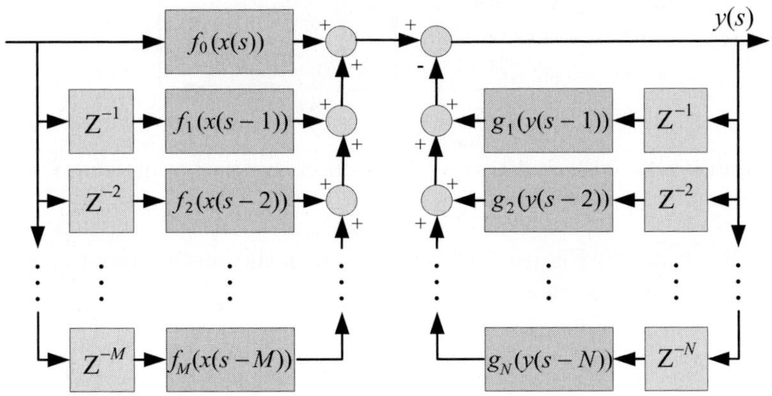

Figure 5.12 Block diagram of the nonlinear autoregressive moving-average (NARMA) model.

is given by

$$y(s) = f_0(x(s)) + \sum_{i=1}^{M} f_i(x(s-i)) - \sum_{j=1}^{N} g_j(y(s-j)), \tag{5.21}$$

$f_i(\cdot)$ and $g_j(\cdot)$ being nonlinear functions that can be implemented by using, for example, polynomials. The present output sample depends on the sum of different static nonlinearities related to the present sample $x(s)$ of the input and both input and output past samples, $x(s-i)$ and $y(s-j)$, $i = 1, 2, \ldots, M$ and $j = 1, 2, \ldots, N$, as shown in Figure 5.12.

If polynomials are used to implement the nonlinear functions $f_i(\cdot)$ and $g_j(\cdot)$, it is possible to rewrite Equation (5.21) as

$$\begin{aligned} y(s) = & \sum_{i=0}^{M} c_i \left[\sum_{k=0}^{P-1} b_k x(s-i) \left| x(s-i) \right|^k \right] \\ & - \sum_{j=1}^{N} d_j \left[\sum_{k=0}^{P-1} a_k y(s-j) \left| y(s-j) \right|^k \right], \end{aligned} \tag{5.22}$$

where P is the order of the memoryless polynomial and M and N are respectively the numbers of the delayed input and output samples used for modelling the PA dynamics.

5.4.2 *Stability test: small-gain theorem*

The small-gain theorem is an input–output stability method based on bounded norms. It was developed by G. Zames in 1966 and it is well explained and applied in [14]; it is capable of determining the stability of nonlinear systems when these are bounded by some kind of norm.

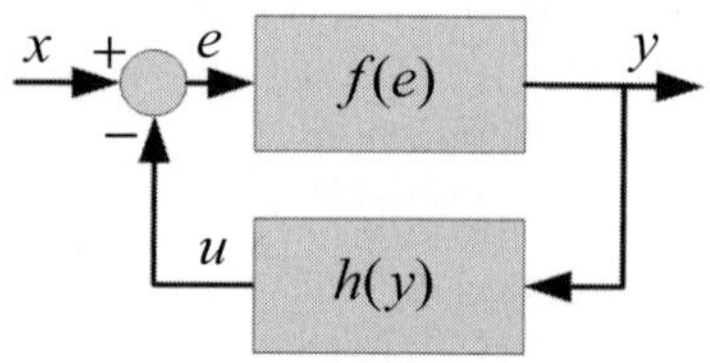

Figure 5.13 Block diagram of a feedback system for applying the small-gain stability method.

Referring to Figure 5.13 and applying the methodology described in [14], the following inequalities are obtained:

$$\begin{aligned} & \|y\|_p \leq \gamma_f \|e\|_p, \\ & \|e\|_p \leq \|x-u\|_p \leq \|x\|_p + \|u\|_p, \\ & \|u\|_p \leq \gamma_g \|y\|_p, \end{aligned} \tag{5.23}$$

where $\|\cdot\|_p$ is the pth-order norm and γ_g, γ_f are bounds of the induced norms (gains of the pth order norm).

Combining the inequalities in Equation (5.23) leads to the following inequality:

$$\|y\|_p \leq \frac{\gamma_f}{1-\gamma_f\gamma_g}\|x\|_p \,. \tag{5.24}$$

In other words, the output norm is bounded and this bound exists if $\gamma_f\gamma_g < 1$; this is a sufficient stability condition [14].

Now it is possible to apply the small-gain stability method to the proposed NARMA model. By defining $|\cdot|$ as the modulus and $\|\cdot\|_2$ as the second-order norm, the following expressions are obtained:

$$|f_i(v)| \leq \alpha_i\,|v| \qquad \rightarrow \qquad \|f_i\|_2 \leq \alpha_i\,\|v\|_2 \,, \tag{5.25}$$

$$|g_j(v)| \leq \gamma_j\,|v| \qquad \rightarrow \qquad \|g_j\|_2 \leq \gamma_j\,\|v\|_2 \,, \tag{5.26}$$

where α_i and γ_j are the bounds of the second-order norm gain of each polynomial term of the nonlinear functions, $f_i(\cdot)$ and $g_j(\cdot)$ respectively and where v is an auxiliary independent variable. As a result, the following inequality is obtained for our particular NARMA structure:

$$\|y\|_2 \leq \alpha_0\,\|x\|_2 + \alpha_1\,\|x\|_2 + \cdots + \alpha_M\,\|x\|_2 + \gamma_1\,\|y\|_2 + \cdots + \gamma_N\,\|y\|_2 \,. \tag{5.27}$$

Finally, starting from relations (5.23) it is possible to obtain the following inequality:

$$\|y\|_2 \leq \frac{\sum_{i=0}^{M}\alpha_i}{1-\sum_{j=1}^{N}\gamma_j}\|x\|_2 \,. \tag{5.28}$$

In conclusion, the stability of the NARMA structure is ensured (and it is a sufficient condition) if the following inequality holds [14]:

$$\sum_{j=1}^{N}\gamma_j \leq 1 \,. \tag{5.29}$$

5.4.3 Example

In order to investigate the PA low-pass complex-envelope behaviour predicted by a NARMA model, input and output discrete complex data were extracted from a three-stage class-AB LDMOS PA.

For the measurements, a 16-QAM root-raised cosine (RRC) filtered (rolloff 0.25) modulated signal was used. This signal provides good power spectral efficiency but lacks a constant envelope and is therefore highly sensitive to PA nonlinearities. The measurements were performed with an RF signal having a bandwidth of 1.25 MHz at a centre frequency of 1.96 GHz. The PA was operating at 2 dB input power backoff (IBO) and had the following nominal characteristics: a frequency range of 1.93–1.96 GHz, maximum output power 48 dBm and gain 36 dB.

The NARMA model considered used five non-consecutive taps (three input delays and two output delays) and yielded a normalised mean-square error (NMSE) less than −30 dB.

In order to ensure the stability of the model the small-gain criterion (see subsection 5.4.2) was checked. In Figure 5.14 the functions are plotted. The main memoryless nonlinear function is f_0, while f_1, f_2, f_3 are the nonlinear functions related to the three delayed samples of the input and g_1, g_2 are the nonlinear functions related to the delayed samples of the output. As can be seen, the sum of the norms of g_1 and g_2 does not exceed the reference bound (i.e. the straight line with unit slope). Therefore, the small-gain criterion is satisfied and so the extracted NARMA model is inherently stable.

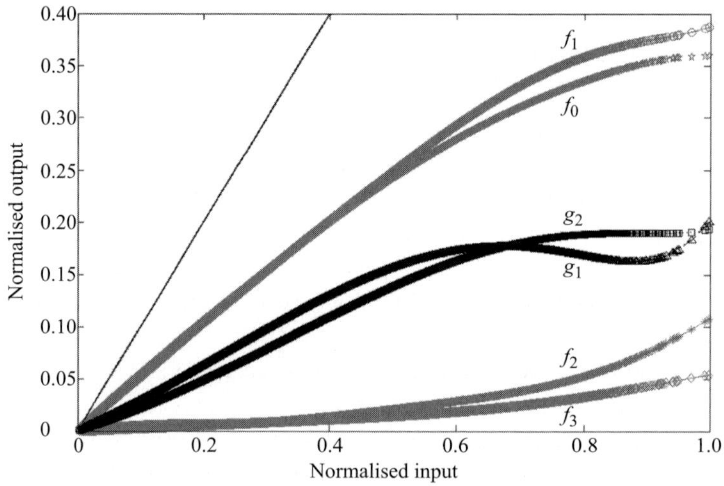

Figure 5.14 Nonlinear functions of the delayed samples in the NARMA model. Above the curves is the reference bound.

In order to highlight the accuracy achieved by the NARMA model and its significant improvement with respect to the memoryless nonlinear model, the output power spectra, the AM–AM and AM–PM characteristics, the in-phase and quadrature components and the error-vector magnitude (EVM) will now be presented. Figure 5.15 shows the output power spectra and Figure 5.16 shows the AM–AM and AM–PM characteristics of the PA measured data, the NARMA model and the memoryless nonlinear model. While the NARMA model is capable of reproducing the AM–AM and AM–PM data dispersion, the memoryless model cannot. Figure 5.17 shows the in-phase and quadrature components of the respective models. It may be observed that the memoryless model cannot perfectly fit the measured output signal since no PA dynamics are considered in this model.

Finally, in order to see the capability of the NARMA model to accurately

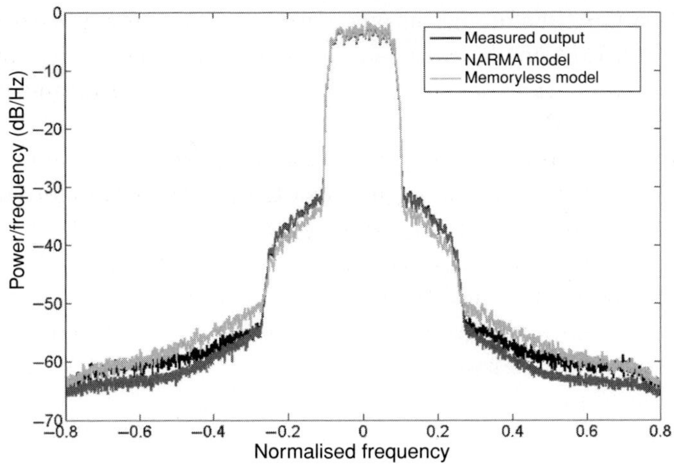

Figure 5.15 Output power spectra of the PA measured data (black), the NARMA model (mid-grey) and the memoryless nonlinear model (light grey).

reproduce in-band distortion, the output data were demodulated in order to plot the constellation diagram and determine the EVM. Figure 5.18 shows the constellation diagrams as measured, as predicted by the NARMA model and as predicted by the memoryless model. The NARMA model's predicted EVM (7.18%) is very close to the measured value (7.51%), whereas the memoryless model's value (4.40%) is some way off.

This example shows how it is possible to obtain inherently stable NARMA models that are useful for developing low-pass complex-envelope dynamic PA models. Moreover, owing to its compromise between complexity and accuracy, the NARMA model can also be included in linearisation schemes such as digital baseband predistortion [15].

5.5 Parallel-cascade Wiener model

The Wiener model was discussed in Section 4.2 as an approach to modelling nonlinear applications with linear memory. The *parallel-cascade Wiener model* is an extension that allows the modelling of nonlinear memory. As explained in Section 4.2, the Wiener model is a cascade connection of a linear time-invariant (LTI) system and a memoryless nonlinear system. In the case of a parallel-cascade Wiener model, the LTI systems and memoryless nonlinear systems are connected in parallel branches. The LTI systems model the memory effects with a change in the envelope frequency. In [16], the following topology was proposed. The first branch is set to the memoryless AM–AM and AM–PM functions. Using two-tone signals, AM–AM and AM–PM curves are extracted for each envelope frequency by measuring IMD products. The error between the memoryless model in the first branch and the measured two-tone data is modelled by adding a parallel cascade of LTI systems

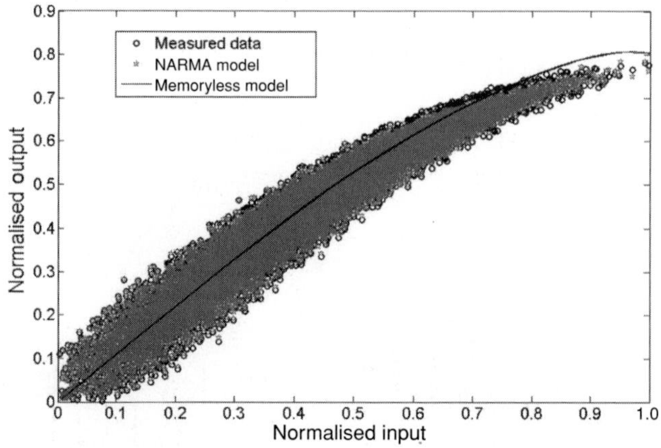

(a)

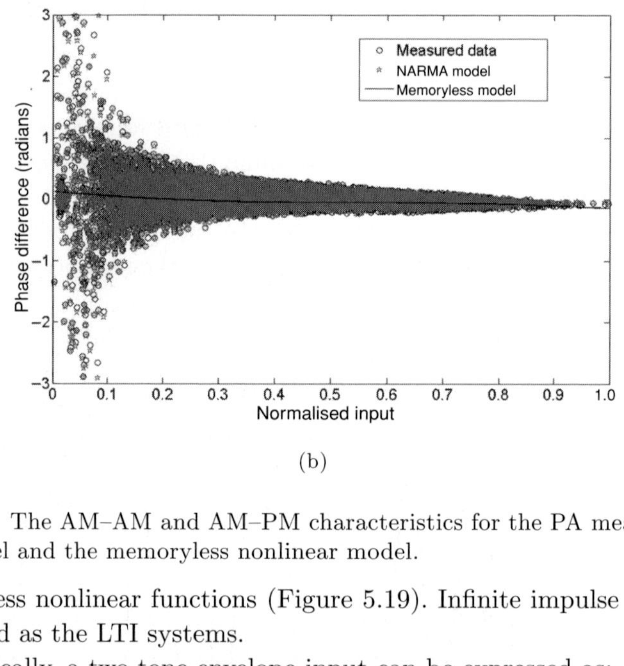

(b)

Figure 5.16 The AM–AM and AM–PM characteristics for the PA measured data, the NARMA model and the memoryless nonlinear model.

and memoryless nonlinear functions (Figure 5.19). Infinite impulse response (IIR) filters are used as the LTI systems.

Mathematically, a two-tone envelope input can be expressed as:

$$x(t) = A \cos \omega_{\mathrm{m}}(t), \tag{5.30}$$

with ω_{m} the envelope frequency. The frequency-dependent complex power series that expresses memory effects is of the following form:

$$\begin{aligned} F(x, \omega_{\mathrm{m}}) &= a_1(\omega_{\mathrm{m}})x + a_3(\omega_{\mathrm{m}})x^3 + \cdots + a_{2n-1}(\omega_{\mathrm{m}})x^{2n-1} \\ &= \sum_{k=1}^{n} a_{2k-1}(\omega_{\mathrm{m}})x^{2k-1}, \end{aligned} \tag{5.31}$$

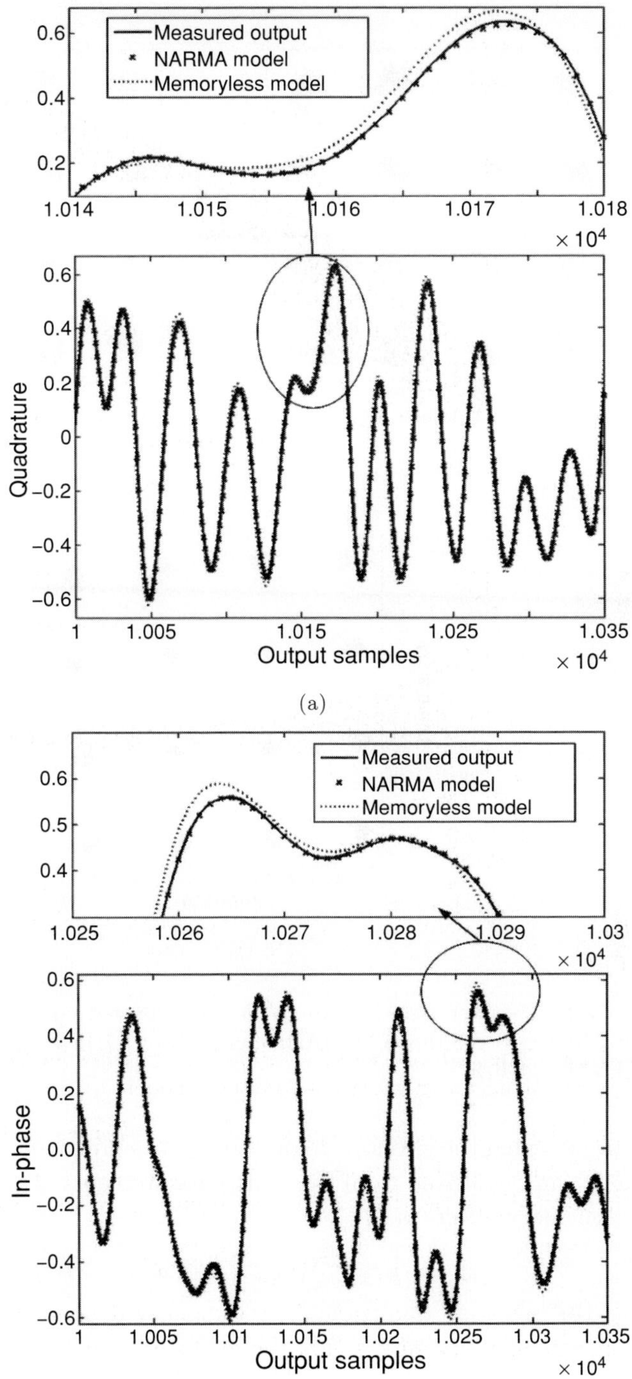

Figure 5.17 (a) In-phase and (b) quadrature components of the PA measured data, the NARMA model and the memoryless nonlinear model. Enlargements are given at the top of (a) and (b). These show that the NARMA model fits the measured output closely.

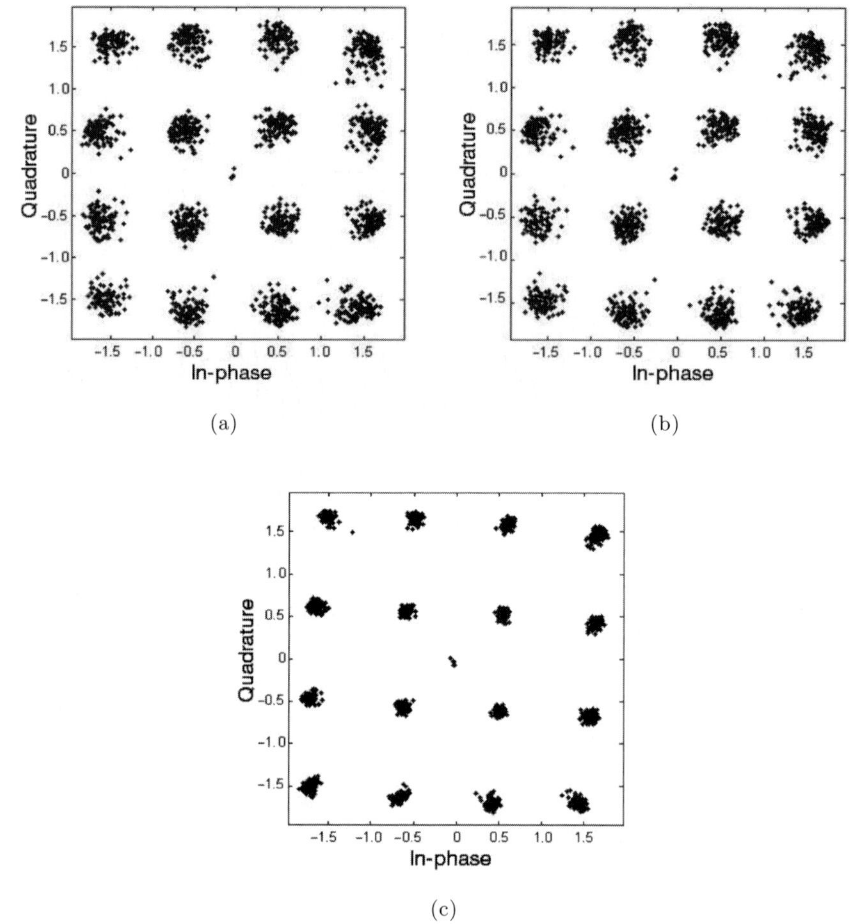

Figure 5.18 Scatter plots for 16-QAM RRC filtered constellation (a) as measured, EVM = 7.51%, (b) for the NARMA model, EVM = 7.18% and (c) for the memoryless model, EVM = 4.40%.

where, by sweeping the envelope frequency $\omega_{\rm m}$, the frequency-dependent coefficients $a_{2k-1}(\omega_{\rm m})$ can be derived from two-tone measurements. Therefore, the output of a PA with memory effects is

$$\omega(t) = |F(x, \omega_{\rm m})| \cos\left[\omega_{\rm c} t + \theta(t) + \angle F(x, \omega_{\rm m})\right]. \tag{5.32}$$

Here the frequency-dependent complex polynomial in Equation (5.31) is realised by a parallel-cascade structure of LTI systems connected in series with memoryless nonlinear systems.

The LTI system has the following characteristic function:

$$H_i(\omega) = |H_i(\omega)|\, e^{j\Omega_i(\omega)}, \tag{5.33}$$

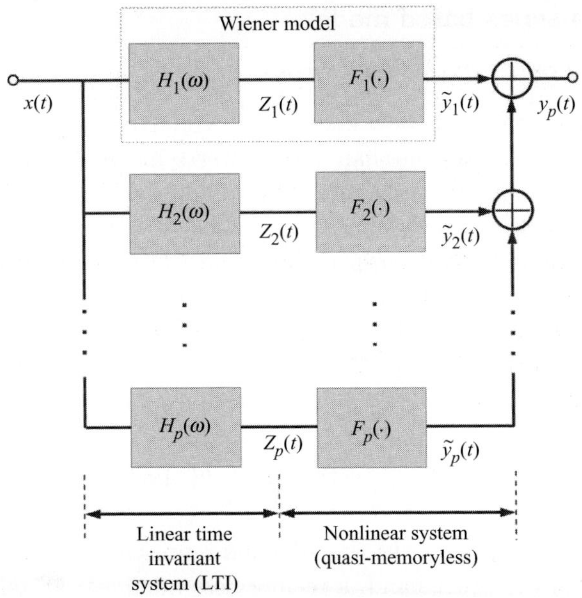

Figure 5.19 PA model for a system with memory using the parallel-cascade Wiener model. (Reprinted with permission from [16], © 2002 IEEE.)

where $F_i(\cdot)$ is a complex polynomial with coefficients $a_{2k-1,i}$, $k = 1, \ldots, n$, $i = 1, \ldots, p$. Thus the output of the parallel-cascade Wiener system in Figure 5.19 is given by

$$\begin{aligned} y_p(t) &= \sum_{i=1}^{p} \tilde{y}_i(t) \\ &= \sum_{i=1}^{p} F_i(Z_i(t)) \\ &= \sum_{i=1}^{p}\sum_{k=1}^{n} a_{2k-1,i}\left\{A\left|H_i(\omega_{\mathrm{m}})\right|\cos[\omega_{\mathrm{m}}t + \Omega_i(\omega_{\mathrm{m}})]\right\}^{2k-1}, \end{aligned} \tag{5.34}$$

where p is the number of parallel branches, $Z_i(t)$ is the output for the LTI system of the ith branch and $\tilde{y}_i(t)$ is the output for the nonlinear system of the ith branch.

In Equation (5.34), values of $a_{2k-1,i}$ and $H_i(\omega_{\mathrm{m}})$ can be determined that minimise the mean-square error ε_i^2, which is defined by

$$\begin{aligned} \varepsilon_i &= F(x(t), \omega_{\mathrm{m}}) - y_i(t) \\ &= F(x(t), \omega_{\mathrm{m}}) - \sum_{s=1}^{i} \tilde{y}_s(t), \end{aligned} \tag{5.35}$$

where $i = 1, \ldots, p$.

This model is simple compared with the general Volterra series model (Section 5.6). It has the capability to compensate for the drawbacks of the Volterra model by quantifying the memory effects in a PA.

5.6 Volterra-series-based models

5.6.1 *Introduction to the Volterra series*

The notion of what is now known as a Volterra series was first introduced in 1887 by the Italian mathematician Vito Volterra in his *Theory of Functionals.* The first major application of Volterra's work to nonlinear circuit analysis was made by the mathematician Norbert Wiener, who used it in a general way to analyse a number of problems including the spectrum of an FM system with a Gaussian noise input. Since then the Volterra series has become one of the most often used tools for nonlinear system modelling and characterisation [17].

It is well known that on the one hand any causal linear system with memory can be described in terms of its impulse response $h(\tau)$ through a convolution integral:

$$y(t) = \int_{-\infty}^{+\infty} h(\tau)x(t-\tau)d\tau, \tag{5.36}$$

where $x(t)$ and $y(t)$ are the time-domain representations of the input and the output signals. On the other hand, a memoryless nonlinear system can be described by a power series:

$$y(t) = \sum_{i=1}^{\infty} a_i[x(t)]^i, \tag{5.37}$$

where the a_i are the coefficients of the power series and $x(t)$ and $y(t)$ are the input and output signals.

The Volterra approach assumes that the response of a nonlinear system with memory, having input $x(t)$ and output $y(t)$, can be expressed, combining the above two equations, as

$$y(t) = \sum_{n=1}^{\infty} y_n(t), \tag{5.38}$$

where y_n is the nth-order term in the response and is given by

$$y_n(t) = \int_{-\infty}^{+\infty} \cdots \int_{-\infty}^{+\infty} \int_{-\infty}^{+\infty} h_n(\tau_1, \ldots, \tau_n) \prod_{p=1}^{n} x(t-\tau_p)d\tau_p, \tag{5.39}$$

and where $h_n(\tau_1, \ldots, \tau_n)$ is the nth-order impulse response. For $n = 1$, $h_n(\tau)$ coincides with the linear impulse response $h(\tau)$ in Equation (5.36). Any higher-order impulse response ($n > 1$) serves to characterise a nonlinearity order, e.g. $h_2(\tau_1, \tau_2)$ and $h_3(\tau_1, \tau_2, \tau_3)$ describe the second-order and third-order nonlinearity of the system respectively. Each kernel transform contains many contributions, namely one for each order of nonlinear coefficient of the power series description up to the order of the kernel. Thus a high-level model of each kernel transform can be constructed in the form of a block diagram that comprises many parallel paths [18].

It is also possible to rewrite the Volterra series in a frequency-domain form:

$$y(t) = \sum_{n=1}^{N} \int_{-\infty}^{+\infty} \cdots \int_{-\infty}^{+\infty} \int_{-\infty}^{+\infty} H(\mathrm{f}_1, \ldots, \mathrm{f}_n) \prod_{p=1}^{n} X(\mathrm{f}_p) e^{j2\pi \mathrm{f}_p \tau_p} \, d\mathrm{f}_p, \tag{5.40}$$

where $X(\mathrm{f})$ is the Fourier transform of the input signal $x(t)$ and $H(\mathrm{f}_1, \ldots, \mathrm{f}_n)$ is the nth-order nonlinear transfer function, which is defined to be the n-dimensional Fourier transform of $h(\tau_1, \ldots, \tau_n)$. Once the system is characterised by its impulse responses or its nonlinear transfer functions, it is possible to determine the system response for input signals defined either in the time or frequency domain.

While basic behavioural models for RF and microwave amplifiers, such as those based on AM–AM and AM–PM characterisation, provide accurate predictions of narrowband systems they are inadequate for wideband systems, since the frequency independence that they intrinsically assume is only valid for a narrow modulation band and cannot take account of memory effects.

Volterra-series-based models can provide an accurate characterisation of microwave PAs since they can take account of both nonlinearities and memory effects. Unfortunately, their effectiveness is limited to weakly nonlinear systems. This is due to the non-convergence of the series for strong nonlinearities, the increase in computational complexity with series order and the difficulties in measuring the higher-order Volterra kernels. Much research has been directed towards overcoming these drawbacks in Volterra-series-based modelling.

The basic methods for Volterra kernel estimation suggest two possible ways to improve the modelling: computing the kernels from the synapses' weights in an associated neural network representation, as shown in [20, 21], or computing them from input–output data, where the choice of input signals and parameter optimisation criteria employed are critical elements of the identification procedure.

Publications such as [22, 23] present enhanced measurement techniques for Volterra kernels but, despite the possibility of measuring higher-order kernels, there is a practical limit to the order in classical Volterra series because of the increase in computational complexity with order. A number of modified approaches are described in the next few subsections.

5.6.2 *Parallel FIR model*

To reduce the computational complexity, a *parallel FIR model* was proposed [24]. It uses a bank of parallel FIR filters to implement a Volterra-based behavioural model. The filters' coefficients may be extracted from circuit-envelope simulations or time-domain measurements. In [24] a discrete-domain finite-memory complex baseband Volterra model is considered:

$$\begin{aligned}\tilde{y}(s) &= \sum_{i=0}^{m_1-1} h_1(i)\tilde{x}(s-i) \\ &+ \sum_{i_1=0}^{m_3-1}\sum_{i_2=i_1}^{m_3-1}\sum_{i_3=i_2}^{m_3-1} h_3(i_1,i_2,i_3)\tilde{x}(s-i_1)\tilde{x}(s-i_2)\tilde{x}^H(s-i_3) \\ &+ \sum_{i_1=0}^{m_5-1}\sum_{i_2=i_1}^{m_5-1}\sum_{i_3=i_2}^{m_5-1}\sum_{i_4=i_3}^{m_5-1}\sum_{i_5=i_4}^{m_5-1} h_5(i_1,i_2,i_3,i_4,i_5)\prod_{j=1}^{5}\tilde{x}(s-i_j) \\ &+ \cdots + \eta(s), \end{aligned} \tag{5.41}$$

where $h_l(i_1, i_2, \ldots, i_l)$ is the lth-order Volterra kernel, m_l represents the memory of the corresponding nonlinearity, $\tilde{x}^H$ represents the conjugate transpose of $\tilde{x}$ and $\eta(s)$ is the unmeasured disturbance. In Equation (5.41) the redundant items associated with kernel symmetry and the even-order kernels, whose effects can be neglected in bandlimited modulation systems, have been removed. As Zhu *et al.* explain in [24], using V-vector algebra results in a Volterra modelling approach which inherits a time-shift-invariance property (each element in a row is a delayed version of the former one) and this allows the implementation of the nonlinear system as a group of parallel linear subsystems such as transversal FIR filters. All the information needed to estimate the convolution, corresponding to the first column of the input data V-vector $\tilde{x}(s)$, can be obtained by defining a set of primary signals and then implementing the convolution for each row of $\tilde{x}(s)$ using an FIR filter. The final output of the Volterra behavioural model is obtained by summing all the filter outputs. As shown in [24], this kind of parallel fast algorithm significantly improves the data processing speed and so reduces the computation time. Figure 5.20 shows the basic configuration required for extraction of the behavioural model parameters.

However, the problems associated with the high complexity of general Volterra models still affect this model since no attempt was made to simplify the model structure. Using such a model to describe a strongly nonlinear device that presents long-term memory effects, as a PA could, may be impractical in some real applications because it involves too many filters.

To avoid this problem, a pruning algorithm that allows the number of coefficients in the series to be reduced without a drastic loss of accuracy was proposed in [25]. This pruning algorithm exploits the fact that in practical situations memory effects presented by real amplifiers decline with time. The simplification consists of forcing to zero, during model extraction, the coefficients that correspond to elements that have the least effect on the output signal, i.e. those with longer time-delay taps in the input vector of the model. Generally, a brute-force pruning method involves evaluating the change in the error resulting from setting individual coefficients to zero. If an unacceptable increase in error is detected then the coefficient is restored, otherwise it is removed. One simple pruned Volterra model is the diagonal Volterra model [26], where all off-diagonal terms are zero (for this reason it is also called the memory polynomial model [27]). Although the reduction in the number of

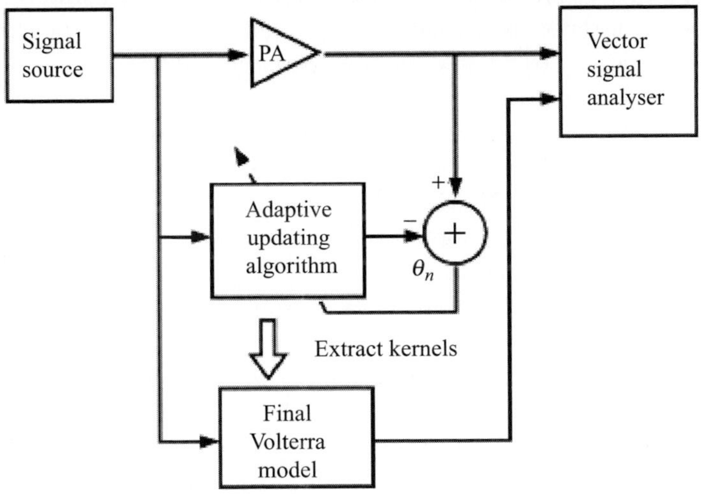

Figure 5.20 Block diagram for extraction of the Volterra model. (Reprinted with permission from [25], © 2004 IEEE.)

model parameters on applying this restriction is very significant, it also has major secondary effects such as decreasing the fidelity of the model since, in some cases the off-diagonal terms are more important than the diagonal terms. To avoid this the diagonal model may be relaxed to a 'near-diagonal' model, which gives some increase in flexibility at the expense of increasing the number of model parameters (with respect to the strictly diagonal case) [25]. In this case not all off-diagonal coefficients are set to zero, only those that are at a distance from the diagonal greater than a fixed limit l. With this 'near-diagonal' structural restriction, the coefficients that are far from the main diagonal in the model are removed; thus the corresponding number of FIR filters in the filter bank can be reduced. The 'near-diagonal' reduction approach dramatically simplifies the structure of the model, reduces the computational complexity of model extraction and allows the modelling of PAs with stronger nonlinearity or longer memory effects. Different values of l can be chosen for different orders, and thus a further reduction in complexity obtained.

5.6.3 Laguerre–Volterra model

Another approach to reducing the number of model parameters is found in the *Laguerre–Volterra model.* In the classical Volterra model, the elements of the Volterra series, i.e. the Volterra kernels, are Dirac impulse responses that have to be estimated as individual elements. However, the use of Dirac impulse responses may be an inefficient description since these functions tend to decay linearly over time. There is a direct relationship between the 'memory length' M and the duration of actual memory in the system and so to accurately describe the output response of the system a sufficiently large M value must be chosen. Otherwise the approximation error would become too large and the dynamic representation of the model would be poor. For this reason, when using the traditional Volterra approach a large number of parameters is often necessary to identify the system. The huge number of parameters in many cases limits the practical usefulness of the classical Volterra model. In the approach proposed in [28], the Dirac impulses in the FIR filter are replaced by more general and complex orthonormal functions $\{\varphi_k(m)\}$, which decay exponentially to zero at a controllable rate. The discrete-time Laguerre function $\{\varphi_k(m)\}$ is defined by its Z-transform according to

$$L_k(z, \lambda) = \frac{\sqrt{1 - |\lambda|^2}}{1 - z^{-1}\lambda} \left(\frac{-\lambda^* + z^{-1}}{1 - z^{-1}\lambda} \right)^k, \qquad k \geq 0, \tag{5.42}$$

where λ is the pole of the Laguerre function ($|\lambda| < 1$) and λ^* represents the complex conjugate. According to this, a linear model based on the Laguerre functions will be of the form

$$y(s) = \sum_{k=0}^{L-1} b_k L_k(z, \lambda) x(s), \tag{5.43}$$

where b_k is the kth regression coefficient and $L_k(z, \lambda)$ is the kth discrete Laguerre function, given by Equation (5.42). Figure 5.21 depicts an implementation of the Laguerre filter.

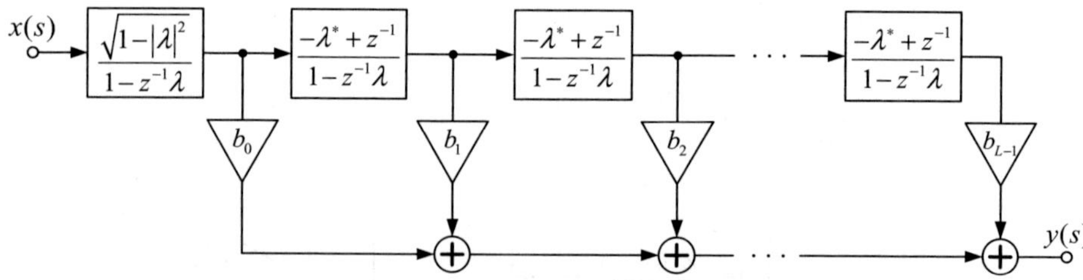

Figure 5.21 Laguerre filter of order L. (Reprinted with permission from [28], © 2005 IEEE.)

The accuracy of the model depends on the number L of Laguerre basis functions. However, an appropriate selection of the basis functions results in an equivalent accuracy with a large reduction in the number of parameters needed in comparison with the classic Volterra model. As described in [28], the implementation of the nonlinear Laguerre model is easily achieved by means of a 'nonlinear combiner' that sums all the weighted product-term combinations; see Figure 5.22. In [28] Zhu and Brazil show how, assuming that the Volterra kernels $h_l(i_1, i_2, \ldots, i_l)$ in Equation (5.39) are absolutely summable on the system memory $[0, M]$, they can be approximated by a complete basis $\{\varphi_k(m)\}$ of Laguerre functions defined over $[0, L]$. The expansions for the first- and third-order kernels are:

$$h_1(i) = \sum_{k=0}^{L-1} c_1(k)\varphi_k(i) \tag{5.44}$$

and

$$h_3(i_1, i_2, i_3) = \sum_{k_1=0}^{L-1} \sum_{k_2=k_1}^{L-1} \sum_{k_3=0}^{L-1} c_3(k_1, k_2, k_3)\varphi_{k_1}(i_1)\varphi_{k_2}(i_2)\varphi^*_{k_3}(i_3). \tag{5.45}$$

These expansions extend to all kernels present in the system. Then the Volterra model becomes

$$\begin{aligned}\tilde{y}(s) &= \sum_{k=0}^{L-1} c_1(k)l_k(s) \\ &+ \sum_{k_1=0}^{L-1} \sum_{k_2=k_1}^{L-1} \sum_{k_3=0}^{L-1} c_3(k_1, k_2, k_3)l_{k_1}(s)l_{k_2}(s)l^*_{k_3}(s) \\ &+ \cdots \end{aligned} \tag{5.46}$$

where

$$l_k(s) = \sum_{m=0}^{M-1} \varphi_k(m)\tilde{x}(s-m) = L_k(z, \lambda)\tilde{x}(s) \tag{5.47}$$

and $c_p(k_1, k_2, \ldots, k_p)$ are the kernel expansion coefficients.

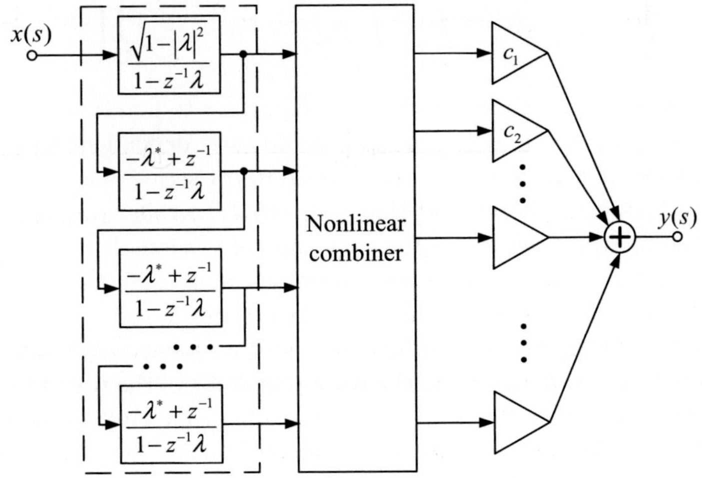

Figure 5.22 Structure of a Laguerre model. (Reprinted with permission from [28], © 2005 IEEE.)

5.6.4 Modified or dynamic Volterra series

To overcome the high complexity of a general Volterra series, a Volterra-like approach, called the *modified Volterra series* [29] or *dynamic Volterra series* [30] has been proposed. This modified series has the important property that it separates the purely static effects from the memory effects, which are intimately mixed in the classical series. If the nonlinear memory duration in the device is short enough with respect to the signal period, this series can be truncated to a single-fold integral, which allows modelling for not only weak but also relatively strong nonlinearities.

In the following paragraphs we review the fundamental theory behind the modified or dynamic Volterra series as presented in [29] and [30]. First, theoretical aspects are explained and both time- and frequency-domain identification procedures are presented. Then an alternative formulation is described.

Model description

By introducing the dynamic deviation

$$e(t,\tau) = x(t-\tau) - x(t), \tag{5.48}$$

a dynamic-deviation-based version of the Volterra series is obtained [29]:

$$y(t) = z_0(t) + \sum_{n=1}^{+\infty} z_n(t), \tag{5.49}$$

where $z_0(t)$ is a purely algebraic function that represents the output of the system for a DC input (i.e. when $e(t,\tau) = 0$) and where the higher-order terms z_n, where

$$z_n(t) = \int_{T_\mathrm{B}}^{T_\mathrm{A}} \cdots \int g_n\left\{x(t), \tau_1, \ldots, \tau_n\right\} \prod_{i=1}^{n} e(t,\tau_i) d\tau_i, \tag{5.50}$$

account for the memory effects. It is important to note that the kernels of the modified Volterra series, $g_n\{\cdot\}$, are now nonlinearly dependent on the input signal $x(t)$. Moreover, it can be shown that the Volterra series is a particular case of Equation (5.50) for $x(t) = 0$ and that the kernels in this equation can be expressed as a function of the kernels of the conventional version of the Volterra series in Equation (5.39). From previous considerations it is clear that both the conventional Volterra series and the modified Volterra series have the same asymptotic convergence properties. However, when for practical reasons only a relatively small number of terms must be considered, the basic properties of the two series are quite different. For instance, when both series are truncated to a single integral, the conventional Volterra model corresponds to a linear convolution (i.e. a purely linear dynamic system), while the modified model is capable of describing not only nonlinear systems without memory through its first term but also some nonlinear dynamic effects, represented by the single-fold convolution integral. The reason is that the kernels in the modified series are nonlinearly dependent on the instantaneous value of the input. Thus, an adequate comparison of the two series should be based on a study of the model accuracy in the presence of a quite limited, practically usable, number of terms. Intuitively, for a periodic input $x(t)$ the terms $x(t-\tau)$ in Equation (5.39) need not necessarily be small, whereas the dynamic deviations $e(t,\tau)$ in Equation (5.48) can be small even in the presence of large values of the input signal $x(t)$ if the memory interval is sufficiently short. Under these conditions, the contributions of the successive products in Equation (5.50) become progressively less important. In other words, the system can be characterised by a small number of terms of the Volterra series only in the presence of a small-amplitude signal, independent of the memory interval, whereas, using the modified Volterra series the output signal can also be represented with a small number of terms in the presence of large-amplitude signals provided that the memory interval is sufficiently short. A detailed study of the convergence properties of the modified Volterra series is included in [31], where it is shown that, for a given signal shape, the truncation error is proportional to the product of the maximum input frequency and the maximum equivalent time duration of the nonlinear effects in the system. In particular, it should be emphasised that, in the dynamic-deviation-based Volterra series, the truncation error for a given system depends not only on the amplitude of the applied signal, as in the conventional Volterra description, but on a tradeoff between its peak-to-peak value and its fundamental frequency for a given 'shape' (or equivalently its bandwidth). The upper limit of the truncation error in the Volterra series is instead dependent only on the signal amplitude and, unfortunately, is not necessarily small in the presence of low-bandwidth signals. The modified series is thus usable in strongly nonlinear operation provided that the memory effects in the system are relatively short with respect to the inverse of the signal frequency. This condition is satisfied, for instance, in electron devices described in a voltage-controlled form (possibly after parasitic de-embedding), or in sample-and-hold ADC devices (after suitable modifications in the system description) [32]. In such cases, the modified Volterra series can be truncated to the first convolution integral, giving

$$y(t) = z_0(t) + \int_{T_B}^{T_A} g_1(x(t), \tau) e(t, \tau) d\tau. \tag{5.51}$$

In order to model telecommunication PAs, which usually exhibit both strong nonlinearity and nonlinear memory effects and which are not usually narrowband systems, a convenient formulation of the modified Volterra series described above was formulated in [33] and [34]. A mathematical formulation of the nonlinear dynamic model was proposed by Mirri *et al.* [33], together with its computer-aided design (CAD) validation of a 2 GHz PHEMT PA for WCDMA applications. An alternative time-domain identification procedure was presented by Dooley *et al.* [35].

In all cases of practical interest a bandlimited input signal $x(t-\tau)$, for any shift τ, can be described as a single carrier, with generic amplitude and phase modulations $|a(t-\tau)|$ and $\angle a(t-\tau)$ with respect to τ, in the following form [33]:

$$x(t-\tau) = 2\,\mathrm{Re}\,\{a(t-\tau)\exp[j2\pi \mathrm{f}_0(t-\tau)]\} \tag{5.52}$$

$$= |a(t-\tau)| \sum_{\substack{i=-1\\ i\neq 0}}^{+1} \exp\{ji[2\pi \mathrm{f}_0(t-\tau) + \angle a(t-\tau)]\} \tag{5.53}$$

where $a(t-\tau)$ is the equivalent complex modulation envelope,

$$a(t-\tau) = |a(t-\tau)|\,\exp[j\angle a(t-\tau)], \tag{5.54}$$

and f_0 is the associated equivalent carrier frequency. On the basis of Equations (5.52)–(5.54), the amplifier response can be conveniently described by a Volterra-like integral series expansion [36] in terms of the dynamic deviations $e(t,\tau)$ of the signal $x(t-\tau)$ from a convenient reference signal $\hat{x}(t,\tau)$:

$$e(t,\tau) = x(t-\tau) - \hat{x}(t,\tau), \tag{5.55}$$

with $e(t,0) = 0$. In this case, the reference signal is selected as an equivalent sinusoid with respect to τ:

$$\hat{x}(t) = 2\,\mathrm{Re}\,\{a(t)\exp[j2\pi \mathrm{f}_0(t-\tau)]\} = |a(t)| \sum_{\substack{i=-1\\ i\neq 0}}^{+1} \exp\{ji[2\pi \mathrm{f}_0(t-\tau) + \angle a(t)]\}, \tag{5.56}$$

whose amplitude and phase coincide with those of the input signal at the instant t at which the output $u(t)$ is evaluated; in fact $\hat{x}(t,0) = x(t)$.

By expressing the functional of the two functions through a convenient series (see Appendix A in [33]), on the hypothesis that the bandwidth of the complex modulation envelope is sufficiently small to make the product over i of the amplitudes of the dynamic deviations $e(t,\tau_i)$ almost negligible in practice, and by considering only the spectral components of the output signal that fall within the operating bandwidth $y_{\mathrm{B}}(t)$ centred around f_0, it can be shown that the output signal can be expressed in terms of the equivalent output complex demodulation envelope $b(t)$ as follows:

$$y_{\mathrm{B}}(t) = 2\,\mathrm{Re}\,\{b(t)\exp[j2\pi \mathrm{f}_0 t]\}, \tag{5.57}$$

where

$$\begin{aligned} b(t) &= a(t)h(\mathrm{f}_0, |a(t)|) + \int_{T_B}^{T_A} h_1(\tau_1)\,[a(t-\tau_1) - a(t)]\exp(-j2\pi \mathrm{f}_0\tau_1)d\tau_1 \\ &+ \int_0^{T_B} g_1(\tau_1, \mathrm{f}_0, |a(t)|)\,[a(t-\tau_1) - a(t)]\exp(-j2\pi \mathrm{f}_0\tau_1)d\tau_1 \\ &+ a^2(t)\int_0^{T_B} g_2(\tau_1, \mathrm{f}_0, |a(t)|)\,[a^*(t-\tau_1) - a^*(t)]\exp(-j2\pi \mathrm{f}_0\tau_1)d\tau_1. \end{aligned} \tag{5.58}$$

According to this equation, the in-band output signal $y_{\mathrm{B}}(t)$ can be computed as the sum of different terms. The first term represents the memoryless nonlinear contribution, the second represents the purely dynamic linear contribution and the last two the purely dynamic nonlinear contributions. The dynamic contributions are evaluated through a convolution integral expressed in terms of the dynamic deviations of the complex modulation envelope of the input signal. In particular, when the input signal is an unmodulated signal carrier, i.e. $a(t)$ is a constant, according to Equation (5.55) each dynamic deviation is identically zero so that the corresponding output in Equation (5.58) is given just by the first term. It can be easily shown that the amplitude and the phase of $H(\mathrm{f}_0, |a(t)|)$ simply correspond to the well-known and widely-used AM–AM and AM–PM amplifier characteristics. This means, in practice, that the AM–AM and AM–PM plots, which are the only data normally provided to characterise the large-signal amplifier response, simply represent a zero-order approximation, with respect to the dynamic deviations of the complex modulation envelope $a(t)$, of the system behaviour. Thus, in the presence of modulated signals the commonly used AM–AM and AM–PM amplifier characterisation is sufficiently accurate only when the bandwidth of the complex modulation envelope $a(t)$ is so small that the amplitudes of the dynamic deviations $a(t-\tau_i) - a(t)$ for each τ_i are almost negligible. In many practical cases, when dealing with large-bandwidth modulated signals, this constraint cannot be met. For better accuracy in such conditions the generalised 'black-box' modelling approach defined by Equation (5.58) can be used, taking into account more terms of the functional series expansion. In fact, even if the series has been truncated to the first-order term ($n = 1$), considerable improvement in accuracy is achieved with respect to the 'coarser' zero-order approximation of the conventional AM–AM and AM–PM characteristics.

Model identification

The model can be identified using either a time- or a frequency-domain approach. The time-domain approach proposed in [35] starts from Equation (5.51) and yields the following formulation:

$$\tilde{y}(t) = z_0(\tilde{x}(t), \tilde{x}^*(t)) + \int_0^T h_1(\tilde{x}(t), \tilde{x}^*(t), \tau_1) e(t, \tau_1) d\tau_1 + \int_0^T h_2(\tilde{x}(t), \tilde{x}^*(t), \tau_1) e^*(t, \tau_1) d\tau_1, \tag{5.59}$$

where $\tilde{x}(t)$ and $\tilde{y}(t)$ are the complex input and output envelopes respectively and $e^*(t, \tau_1)$ is the dynamic deviation of the complex conjugate of the input signal's complex envelope. Figure 5.23 shows a schematic of the model structure.

From Equation (5.59) it is clear that there are three terms to be extracted. The first term, which describes the AM–AM and AM–PM characteristics of the amplifier, can be obtained by measuring the response of the amplifier to a single-tone power sweep at the operating-band centre frequency. In order to represent the characteristics analytically, so that the resulting model is more efficient, they are described by an equation whose parameters are determined by a best-fit

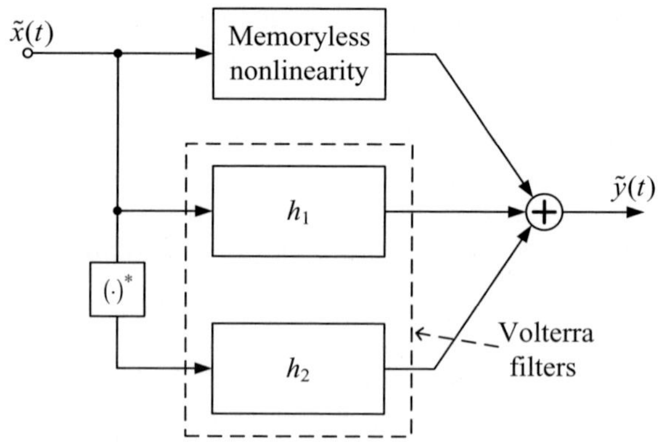

Figure 5.23 Diagram of time-domain model structure. (Reprinted with permission from [35], © 2004 IEEE.)

approximation technique. Having initially characterised the amplifier by its first term only, using the output from this model and the actual output signal from the device for the same input signal, all measured in the time domain, the two remaining kernels can be approximated using adaptive Volterra filters [34]. As with any adaptive filter, the recursive algorithm uses an input signal vector and the desired system response to compute an estimation error. This estimation error is used to update the values of a set of adjustable filter coefficients, which are multiplied by the next input signal vector and used to calculate a subsequent estimation error. The process is continued for a sufficiently large number of iterations until the estimation error falls below a predetermined threshold value.

Alternatively, a frequency-domain approach can be adopted to identify the model described in Equation (5.58) [37]:

$$\begin{aligned} b(t) &= a(t)H(\mathrm{f}_0, |a(t)|) + \int_{-\infty}^{+\infty} [H_1(\mathrm{f}_0+\mathrm{f}) - H_1(\mathrm{f}_0)]A(\mathrm{f})\exp(j2\pi\mathrm{f}t)\,d\mathrm{f} \\ &+ \int_{-\infty}^{+\infty} [G_1(\mathrm{f}_0+\mathrm{f}, \mathrm{f}_0, |a(t)|) - G_1(\mathrm{f}_0, \mathrm{f}_0, |a(t)|)]\,A(\mathrm{f})\exp(j2\pi\mathrm{f}t)\,d\mathrm{f} \\ &+ a^2(t)\int_{-\infty}^{+\infty} [G_2^*(-\mathrm{f}_0-\mathrm{f}, \mathrm{f}_0, |a(t)|) - G_2^*(-\mathrm{f}_0, \mathrm{f}_0, |a(t)|)]\,A^*(\mathrm{f})\exp(-j2\pi\mathrm{f}t)\,d\mathrm{f}. \end{aligned} \tag{5.60}$$

By introducing the Fourier transforms H_1, G_1 and G_2^* of the functions $h_1(\tau_1)$, $g_1(\tau_1, \mathrm{f}_0, |a(t)|)$ and $g_2^*(\tau_1, \mathrm{f}_0, |a(t)|)$ respectively, as well as the new functions

$$\begin{aligned} \hat{H}_1(\mathrm{f}_0, \mathrm{f}) &= H_1(\mathrm{f}_0+\mathrm{f}) - H_1(\mathrm{f}_0), \\ \hat{G}_1(\mathrm{f}_0+\mathrm{f}, \mathrm{f}_0, |a(t)|) &= G_1(\mathrm{f}_0+\mathrm{f}, \mathrm{f}_0, |a(t)|) - G_1(\mathrm{f}_0, \mathrm{f}_0, |a(t)|), \\ \hat{G}_2^*(-\mathrm{f}_0-\mathrm{f}, \mathrm{f}_0, |a(t)|) &= G_2^*(-\mathrm{f}_0-\mathrm{f}, \mathrm{f}_0, |a(t)|) - G_2^*(-\mathrm{f}_0, \mathrm{f}_0, |a(t)|), \end{aligned} \tag{5.61}$$

the final formulation for discrete-spectrum signals becomes

$$b(t) = a(t)H(\mathrm{f}_0, |a(t)|) + \sum_{p=-P}^{P} \hat{H}_1(\mathrm{f}_0, \mathrm{f}_p)A(\mathrm{f}_p)\exp(j2\pi\mathrm{f}_p t)$$
$$+ \sum_{p=-P}^{P} \hat{G}_1(\mathrm{f}_0 + \mathrm{f}_p, \mathrm{f}_0, |a(t)|)A(\mathrm{f}_p)\exp(j2\pi\mathrm{f}_p t)$$
$$+ a^2(t) \sum_{p=-P}^{P} \hat{G}_2^*(-\mathrm{f}_0 - \mathrm{f}_p, \mathrm{f}_0, |a(t)|)A^*(\mathrm{f}_p)\exp(-j2\pi\mathrm{f}_p t). \quad (5.62)$$

The model described by Equation (5.62) can be identified, by using a complex-envelope modulator and demodulator, as the one represented in Figure 5.24.

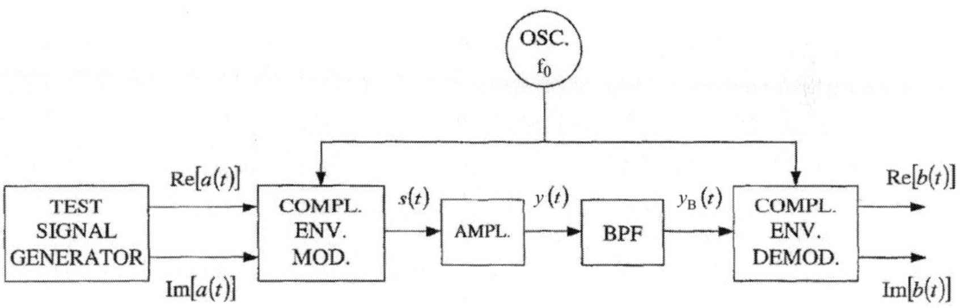

Figure 5.24 Block diagram of frequency-domain identification procedure. (Reprinted with permission from [38], © 1999 IEEE.)

Both the time- and frequency-domain approaches yield accurate modelling results [35, 37].

Alternative model description

In [30], a similar dynamic-deviation-based model was used by Ngoya and Soury to model a 6 W MMIC PA for radar applications and a 10 W C-band HFET-based nonlinear amplifier. By assuming that $\tilde{x}(t)$ and $\tilde{y}(t)$ are the input envelope and output envelope of the PA and following the analysis in [30], the model becomes

$$\tilde{y}(t) = Y_{\mathrm{dc}}(\tilde{x}(t), \tilde{x}^*(t))$$
$$+ \int_0^T h_1(\tilde{x}(t), \tilde{x}^*(t), \lambda)[\tilde{x}(t-\lambda) - \tilde{x}(t)]d\lambda$$
$$+ \int_0^T h_2(\tilde{x}(t), \tilde{x}^*(t), \lambda)[\tilde{x}^*(t-\lambda) - \tilde{x}^*(t)]d\lambda \quad (5.63)$$

or, equivalently, in the frequency domain,

$$\begin{aligned}\tilde{y}(t) &= Y_{\mathrm{dc}}(\tilde{x}(t), \tilde{x}^*(t)) \\ &+ \frac{1}{2\pi}\int_{-BW/2}^{BW/2} H_1(\tilde{x}(t), \tilde{x}^*(t), \Omega)[X(\Omega)e^{j\Omega t}]d\Omega \\ &+ \frac{1}{2\pi}\int_{-BW/2}^{BW/2} H_2(\tilde{x}(t), \tilde{x}^*(t), -\Omega)[X^*(\Omega)e^{-j\Omega t}]d\Omega\,,\end{aligned} \tag{5.64}$$

where BW is the bandwidth of the signal. Accounting for the time invariance of the input–output relation, the kernels in Equation (5.64) take the following form:

$$\begin{aligned}Y_{\mathrm{dc}}(\tilde{x}(t), \tilde{x}^*(t)) &= Y_{\mathrm{dc}}(|\tilde{x}(t)|)e^{j\Phi\tilde{x}(t)}, \\ H_1(\tilde{x}(t), \tilde{x}^*(t), \Omega) &= H_1(|\tilde{x}(t)|, \Omega), \\ H_2(\tilde{x}(t), \tilde{x}^*(t), -\Omega) &= H_2(|\tilde{x}(t)|, \Omega)e^{j2\Phi\tilde{x}(t)}.\end{aligned} \tag{5.65}$$

In the above, $Y_{\mathrm{dc}}(|\tilde{x}(t)|)$ is the static characteristic of the system, i.e. the response to a unmodulated carrier excitation, which corresponds to the AM–AM and AM–PM characteristics. The *synchronous* and *image* Volterra transfer functions of the system are $H_1(\tilde{x}(t), \Omega)$ and $H_2(\tilde{x}(t), -\Omega)$ respectively.

From Equation (5.65) it is clear that it is possible to extract the three kernels by applying at the input a two-tone signal of the form

$$\tilde{x}(t) = |X_0| + \delta X e^{j\Omega t}, \qquad |\delta X| \ll 1. \tag{5.66}$$

The output of the system is then

$$\tilde{y}(t) = Y_0 + \delta Y^+ e^{j\Omega t} + \delta Y^- e^{-j\Omega t}, \tag{5.67}$$

so that

$$\begin{aligned}Y_{\mathrm{dc}}(|X_0|) &= Y_0, \\ H_1(|X_0|, \Omega) &= \frac{\delta Y^+}{\delta X^*} - \frac{1}{2}\left(\frac{\partial Y_0}{\partial |X_0|} - \frac{Y_0}{|X_0|}\right), \\ H_2(|X_0|, -\Omega) &= \frac{\delta Y^-}{\delta X^*} - \frac{1}{2}\left(\frac{\partial Y_0}{\partial |X_0|} - \frac{Y_0}{|X_0|}\right).\end{aligned} \tag{5.68}$$

The extraction procedure is then straightforward and is illustrated in Figure 5.25.

The spacing between the two tones is to be swept throughout the bandwidth, BW, of the system and the input signal magnitude $|X_0|$ is varied from the linear region as far into saturation as required for the system being modelled.

5.6.5 Volterra model from TDNN model

In this final section on Volterra series models, a link with the time-delay neural network (TDNN) models discussed in Section 5.3 is made. It will be shown how a Volterra series model can be generated starting from the weights and bias values of

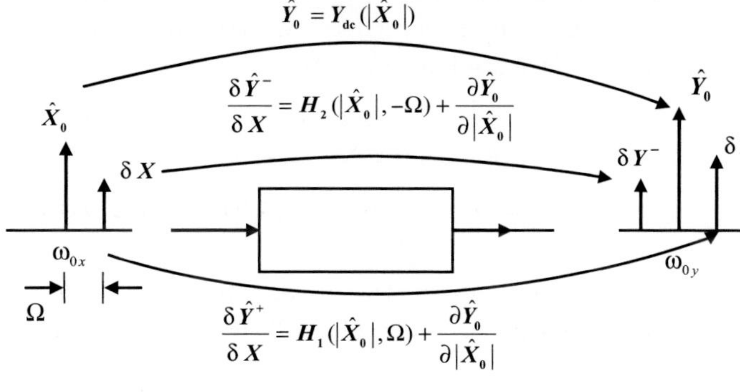

Figure 5.25 Dynamic Volterra kernels extraction setup. (Reprinted with permission from [30], © 2003 IEEE.)

a TDNN model [39]. For clarity, the procedure is detailed for single-input systems, although it can be extended to multiple-input systems.

The single-input TDNN model is described by the following expression:

$$y(s) = w_0^2 + \sum_{j=1}^{N_2} w_j^2 f\left(w_0^1 + \sum_{k=0}^{N_1} w_{kj}^1 x(s-k) \right), \tag{5.69}$$

where $y(t)$ is the output of the TDNN model, $x(s-k)$ with $k = 0, \ldots, N$ represents the input time-delayed samples and w indicates the weight of the links between the various neurons, while the subscripts and superscript refer to the neuron layers and to the link position within a layer respectively.

A Taylor series is expanded around the bias values of the hidden nodes, resulting in

$$y(s) = w_0^2 + \sum_{j=1}^{N_2} w_j^2 f(w_0^1) + \sum_{j=1}^{N_2} w_j^2 \sum_{d=1}^{\infty} \frac{1}{d!} \left. \frac{\partial^d f(x)}{\partial x^d} \right|_{x=w_0^1} \left[\sum_{k=0}^{N_1} w_{kj}^1 x(s-k) \right]^d, \tag{5.70}$$

where $\partial^d f(x)/\partial x^d$ is the dth-order derivative of the hidden-node activation function $f(x)$ with respect to the bias.

Equation (5.70) can be rewritten as follows:

$$\begin{aligned} y(s) = w_0^2 &+ \sum_{j=1}^{N_2} w_j^2 f(w_0^1) + \sum_{j=1}^{N_2} w_j^2 \sum_{d=1}^{\infty} \frac{1}{d!} \left. \frac{\partial^d f(x)}{\partial x^d} \right|_{x=w_0^1} \\ &\times \left[\sum_{k_1=0}^{N_1} w_{k_1 j}^1 x(s-k_1) \sum_{k_2=0}^{N_1} w_{k_2 j}^1 x(s-k_2) \cdots \right. \\ &\times \left. \sum_{k_{d-1}=0}^{N_1} w_{k_{d-1} j}^1 x(s-k_{d-1}) \sum_{k_d=0}^{N_1} w_{k_d j}^1 x(s-k_d) \right]. \end{aligned} \tag{5.71}$$

Observing Equation (5.70), the previous expression can be identified as a discrete Volterra series expansion truncated to the kth order, the kernels of which can be easily found by direct examination to be

$$h_0 = w_0^2 + \sum_{j=1}^{N_2} w_j^2 f(w_0^1), \tag{5.72}$$

$$h_1(k) = \sum_{j=1}^{N_2} \left.\frac{\partial f(x)}{\partial x}\right|_{x=w_0^1} w_j^2 w_{kj}^1, \tag{5.73}$$

$$h_2(k_1,k_2) = \sum_{j=1}^{N_2} \frac{1}{2} \left.\frac{\partial^2 f(x)}{\partial x^2}\right|_{x=w_0^1} w_j^2 w_{k_1 j}^1 w_{k_2 j}^1, \tag{5.74}$$

$$h_d(k_1,k_2,\ldots,k_{d-1},k_d) = \sum_{j=1}^{N_2} \frac{1}{d!} \left.\frac{\partial^d f(x)}{\partial x^d}\right|_{x=w_0^1} w_j^2 w_{k_1 j}^1 \cdots w_{k_{d-1} j}^1 w_{k_d j}^1. \tag{5.75}$$

It should also be noted that for symmetric Volterra kernels it is not necessary to calculate all the possible combinations of $k_1,\ldots,k_d$; in fact it is enough to obtain a kernel as a function of one k_d combination and then multiply it by the number of permutations $k_d!$. For example, when the activation function is a hyperbolic tangent the first three kernels can be derived as:

$$h_0 = w_0^2 + \sum_{j=1}^{N_2} w_j^2 \tanh w_0^1, \tag{5.76}$$

$$h_1(k) = \sum_{j=1}^{N_2} w_j^2 w_{kj}^1 \left(1 - \tanh^2 w_0^1\right), \tag{5.77}$$

$$h_2(k_1,k_2) = \sum_{j=1}^{N_2} w_j^2 w_{k_1 j}^1 w_{k_2 j}^1 \frac{-2\tanh w_0^1 + 2\tanh^3 w_0^1,}{2!}, \tag{5.78}$$

$$h_3(k_1,k_2,k_3) = \sum_{j=1}^{N_2} w_j^2 w_{k_1 j}^1 w_{k_2 j}^1 w_{k_3 j}^1 \frac{-2 + 8\tanh^2 w_0^1 - 6\tanh^4 w_0^1}{3!}. \tag{5.79}$$

The speed and simplicity of the Volterra series extraction may be increased if the system order is known *a priori* and polynomial activation functions can be used for the hidden neurons. In this way the Taylor expansion is avoided and thus also the time-consuming calculation of derivatives of the activation functions.

When polynomial activation functions are used, the TDNN model becomes

$$y(s) = w_0^2 + \sum_{j=1}^{N_2} w_j^2 \left[w_0^1 + \sum_{i=1}^{N_1} w_{ij}^1 x_i(s-k) \right]^M, \tag{5.80}$$

where M refers to the polynomial order. The Volterra-kernel extraction procedure then reduces to distributing the polynomials in such a way that, after common factoring, the kernels are immediately identifiable in the expanded equation:

$$
\begin{aligned}
y(s) = {} & w_0^2 + \sum_{j=1}^{N_2} w_j^2 (w_0^1)^M \\
& + \left[\sum_{j=1}^{N_2} w_j^2 \sum_{i=1}^{N_1} w_{ij}^1 M (w_0^1)^{M-1} \right] x_i(s-k) \\
& + \left[\sum_{j=1}^{N_2} w_j^2 \sum_{i=1}^{N_1} \sum_{k=1}^{N_1} w_{ij}^1 w_{kj}^1 \frac{M(M-1)(w_0^1)^{M-2}}{2!} \right] x_i(s-k) x_i(s-k) \\
& + \left[\sum_{j=1}^{N_2} w_j^2 \sum_{i=1}^{N_1} \sum_{k=1}^{N_1} \sum_{m=1}^{N_1} w_{ij}^1 w_{kj}^1 w_{mj}^1 \frac{M(M-1)(M-2)(w_0^1)^{M-3}}{3!} \right] \\
& \times x_i(s-k) x_i(s-k) x_i(s-k) \\
& + \cdots .
\end{aligned} \tag{5.81}
$$

The new, simpler, expressions for the kernels are then

$$
h_0 = w_0^2 + \sum_{j=1}^{N_2} w_j^2 (w_0^1)^M, \tag{5.82}
$$

$$
h_1(k) = \sum_{j=1}^{N_2} w_j^2 w_{kj}^1 M (w_0^1)^{M-1}, \tag{5.83}
$$

$$
h_2(k_1, k_2) = \sum_{j=1}^{N_2} w_j^2 w_{k_1 j}^1 w_{k_2 j}^1 \frac{M(M-1)(w_0^1)^{M-2}}{2!}, \tag{5.84}
$$

$$
h_3(k_1, k_2, k_3) = \sum_{j=1}^{N_2} w_j^2 w_{k_1 j}^1 w_{k_2 j}^1 w_{k_3 j}^1 \frac{M(M-1)(M-2)(w_0^1)^{M-3}}{3!}, \tag{5.85}
$$

$$
h_d(k_1, k_2, \ldots, k_d) = \sum_{j=1}^{N_2} w_j^2 w_{k_1 j}^1 w_{k_2 j}^1 \cdots w_{k_d j}^1 \frac{M \cdots (M-(d-1))(w_0^1)^{M-d}}{d!}. \tag{5.86}
$$

5.7 State-space-based model

The final modelling approach considered in this chapter is the *state-space-based model* [40]. It belongs to the class of circuit-level models described in Chapter 1. Whereas other methods usually consider a band-pass characteristic around the carrier frequency only, the model formulation used here is such that higher-order harmonics are intrinsically included. The term 'state-space-based' originates from the fact that such models are strongly related to the concept of state space and state equations. An important advantage of this technique is that it is not restricted to the modelling of weakly nonlinear phenomena, unlike methods such as Volterra series analysis. The technique is based directly on large-signal microwave time-domain data. The data can be collected either by simulating an existing PA design or through measurements. The latter approach is more general as it allows the modelling of packaged or chip PAs of which the actual design is unknown. As the method provides a full two-port, not just a single-input–single-output, description the measurements have to be two-port vector large-signal measurements (Figure 2.34).

5.7.1 Model description

It should be noted that the core model can only handle linear memory. Modelling nonlinear memory becomes possible through an extended formulation. This subsection will outline the formulation and demonstrate the modelling approach on a practical example, attention being paid to the experimental design so as to ensure accurate data collection.

Formulation for linear memory

The state-space-based modelling approach essentially depends on nonlinear system identification using techniques developed in nonlinear time-series analysis [41]. An advantage of this technique is that the resulting model is transportable: in other words it is usable over a range of environments and not restricted to a small domain of applicability, for example a single bias condition. The model is described directly by time-differential equations that are reconstructed from measured data. By this means, all the observable dynamics of the device are determined. Finally, this black-box modelling principle is applicable to any device type, regardless of its complexity, because no physical preknowledge is required. Since only the observable dynamics are captured, the model size, especially for circuits, will generally be significantly smaller than when these circuits are represented by separate models for each constituent component. This enables the construction of a compact, accurate and transportable dynamic model [42].

In general, the behaviour of electrical and mechanical devices can be described in terms of their state-space representation:

$$\dot{\mathbf{X}}(t) = F_{\mathrm{a}}(\mathbf{X}(t), \mathbf{U}(t)), \tag{5.87}$$

$$\mathbf{Y}(t) = F_{\mathrm{b}}(\mathbf{X}(t), \mathbf{U}(t)), \tag{5.88}$$

where $\mathbf{X}$ is the vector of state variables, $\mathbf{U}$ is the vector of input variables and $\mathbf{Y}$ is the vector of output variables and the dot above the symbol $\mathbf{X}$ denotes the time derivative.

In the case of a microwave device, it would be logical from the user's point of view to associate the terminal voltages with the inputs and the terminal currents with the outputs:

$$\dot{\mathbf{X}}(t) = F_{\mathrm{a}}(\mathbf{X}(t), \mathbf{V}(t)), \tag{5.89}$$

$$\mathbf{I}(t) = F_{\mathrm{b}}(\mathbf{X}(t), \mathbf{V}(t)) \tag{5.90}$$

where **V** is the vector of the terminal voltages and **I** is the vector of the terminal currents. As an example, in the case of a nonlinear resistor Equation (5.90) becomes $I_1(t) = f_1(V_1(t))$ and in the case of a nonlinear capacitor it becomes $I_1(t) = f_1\left(\dot{V}_1(t), V_1(t)\right)$.

In practice, it is more straightforward to work not with two separate (sets of) equations but with a combined expression. In other words, the behaviour of a two-port microwave device can be described by the following nonlinear ordinary time-differential equations:

$$I_1(t) = f_1\left(V_1(t), V_2(t), \dot{V}_1(t), \dot{V}_2(t), \ddot{V}_1(t), \ldots, \dot{I}_1(t), \dot{I}_2(t), \ldots\right), \qquad (5.91)$$

$$I_2(t) = f_2\left(V_1(t), V_2(t), \dot{V}_1(t), \dot{V}_2(t), \ddot{V}_1(t), \ldots, \dot{I}_1(t), \dot{I}_2(t), \ldots\right), \qquad (5.92)$$

in which the number of dots above the variable indicates the order of the derivative. Note that feedback is also included by means of the time derivatives of the currents. Alternatively, it is also possible to express the terminal voltages as a function of the appropriate independent variables or to consider the relationships between the incident travelling voltage waves a_i as input variables and the scattered travelling voltage waves b_i as output variables.

It should be noted that Equations (5.91) and (5.92) are applicable only to devices that do not exhibit slow, and thus nonlinear, memory effects. An extension to account for such memory effects is presented later.

This modelling approach requires the determination of the number of required independent variables, and subsequently the determination of the functions $f_1(\cdot)$ and $f_2(\cdot)$. This is achieved in three steps.

The first step involves collecting the data, which can originate from either time-domain measurements or simulations. The design of the input excitations is highly important because the set of collected data should adequately cover the state space of interest. The experiment design is detailed in subsection 5.7.2.

The second step is to determine the minimal set of independent variables needed to predict accurately the device dynamics. The initial set of independent variables from which the model is built consists of the measured terminal voltages and currents and their time derivatives as well as the time derivatives of the measured terminal currents, as defined in Equations (5.91) and (5.92). In principle, one could use all the possible independent variables in an arbitrarily fixed order, but this would result in models that are needlessly complex. There exist more rigorous methods for selecting a subset of independent variables from which to construct the model. In practical terms a 'good' (though not necessarily unique) subset of independent variables should have the property that a given response to each such variable is a single-valued function of the variable. The technique for finding this subset of independent variables is the s-ocalled 'false nearest neighbours' method [43, 44].

The third step, determining the functions $f_1(\cdot)$ and $f_2(\cdot)$, is essentially a function-fitting problem. Various types of fitting functions can be used, among which are multivariate polynomials, radial basis functions and artificial neural networks (subsection 5.3.2). As the last has turned out to be the most suitable option for this kind of modelling problem [45], it has been adopted by several research groups.

After constructing the PA model, it can be implemented in a microwave circuit simulator to enable it to be used in system design. The condition for implementation is that the simulator should be able to calculate time derivatives during the simulation and that a variable can be dependent on these time derivatives. A final advantage of this modelling approach is that it usually reduces the simulation time considerably. The reduction ratio is not fixed but depends strongly on the complexity of the original design. For example, when an MMIC amplifier was modelled using this approach a reduction by a factor 10 was obtained [40], but factors as high as 500 were reported when this technique was applied to transistors [46].

Formulation for nonlinear memory

As mentioned above, Equations (5.91) and (5.92) can only describe PAs with linear memory; the formulation has to be extended to include the effects of nonlinear memory [47].

As the derivative is related to the concept of a time delay, Equations (5.91) and (5.92) can be rewritten as

$$I_1(t) = f_1\left(V_1(t), V_2(t), V_1(t-\tau_{\mathrm{rf}}), V_2(t-\tau_{\mathrm{rf}}), \ldots, V_1(t-n\tau_{\mathrm{rf}}), V_2(t-n'\tau_{\mathrm{rf}})\right), \tag{5.93}$$

$$I_2(t) = f_2\left(V_1(t), V_2(t), V_1(t-\tau_{\mathrm{rf}}), V_2(t-\tau_{\mathrm{rf}}), \ldots, V_1(t-n\tau_{\mathrm{rf}}), V_2(t-n'\tau_{\mathrm{rf}})\right), \tag{5.94}$$

where τ_{rf} is a time constant of the order of nanoseconds, which corresponds to behaviour at gigahertz frequencies, and n, n' are integers representing higher-order time derivatives of the voltages. The time derivatives of the currents have been omitted, because inductive effects are often negligible.

Slow-memory effects can be modelled in a similar way. Slow-memory behaviour means that the device characteristics are changing in the kilohertz to megahertz frequency range, which corresponds to time constants of the order of milliseconds to microseconds. The consequence of this is that the terminal currents are dependent not only on what happened nanoseconds earlier (due to fast-memory effects) but also on what happened microseconds or even milliseconds earlier. The modelling equations subsequently become

$$\begin{aligned} I_1(t) =& f_1\left(V_1(t), V_2(t), V_1(t-\tau_{\mathrm{rf}}), V_2(t-\tau_{\mathrm{rf}}), \ldots, V_1(t-n\tau_{\mathrm{rf}}), V_2(t-n'\tau_{\mathrm{rf}}),\right. \\ &\left. V_1(t-\tau_{\mathrm{if}}), V_2(t-\tau_{\mathrm{if}}), \ldots, V_1(t-m\tau_{\mathrm{if}}), V_2(t-m'\tau_{\mathrm{if}})\right), \end{aligned} \tag{5.95}$$

$$\begin{aligned} I_2(t) =& f_2\left(V_1(t), V_2(t), V_1(t-\tau_{\mathrm{rf}}), V_2(t-\tau_{\mathrm{rf}}), \ldots, V_1(t-n\tau_{\mathrm{rf}}), V_2(t-n'\tau_{\mathrm{rf}}),\right. \\ &\left. V_1(t-\tau_{\mathrm{if}}), V_2(t-\tau_{\mathrm{if}}), \ldots, V_1(t-m\tau_{\mathrm{if}}), V_2(t-m'\tau_{\mathrm{if}})\right), \end{aligned} \tag{5.96}$$

where τ_{if} is the millisecond-to-microsecond-level time delay and n, n', m, m' are integers.

The modelling procedure itself remains largely unchanged. The major addition when considering slow-memory effects is that the τ_{if} time-delayed data must also be generated. This is accomplished by post-processing the acquired data. Once all independent and dependent data are available, the functional relationships $f_1(\cdot)$ and $f_2(\cdot)$ may again be represented by an ANN.

5.7.2 *Example*

Experiment design

In the case of a single-tone excitation, the trajectory of $V_2(t)$ versus $V_1(t)$ is a distorted ellipse, as illustrated in Figure 5.26.

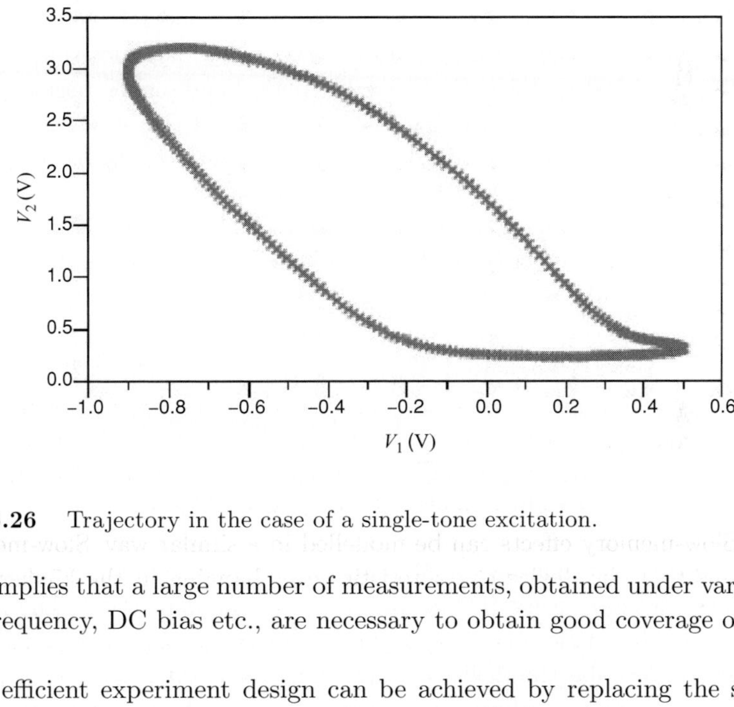

Figure 5.26 Trajectory in the case of a single-tone excitation.

This implies that a large number of measurements, obtained under varying input power, frequency, DC bias etc., are necessary to obtain good coverage of the state space.

More efficient experiment design can be achieved by replacing the single-tone excitation by a multisine [48]. The design of multisine excitations is discussed in Chapter 2.

A multisine is often represented by a complex-envelope, or 'low-pass equivalent' formulation [49]. This means that, instead of having a phasor representation for each tone, the behaviour is summarised by the fact that the amplitude and phase of the RF carrier (and its harmonics) vary as a function of time. In the case of $I_2(t)$, for example, we have

$$I_2(t) = \text{Re}\{A_1(t)e^{j\omega t+P_1(t)} + A_2(t)e^{j2\omega t+P_2(t)} + \cdots + A_h(t)e^{jh\omega t+P_h(t)}\}, \quad (5.97)$$

where $A_i(t)$ and $P_i(t)$ are the amplitude and phase of the ith harmonic and h is the number of harmonics.

Theoretically, the steady-state time-domain representation can still be used for building the model. The data processing would become very cumbersome, however. For a multisine, the intermediate-frequency (IF) envelope period corresponds to $1/\Delta\text{f}$, where Δf is the intercarrier spacing. Thus, one envelope period is several thousand high-frequency periods and hence, in order to meet the Nyquist criterion, the number of required data points should be a multiple of this. For example, an IF period of 1/25 kHz or 40 μs encompasses 40 000 RF periods of 1 ns (considering 1 GHz as the fundamental frequency). This would mean that at least 800 000 data points are required when ten harmonics are considered. The question is whether it is necessary to deal with all these data points, or can the sampling be done more efficiently?

One IF period of a multisine excitation corresponds to a closed curve in the (V_1, V_2) voltage plane, meaning that the start and end points have the same value. If one zooms in from one IF period down to one RF period, it will be noticed that one RF period is not a closed curve, but it is almost one, owing to the large difference between the scale of one RF period and one IF period, which means that the envelope value of $I_2(t)$, as well as that of all the other variables, changes very slowly in time. Consequently, the envelope values can be assumed 'constant' when one is zooming in locally to the RF period scale.

If it can be assumed that $A_i(t)$ and $P_i(t)$ vary slowly and thus can be kept constant with respect to the RF time scale then locally, around one sampling point of the IF period t_{IF}, one can write

$$\begin{aligned} I_2(t)_{\text{IF}} = \text{Re}\Big\{ & A_1(t_{\text{IF}})e^{j\omega t + P_1(t_{\text{IF}})} + A_2(t_{\text{IF}})e^{j2\omega t + P_2(t_{\text{IF}})} \\ & + \cdots + A_h(t_{\text{IF}})e^{jh\omega t + P_h(t_{\text{IF}})} \Big\}. \end{aligned} \tag{5.98}$$

Similar equations can be written down for the other variables (the terminal voltages etc.).

Note that Equation (5.98) reduces to the single-frequency method described above, because every IF 'sample' can be regarded as one independent measurement in the single-tone case. In other words, the data of one RF period could be used to describe the device's behaviour over a small section of the overall coverage area. The procedure consequently consists of 'sampling' the IF period and at each sample point collecting data from one RF period. This collection of RF periods forms the data set used to build the model. When sampling, one can use an equidistant sampling in time [48] or a more involved approach taking into account the level of nonlinearity [50].

Figure 5.27 shows the coverage of the (V_1, V_2) plane obtained by applying a three-tone excitation to an amplifier. If the conventional time-domain representation is used, the numerous data points make up one big black spot. The reason is that a multisine is a collection of phasors that sweep through the complex plane at

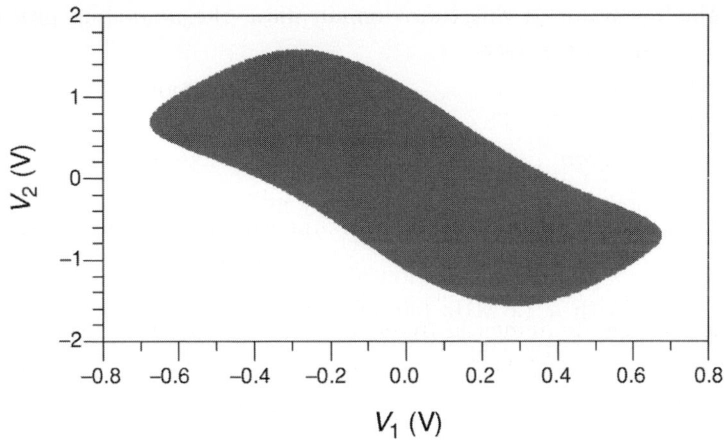

Figure 5.27 Coverage of the (V_1, V_2) plane obtained using one three-tone excitation. (Reprinted with permission from [48], © 2003 IEEE.)

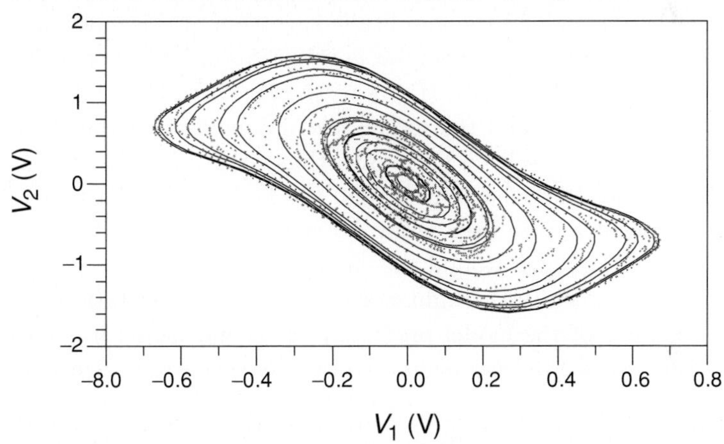

Figure 5.28 The (V_1,V_2) plane as sampled by 16 RF trajectories (black). The grey dots denote the area covered by the three-tone excitation. (Reprinted with permission from [48], © 2003 IEEE.)

slightly different speeds (owing to the small frequency offset), resulting in a large variation in the instantaneous values. Figure 5.28 shows how this space can be sampled efficiently. In this example, 16 IF sampling points are considered, spread equidistantly over time, and the corresponding RF trajectories are plotted. It can be seen that this collection of trajectories provides good coverage of the (V_1, V_2) area of the three-tone excitation. Similar coverage could be obtained by 16 single-tone measurements (or simulations) with varying input powers. As a result, using a multisine excitation reduces the measurement time considerably.

Having obtained this time-domain data, the modelling procedure as described above can be continued.

Model extraction

As an example, the state-space-based modelling method has been applied to an off-the-shelf general-purpose buffer RF amplifier developed for 4.9 GHz wireless applications [51]. The behavioural model was extracted from a 63-tone multisine excitation with a 1.6 MHz bandwidth QPSK-shaped probability density function (PDF). After implementation in the circuit simulator, the model was simulated using a different multisine excitation. The signal was synthesised by the same procedure as the multisine used for the model extraction but using a different set of QPSK-modulated random data.

The set of plots in Figure 5.29 depicts both the time- and frequency-domain simulation results (solid trace) together with the corresponding measurements (circles) of the b_2 scattered travelling voltage wave. The graphs in Figure 5.29(a), (b), (c) show respectively the magnitude and phase of the complex envelope around the carrier frequency and the amplitude spectrum of this envelope.

Similar results for the complex envelope around the second RF harmonic are plotted in Figure 5.29(d), (e), (f) respectively. There is very good agreement between the measurements and the model predictions around the RF carrier frequency as well as around its second harmonic, thus confirming that the model can also accurately predict the behaviour around the higher-order harmonics. This ability is an important advantage over many other behavioural models, which are 'bandwidth limited' to the band around the carrier frequency.

In order to verify the influence of the realistic excitation used for extraction on the accuracy of the model prediction, a second model was created. This time the multisine excitation was composed of only seven tones with equal amplitudes and phases evenly distributed in a 1.6 MHz bandwidth around 4.9 GHz. For clarity, the model based on the QPSK-like multisine will be referred to as model 1, and that based on the seven-tone multisine as model 2.

Model 2 was simulated under the same signal conditions as model 1. The resulting complex envelope of the b_2 travelling voltage wave around the carrier frequency is shown in Figure 5.30. The match between the measured waveform (circles) and those from the model 1 (crosses) and model 2 (triangles) simulations is very good, although it is difficult to quantitatively compare the accuracy of the models' predictions just on the basis of these IQ plots. This is especially true when several input power levels are taken into account.

Therefore, to facilitate quantitative analysis of the simulation results of both

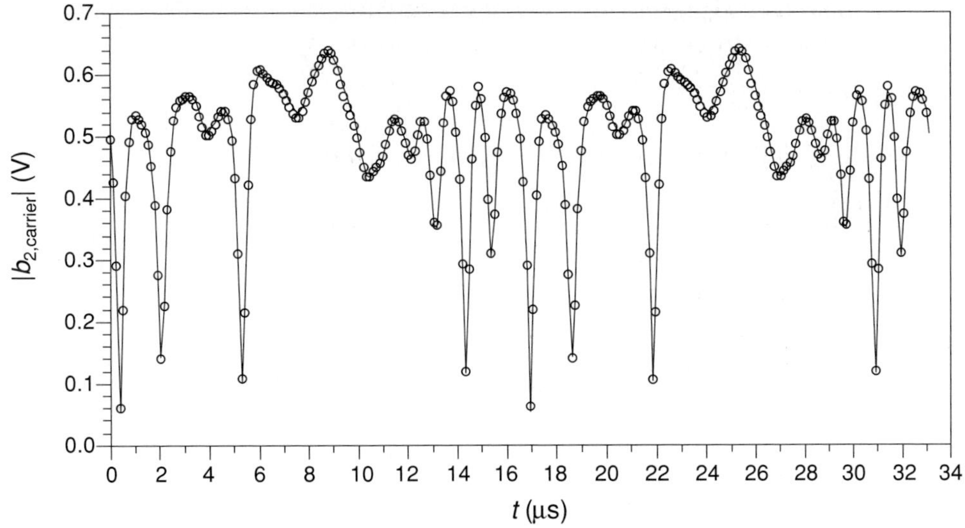

(a)

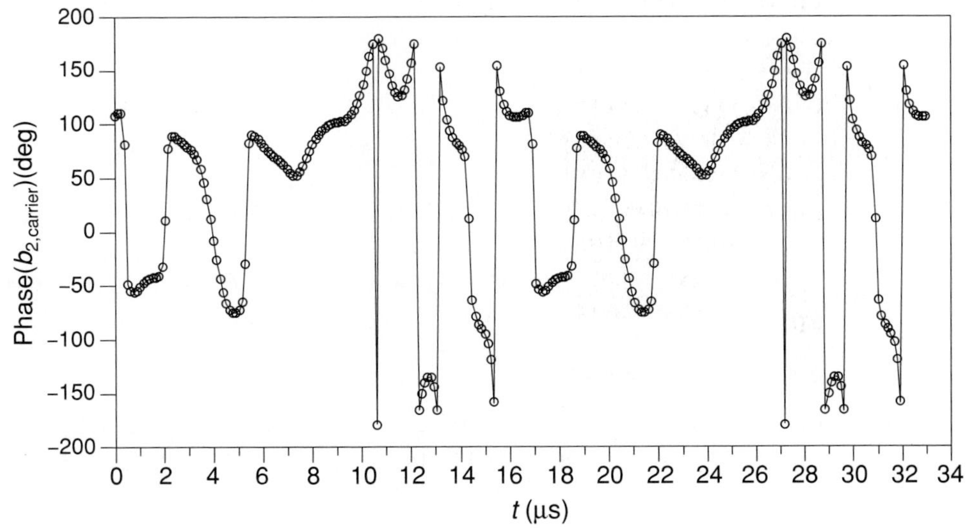

(b)

Figure 5.29 The measured (circles) and simulated (trace or crosses) b_2 scattered travelling voltage wave: (a) magnitude waveform, (b) phase waveform, (c) amplitude spectrum of the complex envelope around the RF carrier frequency, 4.9 GHz, (d) magnitude waveform, (e) phase waveform and (f) amplitude spectrum of the complex envelope around the second RF harmonic. The input power was 6 dBm. (From [51] with permission, © 2005 IEEE.)

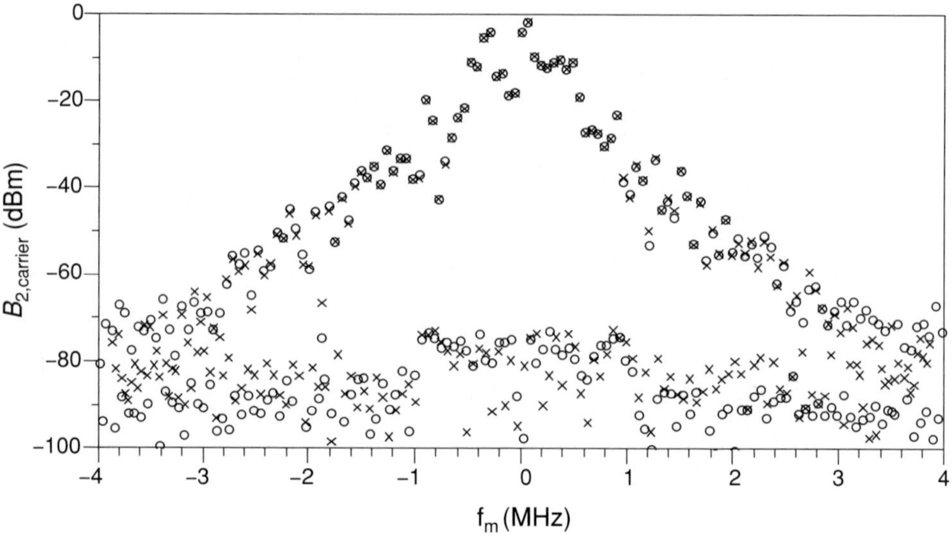

(c)

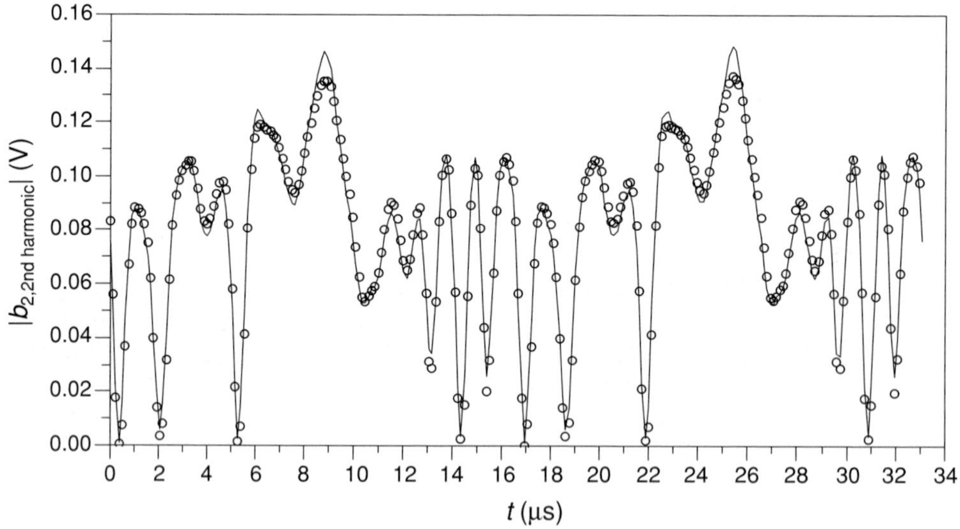

(d)

Figure 5.29 *(cont.)*

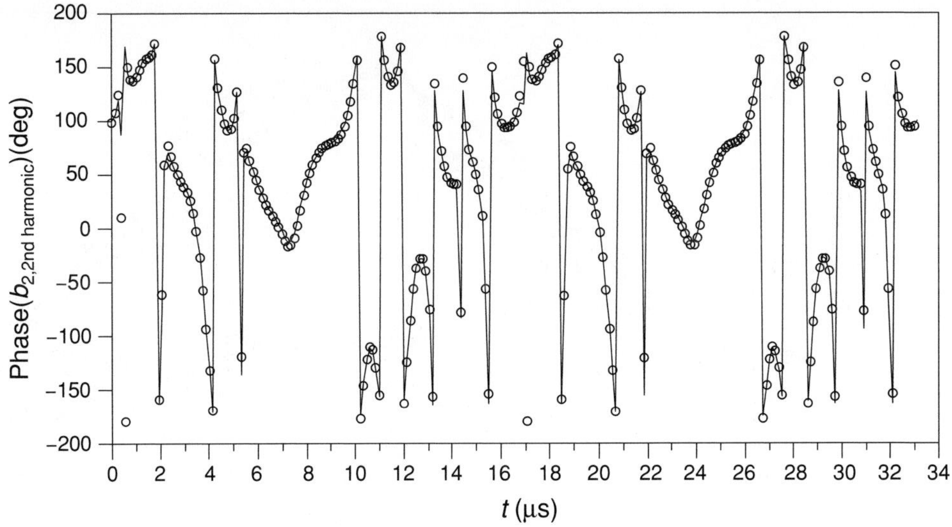

(e)

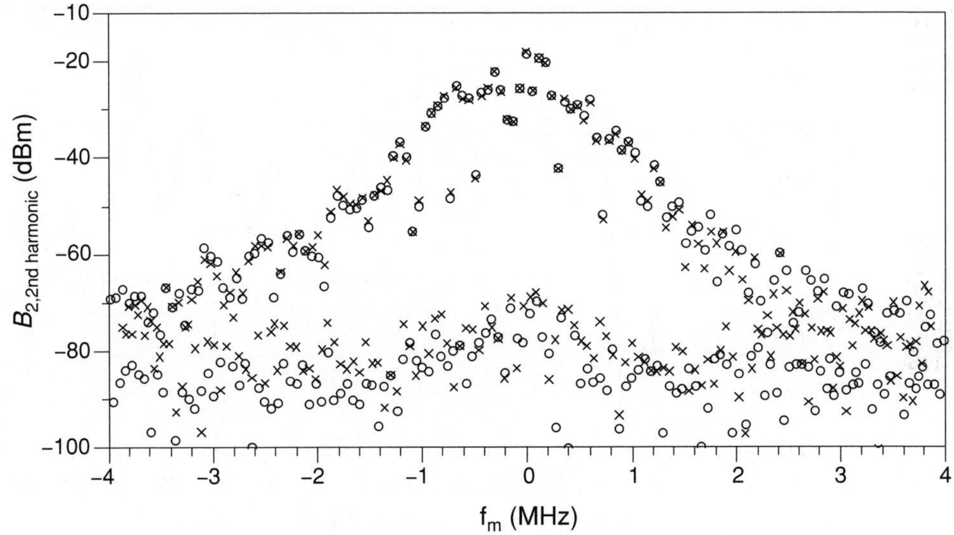

(f)

Figure 5.29 *(cont.)*

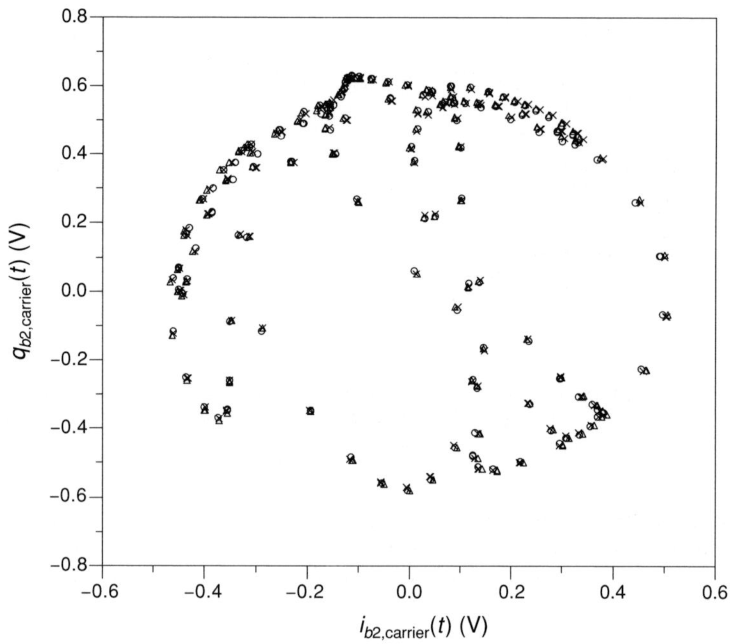

Figure 5.30 Shown is an IQ plot of the measured b_2 wave complex envelope (circles), simulated model 1 (crosses) and simulated model 2 (triangles) at $P_{\text{in}} = 6\,\text{dBm}$ and around 4.9 GHz. (From [51] with permission, © 2005 IEEE.)

behavioural models, an RMS error metric similar to that reported in [52] could be used:

$$e_{k,\text{harm}} = \sqrt{\frac{N^{-1}\sum_{n}|b_{k,\text{harm,sim}}(n) - b_{k,\text{harm,meas}}(n)|^2}{N^{-1}\sum_{n}|b_{k,\text{harm,meas}}(n)|^2}} \tag{5.99}$$

where $b_{k,\text{harm,sim}}$ and $b_{k,\text{harm,meas}}$ represent the simulated and measured complex envelopes at port k and around 'harm' (the harmonic component of the carrier frequency) respectively. The total number of time samples is N and the time-sample index is n.

This error metric was applied to the complex-envelope simulation of the scattered travelling voltage waves b_2 and b_1 for different input power levels, as shown in Figures 5.31 and 5.32 respectively. It can be seen that the RMS error of the model 1 prediction is smaller than that for model 2 at almost all input power levels. This shows the improved accuracy of the model extracted from the 63-tone QPSK-shaped PDF multisine excitation when compared with the seven-tone unshaped multisine. The difference is especially marked at lower power levels.

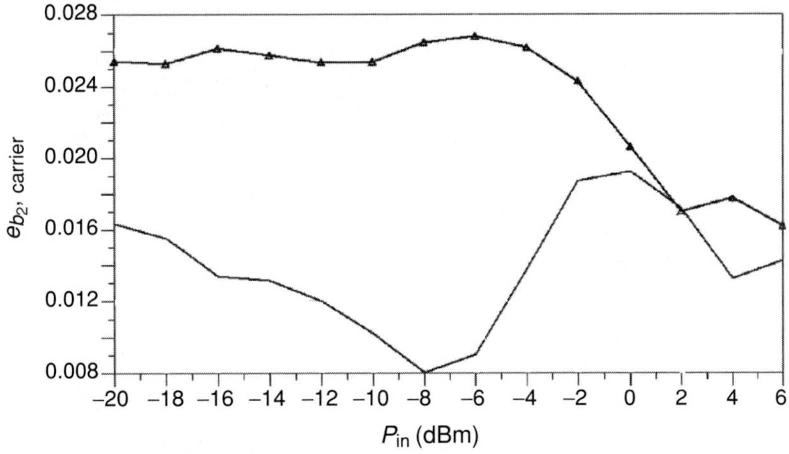

Figure 5.31 The RMS error metric for the model 1 (lower line) and model 2 (upper line with triangles) predictions of the b_2 travelling voltage wave around the carrier frequency, plotted as a function of input power. (From [51] with permission © 2005 IEEE.)

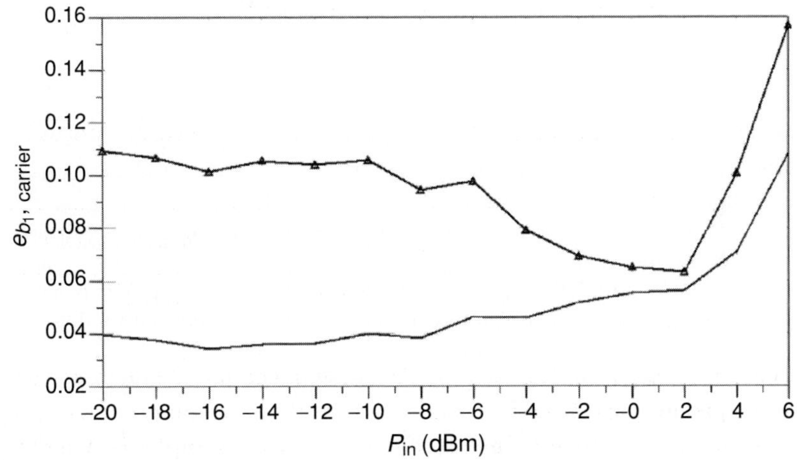

Figure 5.32 The RMS error metric for model 1 (lower line) and model 2 (upper line with triangles) predictions of the b_1 travelling voltage wave around the carrier frequency, plotted as a function of input power. (From [51] with permission, © 2005 IEEE.)

This example illustrates the importance of using metrics to validate and compare models. A detailed discussion on metrics in connection with PA modelling follows in Chapter 6.

References

[1] R. Raich, H. Qian and G. T. Zhou, "Orthogonal polynomials for power amplifier modeling and predistorter design," *IEEE Trans. Vehicular Tech.*, vol. 53, no. 5, pp. 1468–1479, September 2004.

[2] L. Ding, G. T. Zhou, D. R. Morgan *et al.*, "A robust predistorter constructed using memory polynomials," *IEEE Trans. Communications*, vol. 52, no. 1, pp. 159–165, January 2004.

[3] H. Ku and J. S. Kenney, "Behavioral modeling of nonlinear rf power amplifiers considering memory effects," *IEEE Trans. Microw. Theory Tech.*, vol. 51, no. 12, pp. 2495–2504, December 2003.

[4] A. Ahmed, M. O. Abdalla, E. S. Mengistu and G. Kompa, "Power amplifier modeling using memory polynomial with non-uniform delay taps, "in *European Microwave Conf. Dig.*, 2004, pp. 1457–1460.

[5] A. Ahmed, "Analysis, modelling and linearization of nonlinearity and memory effects in power amplifiers used for microwave and mobile communications," doctoral thesis, Department of High Frequency Engineering, University of Kassel, Kassel, Germany, 2005.

[6] Q. J. Zhang, K. C. Gupta and V. K. Devabhaktuni, "Artificial neural networks for rf and microwave design – from theory to practice, "*IEEE Trans. Microw. Theory Tech.*, vol. 51, no. 4, pp. 1339–1350, April 2003.

[7] G. Cybenko, "Approximation by superposition of sigmoidal function, "*Math. Control Signals Systems*, vol. 2, no. 4, pp. 303–314, December 1989.

[8] K. Hornik, M. Stinchcombe and H. White, "Multilayer feedforward networks are universal approximator," *Neural Networks*, vol. 2, no. 5, pp. 359–366, 1989.

[9] Q. J. Zhang and K. C. Gupta, *Neural Networks for RF and Microwave Design*, Artech House, 2000.

[10] V. Devabhaktuni, M. Yagoub and Q. J. Zhang, "A robust algorithm for automatic development of neural network models for microwave applications," *IEEE Trans. Microw. Theory Tech.*, vol. 49, no. 9, pp. 2291–2298, September 2001.

[11] J. Suykens, J. Vandewalle and B. De Moor, *Artificial Neural Network for modeling and control of non-linear systems*, Kluwer, 1996.

[12] M. Ibnkahla and F. Castanie, "Vector neural networks for digital satellite communications," in *IEEE Int. Conf. Communications Dig.*, 1995, pp. 1865–1869.

[13] E. R. Srinidhi, A. Ahmed and G. Kompa, "Power amplifier behavioral modeling strategies using neural network and memory polynomial models," *Microwave Review*, vol. 12, no. 1, pp. 15–20, January 2006.

[14] C. A. Desoer and M. Vidyasagar, *Feedback Systems: Input–Output Properties*, Academic Press, 1975.

[15] P. L. Gilabert, G. Montoro and A. Cesari, "A recursive digital predistorter for linearizing RF power amplifiers with memory effects," in *Asia Pacific Microwave Conf. Dig.*, 2006, pp. 1043–1047.

[16] H. Ku, M. D. Mc Kinley and J. S. Kenney, "Quantifying memory effects in RF power amplifiers," *IEEE Trans. Microw. Theory Tech.*, vol. 50, no. 12, pp. 2843–2849, December 2002.

[17] V. J. Mathews and G. L. Sicuranza, *Polynomial Signal Processing*, John Wiley & Sons, 2000.

[18] P. Dobrovolny, P. Wambacq, G. Vandersteen and H. Dries, "The effective high level modelling of a 5 GHz RF variable gain amplifier," technical report, Interuniversity Microelectronic Centre (IMEC), Leuven, Belgium, 2002.

[19] P. Wambacq, P. Dobrovolny, S. Donnay, M. Engels and I. Bolsens, "Compact

modeling of nonlinear distortion in analog communication circuits," in *Proc. Conf. Design, Automation, and Test in Europe*, 2000.

[20] N. Z. Hakim, J. J. Kaufman, G. Cerf and H. E. Meadows, "Volterra characterisation of neuronal networks," in *Proc. IEEE Int. Conf. Neural Networks*, 1991, pp. 1128–1132.

[21] J. Wray, D. Sanders and G. Green "Neural networks and nonlinear prediction," in *Proc. American Conf. Neural Networks*, 1992.

[22] S. Boyd, Y. S. Tang, and L. O. Chua, "Measuring Volterra kernels," *IEEE Trans. Circuits Syst.*, vol. 30, no. 8, pp. 571–577, August 1983.

[23] G. Palm and T. Poggio, "The Volterra expansion and the Wiener expansion: Validity and pitfalls," *SIAM J. Appl. Math.*, vol. 33, no. 2, pp. 195–216, September 1977.

[24] A. Zhu, M. Wren and T. J. Brazil, "An efficient Volterra-based behavioral model for wideband RF power amplifiers," in *IEEE MTT-S Int. Microwave Symp. Dig.*, 2003, pp. 787–790.

[25] A. Zhu and T. J. Brazil, "Behavioral modeling of RF power amplifiers based on pruned Volterra series," *IEEE Microw. Wireless Compon. Lett.*, vol. 14, no. 12, pp. 563–565, December 2004.

[26] F. J. Doyle III, R. K. Pearson and B. A. Ogunnaike, *Identification and Control using Volterra Models*, Springer-Verlag, 2001.

[27] J. Kim and K. Konstantinou, "Digital predistortion of wideband signals based on power amplifier model with memory," *Electronics Lett.*, vol. 37, no. 23, pp. 1417–1418, December 2001.

[28] A. Zhu and T. J. Brazil, "RF power amplifier behavioral modeling using Volterra expansion with Laguerre functions," in *IEEE MTT-S Int. Microwave Symp. Dig.*, 2005, pp. 963–966.

[29] F. Filicori and G. Vannini, "Mathematical approach to large-signal modelling of electron devices," *Electronics Lett.*, vol. 27, no. 4, pp. 357–359, April 1991.

[30] E. Ngoya and A. Soury, "Modeling memory effects in nonlinear subsystems by dynamic Volterra series," in *Proc. Int. Workshop on Behavioral Modeling and Simulation*, 2003, pp. 28–33.

[31] D. Mirri, G. Iuculano, F. Filicori, G. Pasini, G. Vannini and G. Pellegrini, "A modified Volterra series approach for nonlinear dynamic systems modeling," *IEEE Trans. Circuits Syst.*, vol. 49, no. 8, pp. 1118-1128, August 2002.

[32] D. Mirri, G. Iuculano, F. Filicori, G. Pasini and G. Vannini, "Modeling of non ideal dynamic characteristics in S/H-ADC devices," in *IEEE Instrum. Meas. Tech. Conf. Dig.*, 1995, pp. 27–32.

[33] D. Mirri, F. Filicori, G. Iuculano and G. Pasini, "A nonlinear dynamic model for performance analysis of large-signal amplifiers in communication systems," *IEEE Trans. Instrum. Meas.*, vol. 53, no. 2, pp. 341–350, April 2004.

[34] S. Haykin, *Adaptive Filter Theory*, Prentice-Hall, 1996.

[35] J. Dooley, B. O'Brien and T. J. Brazil, "Behavioural modelling of RF power amplifiers using modified Volterra series in the time domain," in *Proc. High Frequency Postgraduate Student Colloq.*, 2004, pp. 169–174.

[36] D. Mirri, G. Pasini, F. Filicori, G. Iuculano and G. Pellegrini, "Finite memory nonlinear model of a S/H-ADC device," in *Proc. IMEKO Symp. on Development in Digital Meas. Instrum.*, 1998, pp. 873–878.

[37] C. Florian, F. Filicori, D. Mirri, T. Brazil and M. Wren, "CAD identification and validation of a nonlinear dynamic model for performance analysis of large-signal amplifiers," in *IEEE MTT-S Int. Microwave Symp. Dig.*, 2003, pp. 2125–2128.

[38] D. Mirri, F. Filicori, G. Iuculano and G. Pasini, "A non-linear dynamic model for

performance analysis of large-signal amplifiers in communication systems," in *IEEE Instrum. Meas. Tech. Conf. Dig.*, 1999, pp. 193–197.

[39] G. Stegmayer, "Neural networks and Volterra series for modeling new wireless communication devices," in *Proc. Int. Joint Conf. on Neural Networks*, 2007, pp. 489–494.

[40] D. Schreurs, J. Wood, N. Tufillaro, L. Barford and D. Root, "Construction of behavioural models for microwave devices from time-domain large-signal measurements to speed-up high-level design simulations," *Int. J. RF and Microwave Computer Aided Eng.*, vol. 13, no. 1, pp. 54–61, January 2003.

[41] H. Kantz and T. Schreiber, *Nonlinear Time Series Analysis*, Cambridge University Press, 1997.

[42] D. Walker, R. Brown and N. Tufillaro, "Constructing transportable behavioral models for nonlinear electronic devices," *Phys. Lett. A*, vol. 255, nos. 4–6, pp. 236–242, 1999.

[43] M. Kennel, R. Brown and H. Abarbanel, "Determining embedding dimension for phase-space reconstruction using a geometrical construction," *Phys. Rev. A*, vol. 45, no. 6, pp. 3403–3411, March 1992.

[44] J. Wood, D. E. Root and N. B. Tufillaro, "A behavioral modeling approach to nonlinear model-order reduction for RF/microwave ICs and systems," *IEEE Trans. Microw. Theory Tech.*, vol. 52, no. 9, pp. 2274–2284, September 2004.

[45] D. Schreurs, J. Jargon, K. A. Remley, D. DeGroot and K. C. Gupta, "ANN model for HEMTs constructed from large-signal time-domain measurements," in *Automatic RF Techniques Group Conf. Dig.*, June 2002, 6 pp.

[46] D. Schreurs, H. Taher and B. Nauwelaers, "Can the use of black box models significantly reduce simulation time?," in *Proc. European Solid State Circuits Conf. Workshop on robust smart power circuits by design innovation*, 2004, 24 pp.

[47] D. Schreurs, K. A. Remley, M. Myslinski and R. Vandersmissen, "State-space modelling of slow-memory effects based on multisine vector measurements," in *Automatic RF Techniques Group Conf. Dig.*, December 2003, pp. 81–87.

[48] D. Schreurs and K. Remley, "Use of multisine signals for efficient behavioural modelling of RF circuits with short-memory effects," in *Automatic RF Techniques Group Conf. Dig.*, June 2003, pp. 65–72.

[49] M. C. Jeruchim, P. Balaban and K. S. Shanmugan, *Simulation of Communication Systems*, Plenum Press, 1992.

[50] D. Schreurs, K. A. Remley and D. F. Williams, "A metric for assessing the degree of device nonlinearity and improving experimental design," in *IEEE MTT-S Int. Microwave Symp. Dig.*, 2004, pp. 795–798.

[51] M. Myslinski, D. Schreurs, K. A. Remley, M. D. McKinley and B. Nauwelaers, "Large-signal behavioral model of a packaged RF amplifier based on QPSK-like multisine measurements," in *Gallium Arsenide and other Compound Semiconductors Application Symp. Dig.*, 2005, pp. 185–188.

[52] K. A. Remley, J. A. Jargon, D. Schreurs, D. C. DeGroot and K. C. Gupta, "Repeat measurements and metrics for nonlinear model development," in *IEEE MTT-S Int. Microwave Symp. Dig.*, June 2002, pp. 2169–2172.

6 Validation and comparison of PA models

6.1 Introduction

This chapter deals with PA model validation and comparison. In general PAs are complex dynamic systems that combine both short- and long-term memory effects with nonlinear phenomena. In contrast with linear systems with memory, where superposition holds and so any test signal can be used, or nonlinear systems with linear memory, where memory effects can be separately characterised and de-embedded to obtain a simple algebraic descriptive function, nonlinear dynamical systems must be 'locally' modelled and validated. Therefore, test signals and model comparison criteria must be carefully chosen to suit a particular set of typical operating conditions.

This chapter proposes suitable figures and characteristics of merit (i.e. metrics) that enable the performance of different PA models for telecommunication applications to be compared. It is divided into two parts.

In the first part, general figures of merit (FOMs) are presented and the main concepts regarding their applicability are explained. Although most of the proposed metrics can be generalised for sampled and/or stochastic signals, only deterministic continuous-time signals are considered here. Starting from a general time-domain metric, several variants are proposed, each specially suitable for a certain measurement setup.

The second part of the chapter deals with more realistic applications, in that most of the proposed FOMs are formulated for sampled (i.e. discrete-time) signals and are in terms of statistical measures such as the covariance and the power spectral density (PSD). The stochastic-process point of view may be useful for modern measurement instruments and system simulators, where complex telecommunication standards test signals are usually characterised statistically. This part of the chapter also includes an application example, where different FOMs are compared.

The concepts presented here should allow the reader to formulate a suitable FOM for his or her application.

6.2 General-purpose metric

In this section, a general-purpose metric is proposed that is applicable to the various kinds of model addressed in this book (i.e. linear with linear memory, nonlinear without memory and nonlinear with both linear and nonlinear memory). A distinction will be made between the different versions of this proposed FOM according to the type of measurement setup available. First a general time-domain formulation of the metric is given and then the frequency-domain counterparts both for continuous- as well as for discrete-spectrum signals are presented. Also, an approximate version of the metric in the frequency domain, assuming that phase errors are negligible, is considered. This formulation is suitable when only a scalar spectrum analyser is available. After that, the calculation of the metric when the input and output signals are characterised by means of a complex envelope (i.e. an IQ demodulator) is analysed. Finally, a metric for the common case of single-tone AM–AM and AM–PM characteristics is considered.

6.2.1 *Definition*

The PA model scheme under consideration is presented in Figure 6.1. It is the representation of a single-input–single-output PA, with input and output signals defined in terms of power waves on a real reference impedance Z_0 (typically 50 ohm). The input–output relationship is assumed to be describable by means of a nonlinear function of the past values of the input signal up to a past time $\tau = T_{\mathrm{M}}$.

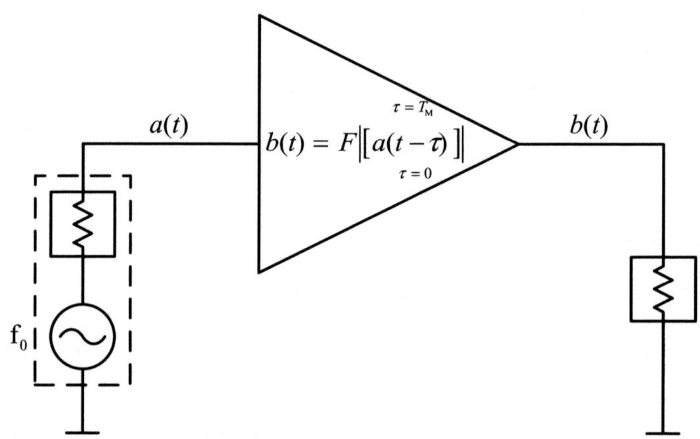

Figure 6.1 PA model scheme.

With reference to Figure 6.1:

- $a(t)$ and $b(t)$ are the input and output real time-domain power waves;
- $F\,\|[\cdot]\|_{\tau=0}^{\tau=T_{\mathrm{M}}}$ is the nonlinear functional which gives the output of the PA in terms of the history of values of the input signal up to T_{M} seconds ago.

Whenever a certain model simulates the signal $b(t)$ at the output of the PA, the modelled magnitude is referred to as $\hat{b}(t)$, to distinguish it from the actual output signal.

Qualitatively, the accuracy of a certain model is its ability to predict the output of the PA given a certain input and set of operating conditions. Accordingly, we define the following FOM,

$$\varepsilon(\mathrm{f}_0, P_{\mathrm{out}}, BW) = \frac{\overline{\Delta b(t)^2}}{\overline{P_{\mathrm{out}}}} = \frac{T^{-1}\left\{\int_0^T [b(t) - \hat{b}(t)]^2 dt\right\}}{T^{-1}\left\{\int_0^T [b(t)]^2 dt\right\}}, \tag{6.1}$$

as the percentage discrepancy between the measured and the modelled PA output signals. The normalisation factor enables the discrepancy to be expressed relative to a certain average output power. The measurement, i.e. observation, time is T; this should be long enough to account for all the significant signal dynamics. As can be observed, this quantity will be dependent on the centre frequency of the PA as well as on the output power level and the bandwidth (BW) of the input signal. These dependences indicate that in order to be able to compare two or more PA models it might be more meaningful to compare the ε versus P_{out} characteristics of the models (at a given bandwidth and central frequency) rather than just comparing the single numbers that result from evaluation of the metric at a fixed output power level.

It might also be of interest to normalise the proposed FOM with respect to the distortion level present in the output signal. The reason for doing so is that it is 'easier' for the model to achieve a small error (i.e. a low metric value) when the distortion is low than when the output signal is highly distorted. Examples of some normalised FOMs are included in the subsections below as appropriate.

6.2.2 *Alternative formulations*

FOM for frequency-domain measurements

The frequency-domain metric proposed, which is equivalent to Equation (6.1) is

$$\varepsilon(\mathrm{f}_0, P_{\mathrm{out}}, BW) = \frac{\int\limits_{\Omega \in O} \left|B(\omega) - \hat{B}(\omega)\right|^2 d\omega}{\int\limits_{\Omega \in O} |B(\omega)|^2 d\omega} \tag{6.2}$$

This metric enables a comparison of the complex spectra of the actual and the modelled output signals, normalised to the actual output signal's spectrum. This quantity should be evaluated for the set O corresponding to the frequency intervals of interest for the performance of the PA.

Discrete-spectrum signals. When one is dealing with periodic or quasi-periodic signals (i.e. single-tone or multitone test signals), as is usually the case in most PA applications, the metric given by Equation (6.2) can be suitably modified to take into account the discrete nature of the spectrum in the following way:

$$\varepsilon(\mathrm{f}_0, P_{\mathrm{out}}, BW) = \frac{\sum\limits_{n \in F} \left| B_n - \hat{B}_n \right|^2}{\sum\limits_{n \in F} \left| B_n \right|^2}, \tag{6.3}$$

where the B_n are complex coefficients of the Fourier series of the output signal. In Equation (6.3) F is the set of frequencies that are present in the spectra of either $b(t)$ or $\hat{b}(t)$ and that fall within the bandwidth of interest, as in Equation (6.2) for the continuous-spectrum case.

Approximated FOM for scalar spectrum-analyser measurements. When only a scalar spectrum analyser is available, the evaluation of Equation (6.3) is not possible since, in the general case, both $B(\omega)$ and $\hat{B}(\omega)$ are complex-valued quantities. However, another FOM that neglects the unknown phase errors can be defined:

$$\varepsilon(\mathrm{f}_0, P_{\mathrm{out}}, BW) = \frac{\int\limits_{\Omega \in O} \left(|B(\omega)| - \left| \hat{B}(\omega) \right| \right)^2 d\omega}{\int\limits_{\Omega \in O} |B(\omega)|^2 \, d\omega}. \tag{6.4}$$

Analogously, for the discrete-spectrum case, the metric becomes

$$\varepsilon(\mathrm{f}_0, P_{\mathrm{out}}, BW) = \frac{\sum\limits_{n \in F} \left(|B_n| - \left| \hat{B}_n \right| \right)^2}{\sum\limits_{n \in F} |B_n|^2}. \tag{6.5}$$

This last formula can be used in the case of conventional two-tone intermodulation measurements.

FOM for time-domain measurements

In many cases PA characterisation is performed in terms of band-pass input and output signals, which are described with a baseband complex modulation envelope applied to a carrier at frequency f_0. In this case, complex I–Q demodulators are employed to measure the in-phase and quadrature components of the input and output waves, as can be seen in Figure 6.2. In this figure, the input and output signals are defined in terms of their equivalent complex-envelope band-pass signals $\mathcal{A}(t)e^{j2\pi \mathrm{f}_0 t}$ and $\mathcal{B}(t)e^{j2\pi \mathrm{f}_0 t}$. Ideal in-phase and quadrature demodulators are included in order to obtain the I (in-phase) and Q (quadrature) components of the signals.

A 'hybrid' (i.e. time–frequency domain) modified version of the metric might

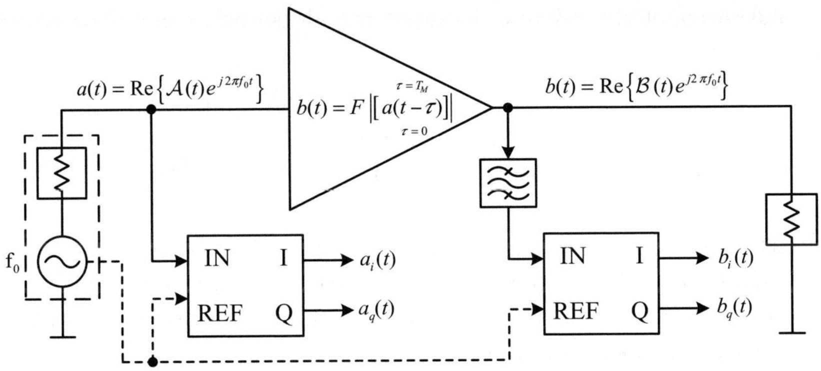

Figure 6.2 Complex-envelope measurement setup.

then be defined as

$$\varepsilon(\mathrm{f}_0, P_{\mathrm{out}}, BW) = \frac{\left(T^{-1}\int_0^T \left|\mathcal{B}(t) - \hat{\mathcal{B}}(t)\right|^2 dt\right)}{\left(T^{-1}\int_0^T |\mathcal{B}(t)|^2 dt\right)}. \tag{6.6}$$

For this metric, and for all the metrics defined in the time domain, it may be necessary to apply a fixed time-delay correction to one of the signals before calculating the difference. In fact, an incorrect input–output delay in the model could lead to a poor FOM that is governed only by a time shift between the measured and the modelled output signals. Accordingly, the corrected metric should be:

$$\varepsilon(\mathrm{f}_0, P_{\mathrm{out}}, BW) = \frac{\left(T^{-1}\int_0^T \left|\mathcal{B}(t) - \hat{\mathcal{B}}(t-\tau_{\mathrm{s}})\right|^2 dt\right)}{\left(T^{-1}\int_0^T |\mathcal{B}(t)|^2 dt\right)} \tag{6.7}$$

where τ_{s} is the time shift that removes the difference $|\mathcal{B}(t) - \hat{\mathcal{B}}(t-\tau_{\mathrm{s}})|$ in small-signal operation.

FOM for single-tone measurements (AM–AM and AM–PM characteristics)

In some cases amplifier characterisation is performed in terms of AM–AM and AM–PM characteristics. In that case the amplifier can be modelled by a describing function

$$\mathcal{H} = \frac{\mathcal{B}(t)}{\mathcal{A}(t)}. \tag{6.8}$$

Under the simplifying assumption of a 'slowly' varying (i.e. narrowband) input signal envelope $\mathcal{A}(t)$, the PA can be assumed to be memoryless with respect to the complex modulation envelopes and so $\mathcal{H}$ will be a function of the central frequency and the absolute value of the input signal. Namely,

$$\mathcal{H}(\mathrm{f}_0, |\mathbf{A}(t)|) = \frac{\mathcal{B}(t)}{\mathcal{A}(t)}. \tag{6.9}$$

In this case the proposed FOM becomes

$$\varepsilon(\mathrm{f}_0, P_{\mathrm{out}}, BW) = \frac{T^{-1}\int_0^T \left|\mathcal{H} - \hat{\mathcal{H}}\right|^2 |\mathcal{A}(t)|^2 \, dt}{T^{-1}\int_0^T |\mathcal{H}|^2 \, |\mathcal{A}(t)|^2 \, dt} \tag{6.10}$$

As can be observed, this metric 'weights' the discrepancies in the measured and predicted describing functions $\mathcal{H}$ by a factor related to the input signal power. Thus, for small amplitudes of the input signal the difference between the describing functions is less relevant than for higher amplitudes.

FOM in the case of high-linearity applications

When evaluating high-linearity PAs, the previously defined metrics can be somewhat misleading in the sense that the power of the error signal (and therefore the metric) may be very small, but the PA model is still unacceptable for the application. In these cases it may be useful to normalise the metric with respect to the level of distortion present in the output signal:

$$\varepsilon(\mathrm{f}_0, P_{\mathrm{out}}, BW) = \frac{1}{T}\int_0^T \left|B(t) - B^{\mathrm{ND}}(t)\right|^2 dt \tag{6.11}$$

in which $B^{\mathrm{ND}}(t)$ is the distortionless version of $B(t)$, i.e., the signal that would result if the PA's behaviour were perfectly linear.

6.3 Figures of merit based on real-world test signals

This section focuses on 'real-world' test signals. Accordingly, several metrics are proposed that are directly applicable for use with stochastic sampled signals.

In the following the input and output sequences of the observed system are identified by $x(s)$, $y(s)$ in the SISO (single-input–single-output) case and by $x_1(s), x_2(s)$ and $y_1(s), y_2(s)$ in the DIDO (dual-input–dual-output) case. The corresponding model outputs are called $\hat{y}(s)$ and $\hat{y}_1(s)$, $\hat{y}_2(s)$, respectively. These sequences describe the behaviour of the observed system and the model in the complex baseband channel (the first-zone contribution [1]). If the model also includes the output sequences at the harmonics, each frequency region must be represented by the corresponding baseband channel. To simplify the notation in this section the DIDO case and the higher-order-zone contribution are not treated separately. Thus a summation over the magnitudes of the complete output sequence corresponds to the following equation in the harmonic DIDO case:

$$\sum_s |y(s)|^2 = \sum_m \sum_s |y_{1,h_m}(s)|^2 + \sum_m \sum_s |y_{2,h_m}(s)|^2, \qquad (6.12)$$

where y_{1,h_m}, y_{2,h_m} denotes the mth-zone contributions of the two DIDO ports. A further simplification is introduced by using the same notation for deterministic sequences and stochastic processes. The FOMs that assume a stochastic input process will be mentioned explicitly.

The length of the sequences used for the validation must be sufficient to avoid their influencing the initial state of the model. Furthermore, the applied stimuli must be persistent to guarantee a comprehensive comparison of the model and the observed system.

6.3.1 Definitions

The general symbol for FOMs is M, subscripted to indicate the metric under consideration.

Error-vector magnitude (EVM)

This metric evaluates the normalised (dimensionless) root-mean-square (RMS) EVM between the considered and the modelled output sequences in the time domain (see [2] and subsection 6.2.1); the FOM is denoted

$$M_{\mathrm{EVM}} = \sqrt{\frac{\sum_s |y(s) - \hat{y}(s)|^2}{\sum_s |y(s)|^2}}, \qquad (6.13)$$

where s runs over the output sequence, as in Equation (6.12).

Power spectral density (PSD)

As for the last metric, the error vector can be evaluated in the frequency domain:

$$e(s) = y(s) - \hat{y}(s),$$

$$M_{\mathrm{PSD}} = \sqrt{\frac{\sum_{\omega_k} |S_e(\omega_k)|^2}{\sum_{\omega_k} |S_y(\omega_k)|^2}}, \qquad (6.14)$$

where $S_y(\omega_k) = \sum_s y(s)\exp(j\omega_k s)$ represents the Fourier transform of the deterministic output sequence and $S_e(\omega_k)$ that of $e(s)$. For random processes M_{PSD} is given by

$$M_{\mathrm{PSD}} = \sqrt{\frac{\sum_{\omega_k} \mathrm{PSD}_e(\omega_k)}{\sum_{\omega_k} \mathrm{PSD}_y(\omega_k)}}, \qquad (6.15)$$

where $\text{PSD}_y(\omega_k) = \sum_s R_{y,y}(s) \exp(j\omega_k s)$ is the Fourier transform of the considered process and $\text{PSD}_e(\omega_k)$ is that of $e(s)$. By choosing the summation range ω_k in a suitable way, the evaluation of M_{PSD} can be performed within a limited bandwidth.

Distortion EVM

The metrics presented up to now require the full output sequence for their evaluation. If the level of nonlinear effects of the system under consideration is low then the power of the distortion generated may well be several orders of magnitude lower than the linear signal's output level. Thus these FOMs will not be able to evaluate the correct modelling of the nonlinear output-signal components. To overcome this disadvantage, a metric must isolate the nonlinear output components and use them for the comparison. This signal separation implies the identification of a linear relationship between the input and output sequences and the cancellation of the corresponding signal. These two steps must be performed for the observed system and for the model.

The linear relationship can be evaluated by minimising the squared inner product of x and $y - f_{\text{L}}(x)$ [3],

$$\min_{f_{\text{L}}} \left|(x, y - f_{\text{L}}(x))\right|^2, \tag{6.16}$$

where f_{L} denotes a linear operator. Here (u, v) denotes the inner product of u and v. Solutions for this minimisation problem use least-squares (LS) techniques. To keep the computational effort for this identification process low the memory length of f_{L} should be chosen to be as short as possible yet still long enough to guarantee that linear effects do not dominate the output of $y - f_{\text{L}}(x)$. Obviously, the same memory length has to be applied for estimation of the linear operator of both the observed system and the model to ensure similar signal cancellation in both cases.

The operator f_{L} so found relates the input sequence to the linear components and to the nonlinear output components that are linearly dependent on the excitation [4]. The residual nonlinear signal components are then used for an EVM calculation:

$$M_{\text{EVM,DIST}} = \sqrt{\frac{\sum\limits_s |z(s) - \hat{z}(s)|^2}{\sum\limits_s |z(s)|^2}}, \tag{6.17}$$

$$z(s) = y(s) - f_{\text{L},z}(x(s)), \qquad \hat{z}(s) = \hat{y}(s) - f_{\text{L},\hat{z}}(x(s)). \tag{6.18}$$

Normalised mean-square error (NMSE)

The statistical equivalent to the EVM is the NMSE, which evaluates the autocovariance of the difference between the observed and the modelled system outputs

[5]; the FOM is denoted

$$M_{\mathrm{NMSE}} = \frac{\mathrm{cov}_{y-\hat{y},y-\hat{y}}(0)}{\mathrm{cov}_{y,y}(0)} \tag{6.19}$$

$$\mathrm{cov}_{x,y}(\tau) = \mathrm{E}\left\{x\left[(s+\tau) - \mathrm{E}x(s)\right]\left[y(s) - \mathrm{E}y(s)\right]^*\right\} \tag{6.20}$$

where E denotes the expectation operator and $*$ the complex conjugate. By the use of the cross covariance, errors introduced by biasing effects will not be included in the NMSE. If the cross covariance is estimated by

$$\mathrm{cov}_{x,y}(0) = \frac{1}{N}\sum_{n=1}^{N}\left[x(n) - \frac{1}{N}\sum_{m=1}^{N} x(m)\right]\left[y(n) - \frac{1}{N}\sum_{m=1}^{N} y(m)\right]^* \tag{6.21}$$

it is easy to show that $M_{\mathrm{NMSE}} = M_{\mathrm{EVM}}^2$ in the zero-mean case.

Variance accounted for (VAF)

The VAF metric is a further statistically based metric, which is closely related to the NMSE [5]:

$$M_{\mathrm{VAF}} = \frac{\mathrm{cov}_{y,y}(0) - \mathrm{cov}_{y-\hat{y},y-\hat{y}}(0)}{\mathrm{cov}_{y,y}(0)} = 1 - M_{\mathrm{NMSE}} \tag{6.22}$$

Coherence function

The coherence function COH relates the cross covariance PSD between the observed and the model output signals to their autocovariance PSD [5]:

$$\mathrm{COH}(\omega_k) = \sqrt{\frac{\left|\mathrm{PSD}_{\mathrm{cov}_{y,\hat{y}}}(\omega_k)\right|^2}{\mathrm{PSD}_{\mathrm{cov}_{y,y}}(\omega_k)\mathrm{PSD}_{\mathrm{cov}_{\hat{y},\hat{y}}}(\omega_k)}} \tag{6.23}$$

The corresponding metric is defined as:

$$M_{\mathrm{COHERENCE}} = \frac{1}{N}\sum_{\omega_k} \mathrm{COH}(\omega_k), \tag{6.24}$$

where N specifies the number of frequency points considered. This metric is not applicable to deterministic sequences. The idea of the coherence function can be explained by assuming a linear relationship between the model and the observed output signal:

$$y(s) = \sum_{q} h(q)\hat{y}(s-q) + v(s), \tag{6.25}$$

where $h(s)$ describes the impulse response function of the linear relationship and $v(s)$ is an additive noise representing the unmodelled signal components of the observed output signal. Assuming that $\hat{y}(s)$ and $v(s)$ are zero-mean stationary random processes, the PSDs are then given by

$$\begin{aligned}\mathrm{PSD}_{\mathrm{cov}_{y,\hat{y}}}(\omega_k) &= H(\omega_k)\mathrm{PSD}_{\mathrm{cov}_{\hat{y},\hat{y}}}(\omega_k),\\ \mathrm{PSD}_{\mathrm{cov}_{y,y}}(\omega_k) &= |H(\omega_k)|^2\,\mathrm{PSD}_{\mathrm{cov}_{\hat{y},\hat{y}}}(\omega_k) + \mathrm{PSD}_{\mathrm{cov}_{v,v}}(\omega_k).\end{aligned} \tag{6.26}$$

The square of the coherence function for this relationship between the observed and modelled output signals is therefore [5]:

$$\begin{aligned}(\mathrm{COH})^2 &= \frac{\left|\mathrm{PSD}_{\mathrm{cov}_{y,\hat{y}}}(\omega_k)\right|^2}{\mathrm{PSD}_{\mathrm{cov}_{y,y}}(\omega_k)\mathrm{PSD}_{\mathrm{cov}_{\hat{y},\hat{y}}}(\omega_k)}\\ &= \frac{\left|H(\omega_k)\mathrm{PSD}_{\mathrm{cov}_{\hat{y},\hat{y}}}(\omega_k)\right|^2}{\mathrm{PSD}_{\mathrm{cov}_{\hat{y},\hat{y}}}(\omega_k)\left[|H(\omega_k)|^2\,\mathrm{PSD}_{\mathrm{cov}_{\hat{y},\hat{y}}}(\omega_k) + \mathrm{PSD}_{\mathrm{cov}_{v,v}}(\omega_k)\right]}\\ &= \frac{1}{1+\dfrac{\mathrm{PSD}_{\mathrm{cov}_{v,v}}(\omega_k)}{|H(\omega_k)|^2\,\mathrm{PSD}_{\mathrm{cov}_{\hat{y},\hat{y}}}(\omega_k)}}\end{aligned} \tag{6.27}$$

Thus the squared coherence function can be interpreted as the fraction of the observed output variance due to the linear part of the relationship, as described by Equation (6.25), to the modelled output signal as a function of the frequency. If significant output signal components are not present in the model or no linear relationship between the two variances can be found then the coherence function will tend to zero.

Adjacent-channel power ratio difference (ΔACPR)

The adjacent-channel power ratio (ACPR) is a measure of the power of the distortion products leaking into the adjacent channels in relation to the signal power in the desired channel [6]:

$$\mathrm{ACPR} = 10\log\left\{\frac{\int_{\omega_{\mathrm{adj}}}\mathrm{PSD}(\omega)d\omega}{\int_{\omega_{\mathrm{ch}}}\mathrm{PSD}(\omega)d\omega}\right\}, \tag{6.28}$$

where ω_{adj} and ω_{ch} specify the frequency bands of an adjacent channel and of the carrier channel respectively. This definition is based on the adjacent-channel leakage ratio (ACLR) for WCDMA modulated signals [6]. The important difference between the ACPR and the ACLR is that the root-raised cosine (RRC) filtering of the considered signal is neglected in Equation (6.28). By comparing the ACPR predicted by the model to the ACPR of the observed output a measure of the accuracy of the modelling of the distortion is found:

$$M_{\Delta\mathrm{ACPR}} = \mathrm{ACPR}_y - \mathrm{ACPR}_{\hat{y}}. \tag{6.29}$$

The disadvantage of this FOM is that only the ratios of the integrated PSDs are compared and not the distortions created in the magnitude and the phase. In contrast with the preceding FOMs, $M_{\Delta\mathrm{ACPR}}$ is usually expressed in dB.

Adjacent-channel error power ratio (ACEPR)

This FOM overcomes the disadvantage of $M_{\Delta\text{ACPR}}$ in that the error between the observed and the modelled signal outputs within the adjacent channel is evaluated and compared with the power of the carrier [7]:

$$e(s) = y(s) - \hat{y}(s),$$
$$M_{\text{ACEPR}} = 10 \log \left\{ \frac{\int\limits_{\omega_{\text{adj}}} |S_e(\omega)|^2 \, d\omega}{\int\limits_{\omega_{\text{ch}}} |S_y(\omega)|^2 \, d\omega} \right\}. \tag{6.30}$$

This calculation of the modelling error due to the adjacent channel can only be applied to a certain realisation of a random process. As with $M_{\Delta\text{ACPR}}$, M_{ACEPR} is usually specified in dB. An example of the use of M_{NMSE}, $M_{\Delta\text{ACPR}}$ and M_{ACEPR} in comparing the performances of several memoryless models is presented in Table 3.2.

6.3.2 *Comparison of the various FOMs*

To compare the metrics presented above the following two-box SISO Wiener model was used:

$$y(s) = \sum_{n=1}^{N} a_{2n-1} f(x(s)) \, |f\,(x(s))|^{2(n-1)}, \tag{6.31}$$

where

$$f(x(s)) = \sum_{q=1}^{Q} h(q)x(s-q) \tag{6.32}$$

and where the linear impulse response length and the maximum order of the nonlinear products were set to $Q = 61$ and $2N - 1 = 7$ respectively.

To visualise the behaviour of the FOMs, the parameters of this model were varied and the responses of the original and altered models were compared. For all comparisons a bandlimited white-noise input signal was used. This excitation signal is characterised by a PAPR of 10 dB. The input and the resulting output signals of the 'reference' model are presented in Figure 6.3 at 11 dB IBO. For the performance comparison of the FOMs the following test cases were considered:

- a delay mismatch of the impulse response function;
- a variation in the linear coefficient a_1;
- a variation in the third-order nonlinear coefficient a_3.

The measures are compared in two groups to account for the linearly and the logarithmically displayed FOMs. All the discussed FOMs except M_{NMSE}, which is directly related to M_{EVM} and M_{VAF}, were included in this evaluation; M_{PSD}

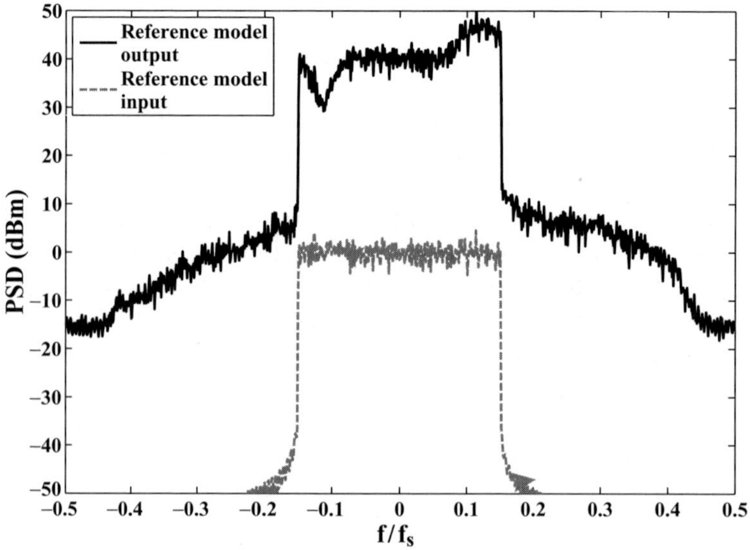

Figure 6.3 Input and output signal of the reference model at 11 dB IBO.

was used to cover on one hand the normalised excitation signal bandwidth $f/f_s = -0.15$ to 0.15 and on the other the distortion spectra $f/f_s = -0.45$ to -0.15 and 0.15 to 0.45. When considering M_{PSD} for the two distortion spectra, the higher value is used. In a similar way, for $M_{\Delta\mathrm{ACPR}}$ and M_{ACEPR} the higher result of the upper and lower adjacent channels is used.

Impulse response delay mismatch

For this comparison the models of the observed (reference) and the altered (model) systems are related by

$$a_{2n-1,\mathrm{model}} = a_{2n-1,\mathrm{reference}}, \qquad h(q)_{\mathrm{model}} = h(q - \Delta\tau)_{\mathrm{reference}}, \tag{6.33}$$

where the parameter $\Delta\tau$ is some fraction of the sampling interval. The delay $\Delta\tau$ was simulated by convolving the reference impulse response with a delayed sinc function. As the mismatch is introduced before the nonlinearity, both the linear and the nonlinear model output components will be time shifted.

Figure 6.4(a) presents the outputs of the linearly scaled FOMs versus delay mismatch. Obviously, the coherence function is independent of $\Delta\tau$ (see the horizontal line at the top of (a)). This behaviour can be explained by the missing signal components that are not linearly related (e.g. unmodelled signal components). The FOMs $M_{\mathrm{PSD,carrier}}$ and M_{EVM} are barely distinguishable; the reason is that most of the signal and the modelling error energy is locatedwithin the carrier bandwidth.

The higher dependence of $M_{\mathrm{PSD,adjacent}}$ compared to that of $M_{\mathrm{PSD,carrier}}$ is caused by the lower signal power at the adjacent channel to which the modelling error is normalised.

The FOMs specified in dB are presented in Figure 6.4(b). In this graph the M_{EVM} and $M_{\mathrm{EVM,DIST}}$ metrics have been added to relate the results to the linearly displayed FOMs in (a). The evaluation of the delay $\Delta\tau$ starts at 0.1 sampling intervals as M_{EVM} and $M_{\mathrm{EVM,DIST}}$ will tend to $-\infty$ dB for $\Delta\tau = 0$. The FOM $M_{\Delta\mathrm{ACPR}}$ shows no dependence on the delay mismatch. The shape of M_{ACEPR} is similar to that of M_{EVM} but, owing to the different normalisation, at a significant lower value. On the basis of these results, a suitable indication of the time-delay mismatch was achieved by using M_{VAF}, M_{EVM}, $M_{\mathrm{EVM,DIST}}$, M_{ACEPR} and the two M_{PSD} metrics.

Linear parameter variation

The behaviour of the metrics under consideration when the linear parameter a_1 varies is shown in Figures 6.5(a), (b). The low sensitivity of $M_{\mathrm{EVM,DIST}}$ to changes in the linear coefficient is due to the subtraction of the estimated linear dependence between the input and output signals. In the case of perfect suppression of the linear contribution this FOM should be independent of linear parameter variations. This complete independence of linear parameter changes is also anticipated for $M_{\mathrm{PSD,adjacent}}$. The departure from the anticipated behaviour can be explained by taking a closer look at the spectra of the modelling error. Figure 6.6 shows the spectra of the scaled input signal and of the modelled and reference output signals. The different PSDs of the model and reference outputs are clearly visible within the carrier frequency range. In the upper and lower adjacent channels the nonlinear distortion dominates the PSD. Additionally, the linear amplified input signal also contributes within this frequency range. Even if the power of this contribution is significantly lower than the power of the nonlinear distortion, it is the dominant source of modelling error within the adjacent channels. By normalising this modelling error to the power of the distortion, the difference between the two models is significantly enhanced. The M_{ACEPR} metric alleviates this problem by normalising the error in the adjacent channel to the power of the carrier.

The $M_{\Delta\mathrm{ACPR}}$ metric shows a change of around 3 dB within the parameter variation that we are considering, this is caused by a rise in the carrier power of the model output signal. Therefore, a linear parameter mismatch is properly visualised by the M_{VAF}, M_{EVM}, $M_{\mathrm{PSD,carrier}}$ and $M_{\Delta\mathrm{ACPR}}$ metrics.

Nonlinear parameter variation

The response of the FOMs to a variation in the magnitude of the third-order nonlinear coefficient a_3 shows a completely different behaviour from the two previous cases. As presented in Figures 6.7(a), (b), the $M_{\mathrm{DIST,EVM}}$, $M_{\mathrm{PSD,adjacent}}$, $M_{\Delta\mathrm{ACPR}}$

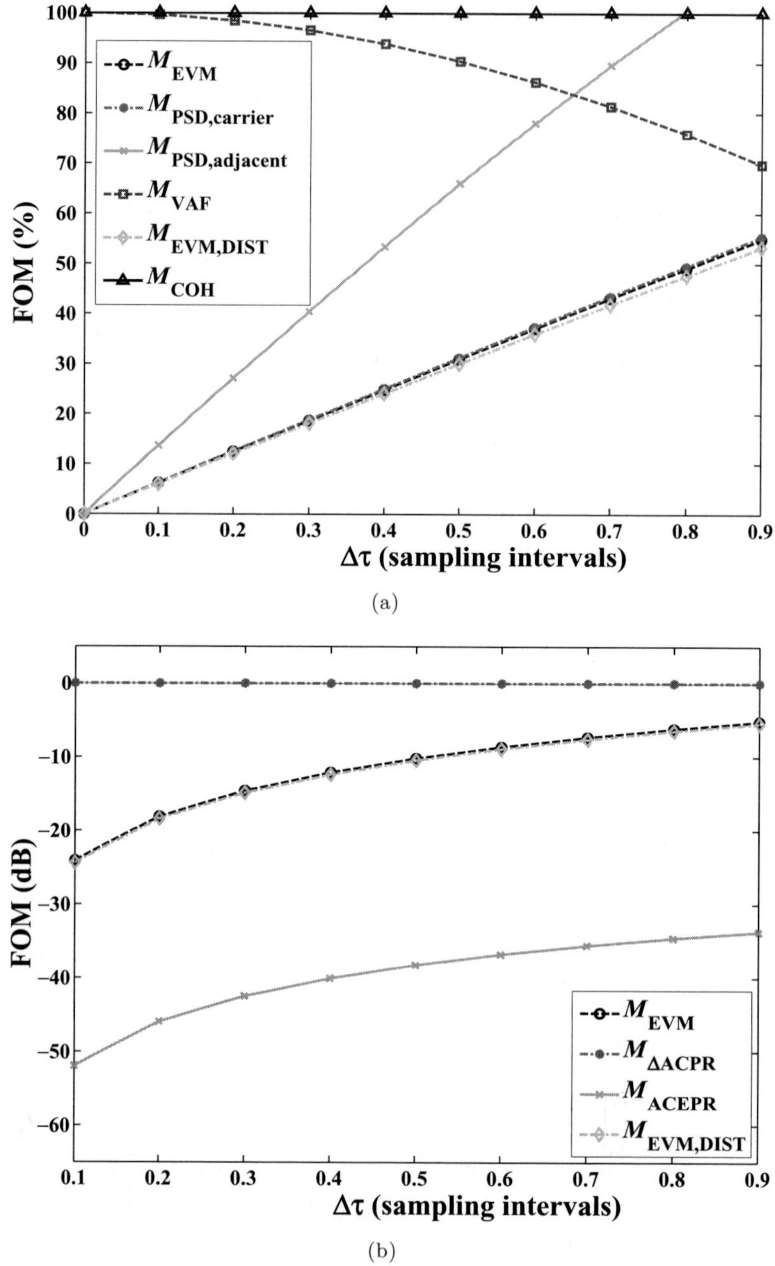

Figure 6.4 Figures of merit shown as functions of $\Delta\tau$ for (a) linearly and (b) logarithmically scaled FOMs.

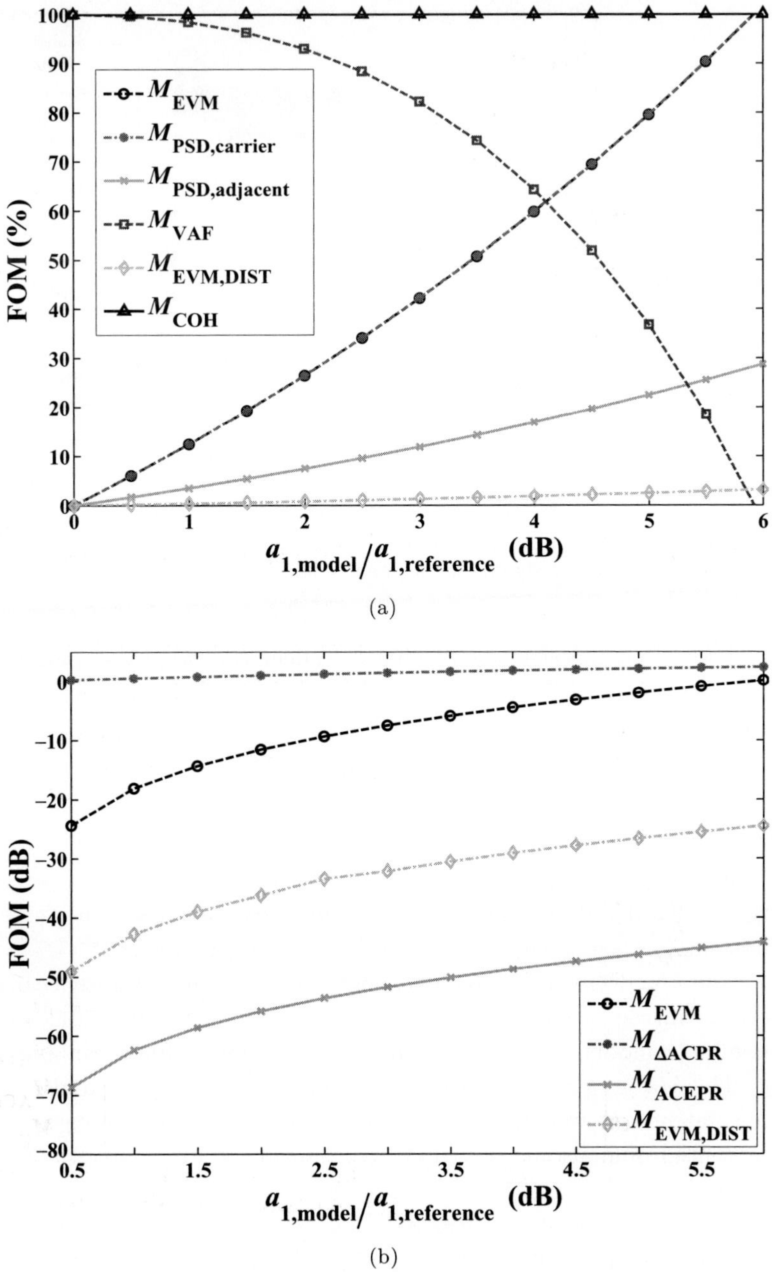

Figure 6.5 The different FOMs for a sweep of the linear coefficient a_1 for (a) the linearly displayed FOMs and (b) the logarithmically displayed FOMs; see Figure 6.4.

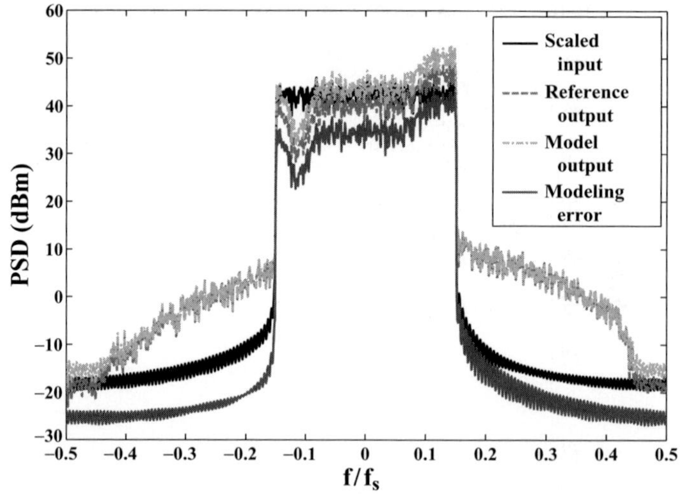

Figure 6.6 The modelling error spectra, the scaled input signal and the reference and the modelled output signals for a linear coefficient mismatch of 4 dB.

and M_{ACEPR} metrics are all able to reflect the nonlinear parameter change. The response of $M_{\Delta\mathrm{ACPR}}$ to changes in a_3 is of the same magnitude as its response to changes in a_1.

Summary

The M_{EVM} and M_{VAF} metrics are capable of identifying a linear mismatch between the observed and modelled systems. Their weakness lies in the detection of nonlinear mismatches, especially in the weakly nonlinear regime.

The M_{PSD} metric showed excellent behaviour in the presence of linear amplitude errors. In the case of a bandlimited excitation signal it can also be used to detect nonlinear modelling errors (assuming a sufficiently high sampling frequency).

The $M_{\Delta\mathrm{ACPR}}$ metric is capable of detecting both linear and nonlinear parameter mismatches. However, owing to the small output value changes (about 3 dB in the presented simulations) this FOM seems unsuitable for detecting marginal modelling errors.

The metrics that are best able to identify nonlinear mismatches are M_{ACEPR} and $M_{\mathrm{DIST,EVM}}$. Of these two, M_{ACEPR} has a higher computational efficiency but demands a bandlimited input signal and a suitable sampling rate. The metric $M_{\mathrm{DIST,EVM}}$ avoids these constraints by modelling the linear input–output relationships of both the observed and the modelled systems.

In the comparisons presented, the selected parameter variations were not detectable from the behaviour of M_{COH}.

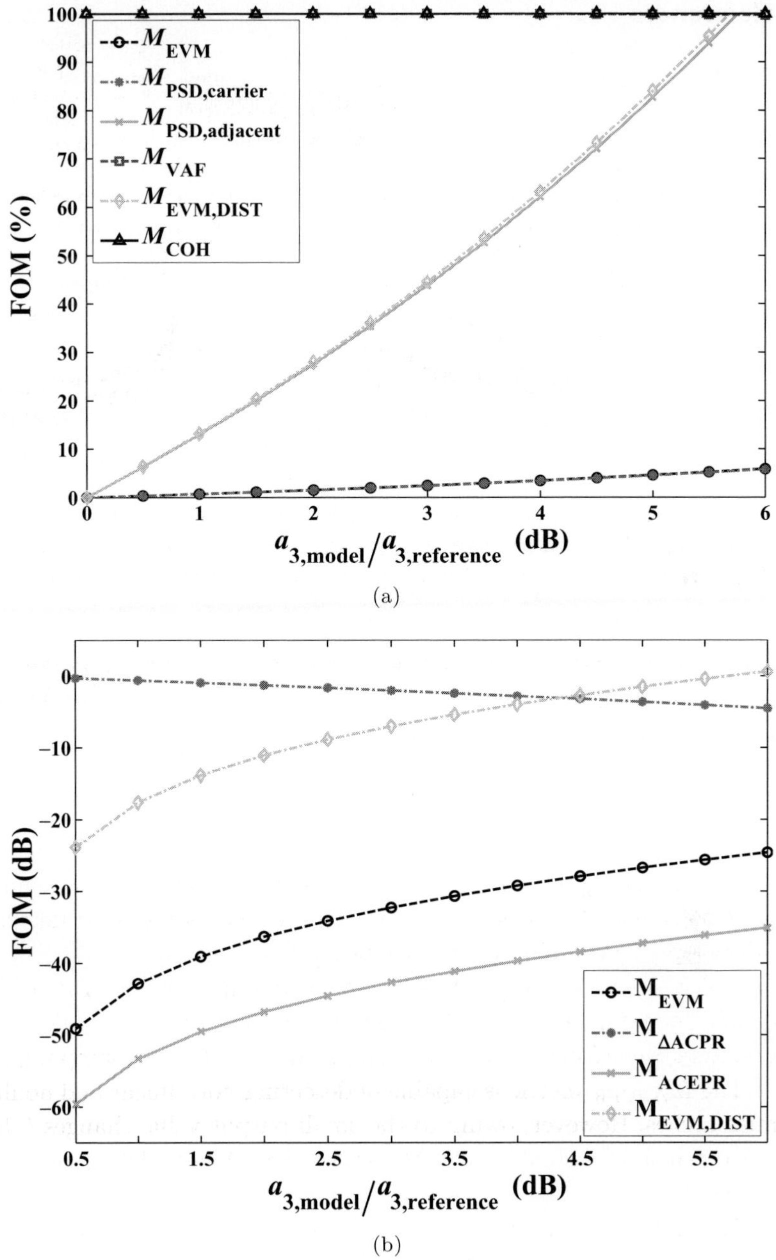

Figure 6.7 The different FOMs for a sweep of the nonlinear coefficient a_3.

References

[1] M. C. Jeruchim, P. Balaban and K. S. Shanmugan, *Simulation of Communication Systems*, second edition, Kluwer/Plenum, 2000.

[2] ETSI, "Digital cellular telecommunications system (phase 2+); radio transmission and reception," 3GPP TS 45.005 version 6.9.0 Release 6, April 2005.

[3] M. E. Gadringer, D. Silveira and G. Magerl, "Dynamic nonlinear model of a cancelation loop with application to the feedforward linearization technique," in *European Microwave Conf. Dig.*, 2006, pp. 1173–1176.

[4] J. C. Pedro and N. B. Carvalho, *Intermodulation Distortion in Microwave and Wireless Circuits*, Artech House, 2003.

[5] D. T. Westwick and R. E. Kearney, *Identification of Nonlinear Physiological Systems*, John Wiley & Sons, 2003.

[6] ETSI, "Universal mobile telecommunications system (UMTS); UTRA (BS) FDD, radio transmission and reception," 3GPP TS 25.104 version 5.2.0 Release 5, March 2002.

[7] M. Isaksson, D. Wisell and D. Rönnow, "A comparative analysis of behavioral models for RF power amplifiers," *IEEE Trans. Microw. Theory Tech.*, vol. 54, no. 1, pp. 348–359, January 2006.

7 Aspects of system simulation

7.1 Introduction

Simulation is widely used in the design and analysis of complex communications systems [1–4]. The aim is to describe the operating characteristics and performance of all or part of a communication link, whether simple or complex, and to mimic through mathematical models all the analogue and digital signal processing activities, whether at baseband, IF or RF frequencies. System-level simulations can be based on time-domain or frequency-domain techniques or on a combination of both. Time-domain models are the norm in system-level digital signal processing (DSP) simulations while the frequency-domain approach is a popular choice for RF circuit-level simulation, even though time-domain simulation is also much used, especially in connection with equivalent baseband envelope techniques such as the circuit envelope. When nonlinearities exist in the system it is difficult to simulate other than in the time domain, although harmonic-balance, mixed frequency-domain and time-domain approaches on statistical techniques may be used.

Well-known commercial dedicated simulation and electronic design automation software systems, such as Applied Wave Research's Microwave Office (AWR-MO) and Virtual System Simulator (AWR-VSS) [5] and Agilent's Advanced Design System (ADS) [6, 7], provide both circuit-level and system-level simulation capability, with the possibility of including nonlinear microwave PA models also. Other robust commercial simulation tools for circuit-level mixed (analogue and digital signal) communication systems are the 'Virtuoso' platform from Cadence [8] and the HSPICE simulator from Synopsis [9]. There is ongoing widespread research in this area especially to increase the capacity to handle more complex systems, to develop library modules for new technologies and to improve simulation efficiency.

Many researchers, individually and in groups, have created their own simulation suites, sometimes for sharing in a research network such as the European Union's TARGET Network of Excellence (Top Amplifier Research Groups in a European Team), where packages are shared for PA modelling and PA linearisation research [10]. These may be created on the commercial platforms mentioned previously or may use other applications such as MATLAB, Simulink or Maple or non-commercial packages such as: Ptolemy Classic II (University of California, Berkeley) [11, 12]; TOPSIM (Politecnico di Torino, Italy, developed with the support of the European Space Agency) [13], which parallels the development of the SYSTID platform by Hughes Aircraft for NASA and COMSAT [3, 14]; MITSYN at MIT, Boston [15]; the IT++ package (Chalmers University, Sweden) [16]. The last is an example of a DSP library for digital-communication-system simulations that is kept up to date; it has found acceptance in some European research networks, e.g. NEWCOM (Network of Excellence in Wireless Communications [50]). Some researchers develop simulation suites using their favourite programming language and in that way, through in-depth knowledge of how modules and functions are simulated, are able to maintain full confidence and understanding of the strengths and limitations of their simulation results.

Simulations can move beyond general signal processing activities to the simulation of network and protocol architectures and complex networking communication activities over logical communication channels. This is a major and distinct field and is sometimes referred to as communication-network simulation. Such simulators tend to use event-driven 'model of computation' approaches, [3, 11, 17, 18]. Examples include the USA's DARPA and NSF sponsored discrete-event network simulator NS-2, targeted at networking research [19–21] and commercial fixed and wireless network simulation software products from OPNET [22] and OMNeT++ [23].

In Europe, a notable product in this field is Telelogic's Rhapsody [24]. Using the Universal Modelling Language (UML), it is a system design, analysis and implementation simulation platform with specific strengths for embedded real-time systems and software engineering for telecommunications protocol-architecture development.

Other platforms include a scalable wireless ad hoc network simulator (SWANS) built atop the 'Java in simulation time' (JiST) platform [25]. For other specification and description language (SDL) simulators [27, 28] and tree and tabular combined notation (TTCN) simulators [29, 30], the discussions in [26] on the model-based design, development, and validation of real-time mobile communication systems is suggested.

From an engineering point of view, these fields – of network simulation and of the simulation of physical communications systems and subsystems – are quite distinct and distant from one another while being mutually dependent when designing total telecommunications services.

In this chapter we specifically address aspects of physical communications system simulation, especially within the context of expanding PA behavioural modelling into full transmitter system and communications link simulations, from transmitter data input to receiver data output.

7.2 Some relevant simulation terminology

The terms 'modelling', 'design', 'executable models', 'constructive models' and 'embedded systems' used in this chapter are based on the thinking in Lee *et al.*'s paper [11].

Modelling is the act of representing a system or subsystem formally. A model might be mathematical and usually is in telecommunication simulations. In such cases, it can be viewed as a set of assertions about properties of the system such as its functionality at different levels of abstraction, possibly down to physical dimensions and physical properties. Much of this book is focused on PA modelling at system level, which corresponds to a high level of abstraction. Circuit-level modelling corresponds to a lower level of abstraction and would include physical electrical properties. A model can also be constructive in that it defines a computational procedure that mimics a set of properties of the system. Constructive models are often used to describe the response of a system to a stimulus from outside. This is the approach in RF PA behavioural modelling.

The design of a system or a subsystem in a simulation context usually involves defining one or more models of the system and refining the models until the desired functionality is obtained within a set of constraints.

Design and modelling, while distinct, are closely coupled. One may design models for a system but not be interested in designing the system through modelling. In the simulation of composite systems some of the models involved may under some circumstances be fixed. For instance, in the process of designing of a subsystem, other subsystems (i.e. their models), constraints, or externally imposed behaviours, are constant and unchanging. An example could be the design, by modelling, of a linearisation scheme for a particular memoryless nonlinear PA for a particular interface: once the signal-representation model and the nonlinear PA model are satisfactorily constructed, they remain fixed in the process of designing the linearisation scheme. Another non-telecommunications example which may delineate this concept better would be the design using modelling of an electronic control subsystem for an electromechanical system: the mechanical subsystem itself, or its model, may not be under design and so would be fixed.

Constructive models may evolve to be 'executable models' in that they cease to be merely a model and become a system or subsystem to be embedded into the real environment. This is frequently the case with embedded software and the development of DSP algorithms. In such cases, the distinction between designing a model and designing a system through modelling may become blurred.

7.3 Analogue-signal behavioural simulators for wireless communication systems

The evolution of the various types of communication-system simulators parallels the natural research, design and implementation divisions within communications systems, i.e. analogue (RF, IF and baseband) and digital baseband. The RF or IF front-end, which involves RF carrier components such as PAs, low-noise amplifiers (LNAs), modulators and demodulators, mixers, filters, couplers, circulators, multiplexers and demultiplexers, up- or downconverters and wireless channels, is predominantly analogue, while the baseband portion consists of signal processing operations that can be implemented using analogue or digital signal processing or both. The interfaces between the analogue and digital sections are provided by analogue-to-digital converters (ADCs) and digital-to-analogue converters (DACs). Baseband components include filters, channels, detector decision circuits and line coding as well as certain aspects of modulators or demodulators and multiplexers or demultiplexers. For digital communication systems, there are also the digital logic components such as bit-and-octet synchronisers, encoders, decoders, interleavers, de-interleavers, frame synchronisers etc. This area partly overlaps with the physical-layer protocols of the communication-network architecture.

Historically, system simulators tend to address one or other domain alone, with the goal of analysing some particular problem or seeking new or improved designs for some specific signal processing operations within the domain. Isolated-system simulations like this may be justified in many circumstances and for many purposes. Once, commercial simulators were used mainly either for pure DSP simulations (these involve the more algorithmic, system-oriented and time-domain-based environments, e.g. digital filters and equalisers) or for analogue-signal-based simulations (these are more hardware-oriented, e.g. they are used for microwave and RF system simulation or for circuit design for printed circuit boards (PCBs) or integrated circuits (ICs)). However, they have evolved into devices that can integrate both types of simulation into mixed-signal simulation platforms.

The two broad categories of analogue-signal-based simulations are system-level and circuit-level simulations. The first category could be generally described as treating communication subsystems as black-box modules defined and characterised by input–output relationships or behavioural functions and in which signals are passed from one module to the next. Early generations of system-level simulation tools [3, 4, 14, 31] addressed modelling, design and analysis at a mathematical level. The downside of such simulations is that they are at one remove from real-system implementation details, e.g. real-circuit synthesis. Nonetheless this abstraction, which is a characteristic of system-level simulations and techniques, reduces subsystem complexity and improves computational efficiency for any particular subsystem; thus the possibility exists of cascading greater numbers and a greater variety of subsystems. This simulation philosophy has enabled the simulation of ever more complex communication systems employing ever-increasingly complex signal and air-interface structures, e.g. spectrally efficient complex OFDM-modulated signals and complex combinations of multiple air-interface signals with high peak-to-average power ratios (PAPRs), which need to be processed through the full gamut of telecommunications systems including, perhaps, common nonlinear power amplification. As a result system-level simulations have served and will continue to serve the telecommunications research and design engineering community very well.

In the second category, circuit-level simulations, subsystems are described by a full representation of their physical circuits, each component being modelled and equivalent-circuit models being used for active devices and other complex components. Those circuit-level simulation tools that have been evolved in parallel with system-level tools have been focused on the physical electronic circuit-level design and ways have been developed to use them in computer aided design (CAD) circuit synthesis, analysis and implementation. The closer models come to reflecting the physical circuit the better, since then a greater accuracy in simulations may be expected. Ultimately, an important output often becomes an actual circuit design element.

Being pedantic, one could say that in system-level simulation, if the system is decomposed into ever smaller subsystems (components) then eventually the simulation would be transformed into a circuit-level simulation. Nonetheless, this serves to highlight the 'abstract' nature of system-level simulation in relation to circuit-level simulation. As one abstracts, one loses the power to model certain effects and certain aspects of the system. For example, in communication systems that include nonlinear PAs, modelling for power efficiency and similar aspects can really only be handled by circuit-level simulators; for a given circuit configuration the associated power-efficiency characteristics are a 'given' in the equivalent system-level simulators. This example should illustrate how circuit-level and system-level simulators each have their own strengths, weaknesses and modelling domains. Another example is the way in which circuit-level simulation handles terminations, impedance matching, reflections, parasitic capacitances, transients etc. directly.

These are important issues in real RF system design and are not easily catered for in system-level simulation. In all real-circuit implementations, as the signal moves from one stage to the next it will encounter some finite level of mismatch, no matter how small (i.e. the input reflection coefficient will never be exactly zero), or some finite parasitic impedance giving rise to effects not seen in the system-level simulation. Experienced designers, when using system-level simulations, do not forget this consequence of abstraction from the physical circuit, i.e. the absence in the models of certain circuit-level effects when considering the accuracy of the simulation results. There are ways, 'work-arounds', to include some of these effects, or their equivalents, in system-level simulations through appropriately designed simulation modules. For instance, a partial solution when modelling mismatch effects is to insert suitably designed filters. In circuit-level simulators it is a normal part of good simulation-model design to cater for these effects in ways that reflect what happens in the real circuit.

From another perspective, design by system-level simulation is analogous to the situation where a telecommunications system designer seeks hardware, components and subsystems that, when combined, will yield the desired system behaviour, particularly at the system output and at key points in the telecommunications chain. Circuit-level simulation, however, is analogous to the situation where component or circuit designers seek to ensure that their designs meet certain basic stand-alone input–output characteristics. They are different, if overlapping, fields.

For narrowband high-powered nonlinear PA systems, distortion is traditionally and frequently modelled using the memoryless AM–AM/AM–PM nonlinear characteristics, see Chapter 3. However, modern advanced high-power solid state PAs (SSPAs) manifesting, for example, dynamic short- and long-term memory effects require more sophisticated models. In Chapters 4 and 5 PA models with linear and nonlinear memory effects were considered. The models presented in these chapters are really geared for system-level simulation and can be realised through mathematical modelling simulator tools such as MATLAB.

7.4 Figure of merit considerations in behavioural simulations

The aim of both circuit-level and system-level simulation is to extract FOMs that can be linked to the overall system performance – directly in system-level simulation and indirectly in circuit-level simulation. When considering systems that include nonlinear PAs, the simulation needs to be able to handle the competing requirements of full transmission-path linearity and good PA power efficiency. One part of this trade-off, power-efficiency modelling and related questions, can only be handled in circuit-level simulators, as mentioned above. Linearity can be handled by both, but in different ways and to different degrees.

Time-varying subsystem characteristics, such as some aspects of thermally-dependent system behaviour or temporal effects due to aging, may also be modelled at a system level. These may be viewed as long-term memory effects and, naturally, the models will only be as good as the characterisation of this time-varying behaviour.

In all simulations, design and performance analysis is achieved through deriving and using quality objectives and FOMs. The principles underscoring the algorithms for some key FOMs in complex nonlinear PA behavioural analysis have already been treated in detail in Chapter 6. A list of quality objectives and FOM measures that may be useful and that can be extracted, both from simulations as well as from measurements at a transmitter output or at any probe point in the overall communications chain, when nonlinear PA systems are present (with or without linearisation schemes) is as follows:

- the two-tone and three-tone behaviour, i.e. intermodulation and harmonic generation behaviour, including the carrier-to-intermodulation ratio C/I, the carrier-to-third-order intermodulation product (IMP) ratio (C3IM), the total IMP distortion and the second- and third-order intercept points (IP_2 and IP_3);
- the power-added efficiency (PAE);
- the noise–power ratio (NPR);
- the modulation fidelity, as represented by e.g. the error-vector magnitude (EVM) or NMSE;
- the PAPR and complementary cumulative distribution functions (CCDFs) of RF envelopes;
- the percentage linearisation (PL) (when PA linearisers are present) [32];
- the bit and symbol error rates (BER and SER);
- co-channel and interchannel interference (nonlinear and linear distortion);
- adjacent-channel leakage and power ratios (ACLR and ACPR);
- spectral regrowth and occupied bandwidth.

Usually, modern software circuit- and system-level simulation tools will be capable of evaluating most of, if not all, these metrics.

7.5 Circuit-level techniques

Circuit-level techniques tend to be mainly devoted to narrowband systems and are largely based on the harmonic balance (HB) [6, 33–35] and circuit-envelope [36, 37] simulation techniques. The FOMs extracted are mainly limited to multi-tone tests, usually two-tone and three-tone – especially for HB, such as C/I ratios, IP_2, IP_3 and harmonic signal generation. The common categories of circuit-level simulators are:

1. time-domain;
2. small-signal frequency-domain;
3. time-invariant harmonic balance;
4. time-varying harmonic balance (circuit-envelope).

The best-known commercial general-purpose time-domain mixed-signal circuit-level simulator product is Spice, a product from MentorGraphics [38], and variants thereof, e.g. the Cadence PSPICE [8] or Synopsis HSPICE products [9], which can perform circuit behaviour design using basic DC, AC, noise or transient analysis. Transient analysis is inherently catered for and, as might be expected in discrete mathematical solutions of differential equations, iterations are normal for each time step to find the waveform values within that time step or instant. The signal path can contain thousands of active elements: besides the PA, components such as crystal voltage-controlled oscillators (VCOs) and other high-Q circuits, transmission lines, surface-acoustic-wave (SAW) devices and filters and other distributed components. Regardless of the complexity of the modelling equations for each circuit component, the sampling must represent signals over the full band from DC to the highest significant harmonic. As operating frequencies go ever higher, so do the sampling frequency and the computing resources required to simulate behaviour over a given number of information symbols. This, together with the number of time-step iterations, means that the computing time required to represent the passage of even a short signal through the circuit can become significant. Nonetheless, this may be the only option for analysing the transient behaviour of circuits, including RF circuits; see also Chapter 3. This Spice type of simulator is not efficient for computing directly the higher-level FOMs listed in Section 7.4, such as the BER of real-world signals using complex modulation formats.

7.5.1 Harmonic-balance simulation

For steady-state analysis and design by simulation, however, the HB technique provides a 'work-around' for the full-time-domain simulation weakness mentioned above. It is a highly accurate frequency-domain circuit simulation approach. It has the capacity to compute steady-state solutions of nonlinear circuits and systems by calculating the Fourier coefficients of the output and capturing the steady-state voltage and capacitor charge waveforms and, through time differentiation of a Fourier series, to obtain accurate steady-state capacitor current waveforms. It has, for quite some time, been used to predict the response of nonlinear PAs to spectrally simple input waveforms consisting of a few sinusoids, e.g. two or three pure tones. The HB solution amounts to a sum of the steady-state sinusoids, the input frequencies in addition to any significant harmonics and IMPs, from which, with limited analysis, simple FOMs such as the C/I ratio may be calculated.

Though it is capable of handling the first set of quality objectives listed in Section 7.4, HB simulation is not appropriate for calculation of the others. There are, nonetheless, ongoing research efforts to establish relationships between PA behavioural responses, as determined by HB, to simple inputs and also to more complex inputs. Even if the IMPs found in a two- or three-tone test are related to the distortion products seen in actual complex digitally modulated signals (signals with complex envelopes, e.g. multicarrier modulated signals and OFDM signals that are effectively in the frequency domain being continuous over wide bandwidths), as yet there is no simple *a priori* relation between these results and system-level specifications such as ACPR, NPR and EVM. For this reason, circuit-level methods for directly simulating the nonlinear response of an amplifier to realistic digitally modulated RF signals are much in demand.

7.5.2 *Circuit-envelope simulation*

Time-varying HB, or circuit-envelope, simulation [36, 37], is an approach that has the capacity to address some of the system FOMs listed in Section 7.4, with varying degrees of success. It effectively applies time-domain techniques on top of the frequency-domain HB solution. It has grown in popularity and is being continually refined and developed.

When combined with the high-level transfer functions of the nonlinear PA, assumed to have minimal dispersion over the envelope bandwidth so that the quasi-steady-state AM–AM and AM–PM responses and quadrature PA model are possible, see Chapter 3, then the PA output power and distortion behaviour when it is amplifying complex input signals may be predicted [39]. For instance, the zonal-output solution (i.e. the solution in that part of the output band corresponding to the input band) for a digitally modulated input signal can be represented as a sum of the RF frequency components, each with a finite-time, time-varying, complex envelope:

$$y(t) = \mathrm{Re}\left(\sum_{k=0}^{N} \tilde{y}_k(t)e^{j\omega_k t}\right), \tag{7.1}$$

where the variables $\tilde{y}_k(t)$, constant parameters in time-invariant HB simulations, represent an arbitrary modulation spectrum around each harmonic output component ω_k, i.e. the envelope of that component. They are solutions of the nonlinear ordinary differential equation for the current at a designated node. The instantaneous amplitude and phase (or quadrature I and Q) modulation information on each component is thus accessible. With this, the possibility of calculating FOMs such as ACPR and EVM is created. As an example, the output ACPR may be found by Fourier transformation of each of these complex time-varying envelopes, integrating over the in-band-channel and adjacent-channel bandwidths and hence calculating the appropriate ratio. Unlike circuit time-domain simulations, the determining bandwidth for the choice of sampling frequency (or time step) is the modulation bandwidth and not the highest RF component. This sampling question is treated in more detail in subsection 7.8.1 below.

By processing the complex modulation information in the time domain, while efficiently handling the RF carriers in the frequency domain, efficient circuit-envelope simulations of digitally modulated RF signals passing through a nonlinear PA are possible.

At circuit level, thermal models constructed through thermal equivalent circuits are possible. These may be integrated into the electrical model and have an important role in predicting performance after circuit packaging [40].

7.5.3 *A mixed-signal high-frequency IC circuit-level simulation*

High-level simulations can be very helpful in complex mixed-signal telecommunication integrated circuit (IC) design [41–43]. Since 1999 the IEEE have agreed an analogue and mixed-signal extension to their standard Very high speed IC Hardware Description Language, IEEE 1076.1-1999 VHDL-AMS [44], which has achieved wide acceptance in the industry. Determining an optimal analogue–digital partitioning through high-level simulations is a first step. Then the functional and behavioural design of the analogue and digital parts can be performed at a high level, with gradual iteration down to the full circuit design of each subsystem via autonomous mixed-signal mixed-mode simulation of the two parts using appropriate tools, e.g. the Cadence and MentorGraphics tools [8, 38]. Designing the integration of the two parts onto the same chip requires the modelling to take account of the switching noise originating in the very fast digital switching logic, which propagates via the chip substrate to the analogue part, causing various types and levels of interference [45, 46]. This is a mixed-signal circuit-integration design problem and, since realisable and optimised IC layout is the goal, it is an important challenge for circuit-level simulators.

Developing suitable models naturally relies on proper characterisation of the problem. An example of a model-and-measurement-validation methodology for predicting noise voltage at the gate level for a low-ohmic IC substrate with an epitaxial layer is that proposed by van Heijningen *et al.* [46].

7.6 System-level techniques

In system-level simulation, models can be of individual systems or subsystems, such as mixers, modulators, amplifiers or filters, through to full telecommunication chains in which all system and subsystem models are included. Techniques vary from ensuring that a subsystem model properly reflects all the real characteristics of that subsystem to ideal-subsystem models for which the subsystem behavioural impairments and/or non-idealities have been translated to another part of the simulation chain, possibly being merged with behavioural impairment models of other subsystems. Sometimes separate 'simulation-only' blocks are added in which these non-idealities and impairments are modelled in such a way as to allow independent control.

Thus, typically, subsystem additive thermal noise will often not be present in the subsystem model but be added in correct and controlled proportions to the signal(s) at an appropriate juncture of the simulation chain. In some approaches, such noise is not added to the signal at all but comes into play in post-processing or extra-processing performance-evaluation stages, at which FOMs such as the signal-to-noise ratio (SNR) are being calculated.

Signals may be 'probed' at any chosen probe point as they traverse the communication-system simulation blocks and are then passed to extra-processing modules for intermediate FOM evaluations.

A system-level simulation is typically described by a block diagram of interconnected subsystems, e.g. Figure 7.1. This example, which is described in more detail in Section 7.9, includes co-simulation [41], i.e. an integration of two 'models of computation' with a digital-logic-system simulation on the left and an analogue-signal-processing simulation on the right. Such a case would typically be implemented following the complex-baseband-envelope approach (see Chapter 3 and Section 7.1). The term analogue signal is used here in its more general understanding and includes both continuous-time analogue signals and continuous-time digital signals.

The accuracy of the simulation results relies on the accuracy of each subsystem-block model and on the signal representation approach. As the models represent real subsystem circuits at a system level this accuracy will be compromised in various ways, simply because the actual physical circuit (e.g. a transistor's equivalent circuit or an RF interdigital or SAW filter) is not modelled. For linear systems such as filters, a model using a standard digital filter synthesis technique may be created on the basis of the impulse response functions. As this itself may become the 'executable model', e.g. a low-pass filter in the real system, its model accuracy is 'perfect'. Otherwise, where the model is mapping the actual filter response function, e.g. of an RF or IF filter onto an equivalent digital filter it will be an approximation of the real filter. As such models may not be realised in an actual system by digital signal-processing engines implementing the model algorithms but by RF or IF components, the real subsystem responses will be different from those of the models. In finite-time window simulations, filters and other linear systems may of

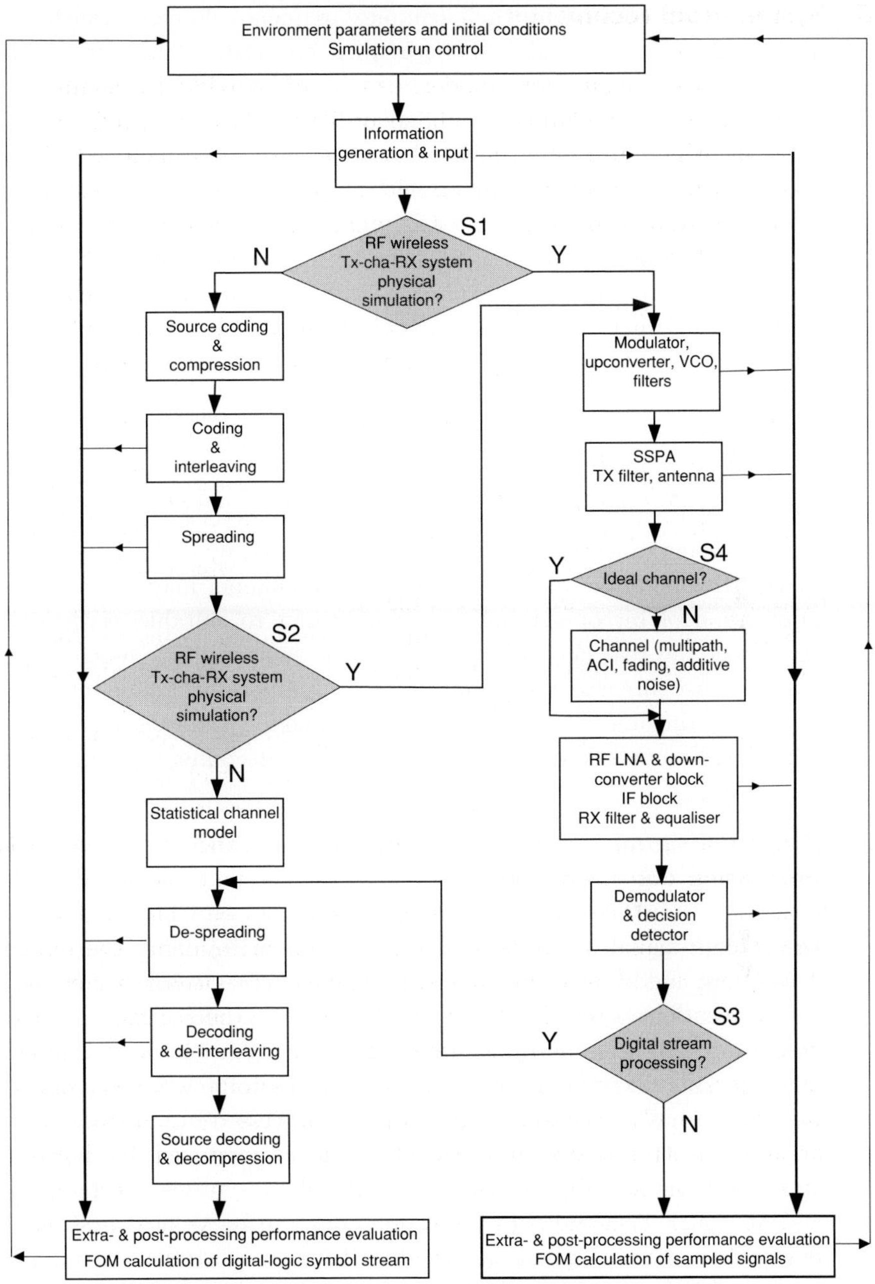

Figure 7.1 Execution flow for a system-level co-simulation of a communication link that combines digital-logic (left-hand side) and analogue-signal (right-hand side) models of computation.

course be represented by their complex frequency-domain transfer function with the simulation process also in the frequency domain. Time-varying linear systems can be described by means of a cascade of delay elements in combination with a multiplication operation using time-varying coefficients. Such an arrangement is sometimes used to model linear devices with memory. Nonlinear systems pose a much more challenging problem, and a wide range of nonlinear PA models have been presented and discussed in Chapters 3 to 5.

7.7 Digital-logic simulation

Logical simulation, also called 'purely digital' or digital-logic simulation, involves system-level simulations of digital portions of the communication system, where all signals are reducible to their binary logical two-state (0 and 1) digital-stream form and all processing is applied at this level; see e.g. the simulation chain in the left arm of Figure 7.1. This figure is more fully described in Section 7.9. The real physical communications system may be modelled by a 'digital-channel' simulation block. Distortion or information loss in the channel may be implemented by the application of bit or symbol errors according to suitable statistical behavioural models of the real channel, e.g. via bit or symbol random-error and burst-error statistics. In this way the performance of symbol and bit de-interleaving, nested-error-detecting and error-correcting channel-coding schemes may be analysed, along with data compression and source encoding or decoding schemes.

7.8 Analogue signal – representation, sampling and processing considerations

In analogue-signal simulation the focus is on maintaining the integrity of signal waveforms and thus mirroring with minimal error what occurs in the physical communications system being modelled. This requires that the analogue signal waveforms are properly represented through suitable sampling. The simulation will attempt to track faithfully the analogue signals (often carrying digital or digitised information) as they pass through, and are processed by, a range of communication system and subsystem functional modules, as portrayed in the right arm of the simulation chain in Figure 7.1.

The signals considered in the simulations can be either wholly deterministic or partly deterministic [2] with a 'pseudo-random' part (100% randomness is generally not possible in simulations). These random-variable-type signals may be used to represent the wanted information as modulated onto carriers and may also be used in different ways to simulate all types of noise contributions such as thermal and shot noise from components, impulsive, burst, intermittent and spurious noise, channel noise, dynamic multipath and fading effects, atmospheric effects, dynamic adjacent-channel interference and so forth. In this way approximations to the output-signal statistical properties may be modelled and evaluated.

When handling analogue signals, two important issues facing simulation-system designers are the sampling rate and the use of continuous-in-time or finite-time-window simulation modes.

7.8.1 *Sampling rate*

The choice of a suitable sampling rate is a key factor, particularly when dealing with aspects such as complex wideband modulation schemes, multicarrier systems, multiple air-interface systems or nonlinear systems. An appendix to this book provides extracts of the details of a number of modern wideband air interfaces. The maximum signal bandwidth tends to dominate the simulation sampling time chosen. Where a direct RF simulation is involved (for instance in a circuit-level simulation CAD platform) and the system is presumed to be linear then, by virtue of the Nyquist theorem, this would correspond to a minimum of twice the highest RF frequency to be represented, f_{s1} in Figure 3.1.

The simulations required to model transient responses accurately are clearly quite different from those modelling steady-state conditions only. If both must be represented in a single simulation then the fastest transient bandwidth will usually determine the choice of sampling rate, for all parts of the simulation.

The inclusion of nonlinear PAs, or any nonlinear devices, in the communications chain demands that great care and attention be paid to sampling rates, as the generation of harmonics and IMPs can quite easily cause aliasing impairments that are not present in the real system. The sampling rate has to be high enough to avoid this type of simulation error altogether or only allow it to occur at levels below the expected noise level. To represent the signal accurately, sometimes it may be necessary to sample at five or six times the Nyquist rate.

As discussed in earlier chapters, see e.g. the 'signal decomposability' attribute of some models discussed in Chapter 3, ways have been developed of implementing behavioural models of PAs wherein the generation of harmonics and IMPs is controllable. Where this 'decomposability' attribute of a nonlinear model holds it can be exploited to avoid having to use an excessively high sampling rate.

Often, however, single-sided frequency components only may be considered, e.g. complex or analytic signal representations such as the Hilbert transform of the real signal. In this case the Nyquist theorem is satisfied by sampling at (or above) the highest RF frequency, i.e. $\frac{1}{2}f_{s1}$. However, each sample is a complex number so in reality the signal-array resources, which translate to computer memory resources, are the same.

In order to reduce the sampling rate, and thus the simulation time, in most system-level simulations the further step is taken of not considering the RF band-pass modulated signals directly but, rather, representing only the relatively slowly varying complex-envelope part of the signal. The carrier frequency is then simply an 'environment parameter'. This is the normal practice in modern system-level simulation platforms for communication systems. In these, complex equivalent baseband signals are used and the minimum Nyquist sampling rate reduces to the maximum bandwidth encountered, f_{s2} in Figure 3.1, each sample being a complex number.

7.8.2 *Multirate sampling*

Nonetheless, imposing a single high sampling rate across all signals in a system may be significantly inefficient from a computational viewpoint, as all simulation-processing actions must be applied to every time sample. For reasons of efficiency or even simulation feasibility, different sampling rates may be used in different parts of a simulation that are dictated by the signal bandwidths in those parts of the system; this is multirate simulation. For instance, when two signals having widely different bandwidths are being combined, e.g. one signal with bandwidth 128 kHz and the other with bandwidth 120 MHz in a wireless air interface, and where interchannel or multipath interference is being investigated, uniform sampling – although an attractive approach by reason of its relative simplicity – would result in vast oversampling of the 128 kHz signal.

At the signal merging or combining points, or where separate simulation parts are integrated, sampling-conversion algorithms are required to provide an appropriate new sampling regime by interpolation or signal-sample dropping, known as 'decimation' (the word is used in a general sense; it does not imply a fixed one-in-ten removal)[31].

7.8.3 *Simulation time-domain modes: continuous-in-time and finite-time-window*

Continuous-in-time simulations mirror the continuous processing of signals through a communications system. Realising this in the simulation of complex systems is quite a challenge. In the real system, each component autonomously processes its input to obtain its output continuously over time. In the simulation, all the processing of all modules must be carried out within a sample time interval (a 'tick' or time step of the simulation clock specific to that simulation), each module taking the input signal sample at that instant, processing it and delivering its output. This is also referred to as 'time-driven simulation' [3]. On a single computer this will be done sequentially, complete knowledge of the state of each simulated component being maintained from sample instant to sample instant. Thus the simulation time will be much slower than in real time. It may be speeded up by using multiple concatenated processors since aspects of the system being simulated may lend themselves to parallel processing, which, while adding computational complexity, can speed up the simulation time. All communication component and subsystem models are time-domain models only, and the inclusion of nonlinear PAs or other such devices intuitively demands time-domain simulation.

An attractive advantage of this open continuous-in-time simulation approach is that it allows for the bit-error performance (in digital communication signals) to be calculated by actual counting of the errors that occur, as generally happens in practice. However, for small error rates long simulation times are necessary when the goal is to establish or validate error performance statistically.

Gathering blocks of signal samples at any 'ideal-probe' point to be fed into post-processing or extra-processing performance-analysis algorithms is quite normal, e.g.

for the calculation of error probabilities on the basis of the SNR or the bit or symbol energy-to-noise-density ratio. This extra-processing may be carried out on other processors, for example through a distributed computer-system infrastructure. This is analogous to the use of multiple instruments such as spectrum analysers, vector analysers or waveform analysers for probing communications systems at desired probe points using 'ideal probes'. Continuous-in-time simulations are also useful for investigating system sensitivity by varying a particular parameter or coefficient while observing the impact of these variations on performance FOMs. Examples would be varying the parameters in a PA adaptive feedback linearisation scheme or varying the fading or multipath behaviour of the RF transmission channel.

An alternative approach is to use a finite-time-window or finite-time-block scheme. Here a finite length of signal is processed by each simulation module and the output block is passed onto the next module and so on sequentially to the end of the system; then the next block is processed. This approach is also known as array processing, vector processing or block processing. Both time-domain and frequency-domain model implementations are possible when considering linear subsystems (e.g. filters) that use a fast Fourier transform (FFT) to move between the domains. Care is required to take account of the windowing effect in the transition from one finite-time block to the next, and algorithms are needed to handle and compensate for or remove this.

In finite-time-window simulations, performance analysis by post-processing is normal. However, the extraction of error probabilities by error counting, as is possible for continuous-in-time simulations, is not usual because of the overhead involved in the extra-processing algorithms required at the transitions between blocks. Also, the finite-time-window approach cannot simulate instantaneous intermodule (system or subsystem) feedback; such as would be required to simulate dynamic digital continuously adaptive predistortion linearisers, although it can be used with care in feedforward systems. Many existing commercial and academic system-level and circuit-level simulators have the capacity to handle both modes.

It is also possible to use special mixed frequency- and time-domain (MFTD) signal representation techniques for modelling highly nonlinear PAs at system level. This technique may be further enhanced, especially for simulation efficiency, by mixing real-signal simulation with statistical-signal representations; this may be used to generate types of sampled noise signals that reflect signal statistical properties sufficiently well without having to generate the signals themselves. An example of this is the MFTD-Stat signal representation technique [47]; the 'Stat' part refers to the generation of a complex signal reflecting all the interfering properties of the third-, fifth- and seventh-order IMPs (or various combinations of these) generated by the nonlinear amplification of an OFDM signal without the actual generation of the vast numbers of IMPs that would make up these signals.

Statistical techniques are popularly used in simulations to evaluate the properties of digital communication systems; an example is the Monte Carlo technique [1, 48, 49]. A sequence of signal frames, i.e. finite independent signal durations, is defined for continuous-in-time simulation and a sequence of window lengths is defined for finite-time-window simulations. Pseudo-random number generators are used to create signals that mimic information-carrying signals and also to generate the various in-band and out-of-band noise sources and spurious signals to be added to the information signals as appropriate. The experimental SER or BER, i.e. the number of errors divided by the total number of symbols or bits processed, and also the SNR, E_b/N_0 and such FOMs, are then evaluated in these frames and estimates of their means and deviations over a statistically representative number of sample frames are obtained to produce a good picture of the behavioural performance. In the finite-time-window method, typically simulation runs will use different pseudo-random number seeds for each random variable and in each finite-time window. The Monte Carlo simulation technique enhances the reliability of the performance estimates in a statistical sense.

7.9 Heterogeneous simulation

Simultaneously simulating the whole of a communications system that extends over more than one domain, as is necessary with digital-logic and analogue simulations or, within analogue, RF and/or baseband-equivalent and DSP system-level and/or circuit-level simulations, requires heterogeneous simulation. The simulation chain on the right arm in Figure 7.1 is an example of where such a mix of analogue RF and DSP portions might occur. This heterogeneous simulation approach can be particularly useful in establishing full end-to-end performance behaviour in complex environments made up of multiple communications interfaces and sources of impairment. There is much ongoing research on widening the range of heterogeneity achievable in complex system simulation and on the development of the corresponding formal software design techniques [1, 3, 17].

7.9.1 Analogue and digital-logic co-simulation

Combining analogue-logic and digital-logic simulation types is a form of heterogeneous simulation; Figure 7.1 presents a typical block diagram illustrating how this may be achieved. The right arm contains a core set of analogue simulation blocks and the left arm the digital-logic blocks. The simulation mode is controlled by setting the three 'switches', i.e. the diamond-shaped decision blocks S1, S2 and S3; block S4 determines whether the channel is ideal, i.e. whether there are channel impairments present. The simulation mode may be (i) purely analogue, if S1, S2 and S3 are set to (Y, –, N) respectively or (ii) purely digital if the setting is (N, N, –) or (iii) a mixture of both analogue and digital if the setting is (N, Y, N). For instance, in the last case the digital-logic information stream is first processed through the source encoder and compressor and the channel encoder and interleaver, followed by spreading (upper left arm); a sampled version is then created for processing through the right arm's analogue blocks. This may or may not include various sections of the transmission path, depending on the goal of the simulation. Shown in the figure is an option of switching in a channel simulator, which will include simulated channel losses and impairments such as multipath or adjacent-channel interference (ACI), noise etc. Omitting the channel simulator is the same as executing a transmitter receiver loopback test. Hardware systems usually have a number of loopback test options. In any case, in the figure, the signal going through the analogue arm eventually passes through the demodulator block, which would include a decision-circuit simulation module, such as a matched filter, that outputs the detected digital data stream. At this point the digital data are returned to the left side for channel and source decoding etc. Clearly, the logical–digital simulation part does not really add knowledge to the analogue (e.g. RF) design and analysis exercise, apart, perhaps, from varying the specifications that the analogue system must satisfy. However, the analogue part will be seen by the logical–digital part as a model of the full communications channel that is more realistic than a statistical channel model. Also the 'models of computation' for the left and right arms are different. The analogue part will use a continuous-in-time or finite-time-window computational approach, whereas the logical–digital part will probably use a data-flow, or synchronous data-flow, model of computation [3, 17, 18].

While this simulation flow diagram shows only a selection of the core components in a single-channel communications system, the approach could be used for simulating quite sophisticated communication channels and systems. It could, for instance, be the core of a simulation system for the design and performance analysis of an advanced modem, whether wireless or wired. Not only would it enable the design of various components and subsystems that satisfy full system performance requirements and specifications, but some of the simulation module algorithms may be transferred across directly into the real system. For example, the same algorithms as those used to model a modulator or demodulator and filters may be directly transferable onto DSP platforms (using application-specific integrated circuits (ASICs), ever-higher-density field programmable gate arrays (FPGAs) or programmable DSP chips) and thus implement in the real system exactly these same modulator/demodulators and filters. In this sense the boundary between some simulation activities and real-system implementation design can become blurred – this is the transfer from a 'constructive' to an 'executable' model, mentioned in Section 7.3.

7.9.2 Complete-system simulation

The gaps between some of these 'models of computation' can be significant. Some new-generation electronic system design automation (ESDA) tools incorporate system simulation support that seeks to bridge these gaps using various levels of ingenuity, complexity and heterogeneity. This occurs particularly as one moves down from system level to circuit level, mixing 'models of computation' in an effort to create an integrated and comprehensive communications system design-and-analysis process. In some of these simulators, RF portions of the system such as PAs or mixers can be described by a standard circuit-level simulation (e.g. a circuit envelope) using in-built or user-defined component-simulation models and then can be included in the system simulation in a seamless manner. An interesting review and discussion of this theme may be found in Lee *et al.* [11]. Typical examples of commercial simulators implementing complete-system simulation, including a description of the RF portion of the communication chain, are the AWR-VSS [5] and Agilent's advanced design system (ADS) implementation of the University of California's Ptolemy platform [7].

References

[1] M. C. Jeruchim, P. Balaban and K. S. Shanmugan, *Simulation of Communication Systems*, second edition, Kluwer/Plenum, 2000.

[2] W. H. Tranter and K. L. Kosbar, "Simulation of communication systems," *IEEE Communications Mag.*, vol. 32, no. 7, pp. 26–35, July 1994.

[3] K. S. Shanmugan, "Simulation and implementation tools for signal processing and communication systems," *IEEE Communications Magazine*, vol. 32, no. 7, pp. 36–40, July 1994.

[4] G. Benelli, V. Cappellini and E. Del Re, "Simulation system for analog and digital transmissions," *IEEE J. Selected Areas In Communications*, vol. SAC-2, no. 1, pp. 77–88, January 1984.

[5] AWR, "VSS users manual," Applied Wave Research, 1960, E. Grand Avenue, Suite 430, El Segundo CA 90245.

[6] Agilent-ADS, URL, 2008: http://www.tm.agilent.com/tmo/hpeesof/products/ads/adsoview.html.

[7] Agilent, "ADS Ptolemy Simulation," 2005, Agilent Technologies, 395 Page Mill Road, Palo Alto, CA 94304. URL, 2008: http://eesof.tm.agilent.com/docs/adsdoc2005A/pdf/ptolemy.pdf.

[8] Cadence, URL, 2008: http://www.cadence.com.

[9] Synopsis, URL, 2008: http://www.synopsys.com/products/mixedsignal/hspice/hspice.html.

[10] M. O'Droma and A. A.Goacher, "Final report on linearisation evaluation map," EU NOE TARGET Deliverable D1.3.2.7 (WP 2.2.E.1), vol. 2, 2006.

[11] E. A. Lee *et al.* "Overview of the Ptolemy project," March 2001. UC Berkeley, USA. URL, 2008: http://ptolemy.eecs.berkeley.edu/publications.

[12] J. Buck, Ha Soonhoi, E. A. Lee and D. G. Messerschmitt, "Ptolemy: a framework for simulating and prototyping heterogeneous systems," *Int. J. Computer Simulation, special issue on simulation software development*, pp. 1–34, August 1992.

[13] Dipartimento di Elettronica, Politecnico di Torino,"TOPSIM III – simulation package for communication systems – user's manual," Torino, Italy.

[14] M. Fashano and A. L. Strodbeck, "Communication systems simulation using SYSTID," *IEEE J. SAC*, vol. 2, no. 1, pp. 8–29, January 1984.

[15] W. Henke, "MITSYN – an interactive dialogue language for time signal processing," *MIT Res. Lab. Electron. Memo. RLETM-1*, February 1975.

[16] URL, 2008: http://itpp.sourceforge.net.

[17] J. Eker, J. W. Janneck, E. A. Lee *et al.*, "Taming heterogeneity – the Ptolemy approach," *Proc. IEEE*, vol. 91, no. 1, pp. 127–144, January 2003.

[18] E. A. Lee, and D. G. Messerschmitt, "Synchronous data flow," *Proc. IEEE*, vol. 75, no. 9, pp. 1235–1245, September 1987.

[19] URL, 2008: http://www.isi.edu/nsnam/ns.
[20] M. Ali, M. Welzl, A. Adnan and F. Nadeem, "Using the NS-2 network simulator for evaluating network on chips (NoC)", in *Proc. 2nd IEEE Int. Conf. on Emerging Technologies, ICET*, Peshawar, Pakistan, November 2006, pp. 506–512.
[21] R. Lemaire, F. Clermidy, Y. Durand, D. Lattard and A. A. Jerraya, "Performance evaluation of a NoC based design for MC-CDMA telecommunications using NS-2", in *Proc. 6th IEEE Int. Workshop on Rapid System Prototyping (RSP'05)*, 2006.
[22] OPNET Technologies, URL, 2008: http://www.opnet.com.
[23] S. Wang, K. Z. Liu and F. P. Hu, "Simulation of wireless sensor networks localization with OMNeT," in *Proc. 2nd IEEE Int. Conf. on Mobile Technology, Applications and Systems*, November 2005, pp. 1–6.
[24] Telelogic, "Rhapsody – model driven development for systems engineering, software development and test of embedded, real-time applications or technical systems," URL, 2008: http://www.telelogic.com/products/rhapsody/index.cfm.
[25] "SWANS – scalable wireless ad hoc network simulator user guide," URL, 2008: http://jist.ece.cornell.edu.
[26] M. Jiang, "An integrated requirements specification and validation framework for model-based systems," *Proc. Software Eng. Appl.*, vol. 514, no. 029, November 2006.
[27] D. Amyot and A. W. Williams. "System analysis and modeling," in *Proc. 4th Int. SDL and MSC Workshop (SAM 2004), Ottawa, Canada*, Springer, 2005. ISBN 3540245618. 301 pp.
[28] ITU-T Recommendation Z.100 (11/99), *Specification and Description Language (SDL)*, International Telecommunications Union, 1999.
[29] URL, 2008: http://www.itu.int/ITU-T/studygroups/com07/ttcn.html.
[30] S. Schulz and T. Vassiliou-Gioles, "Implementation of TTCN-3 test systems using the TRI," in *Proc. 14th Int. Conf. on Testing of Communicating Systems (TestCom), Berlin*, April 2002, pp. 425–441.
[31] J. O'Flaherty, "Developments in the simulation of communications satellite systems," Ph.D. thesis, National University of Ireland, 1982.
[32] M. S. O'Droma, N. Mgebrishvili and A. Goacher, "New percentage linearisation measures of the degree of linearisation of HPA nonlinearity," *IEEE Communications Lett.*, vol. 8, no. 4, pp. 214–216, April 2004.
[33] Agilent Technologies, "A comprehensive guide to harmonic balance for ADS", May 2003. URL, 2008: http://www.agilent.com.
[34] S. Maas, *Nonlinear Microwave and RF Circuits*, second edition, Artech House, 2003.
[35] K. Kundert, J. White and A. Sangiovanni-Vincentelli, *Steady State Methods for Simulating Analog and Microwave Circuits*, Kluwer, 1990.
[36] How-Siang Yap, "Designing to digital wireless specifications using circuit envelope simulation," HP-EEsat Division, Hewlett Packard. URL, 2008: http://eesof.viewmark.com/pdf/ckt_env.pdf. Also in *Proc. Asia Pacific Microwave Conf.*, December 1997, pp. 173–176.
[37] E. Ngoya and R. Larcheveque, "Envelope transient analysis: a new method for transient and steady-state analysis of microwave communication circuits and systems," in *Proc. IEEE MTT Symp. (IMS)*, June 1996, pp. 1365–1368.
[38] MentorGraphics. URL, 2008: http://www.mentor.com.
[39] J. Staudinger and G. Norris, "The effect of harmonic load terminations on RF power amplifier linearity for sinusoidal and $\pi/4DQPSK$ stimuli," in *Wireless Applications Dig., IEEE MTT-S Symp. on Technologies*, February 1997, pp. 23–28.

[40] J. Palacin, M. Salleras, J. Samitier and S. Marco, "Dynamic compact thermal models with multiple power sources: application to an ultrathin chip stacking technology," *IEEE Trans. Advanced Packaging*, vol. 28, no. 4, pp. 694–703, November 2005.
[41] J. L. Pino and K. Kalbasi, "Cosimulating synchronous DSP applications with analog RF circuits," in *Proc. 32nd Asilomar Conf. on Signals, Systems & Computers*, November 1998, vol. 2, pp. 1710–1714.
[42] R. Ahola, A. Aktas, J. Wilson, K. R. Rao, F. Jonsson, I. Hyyrylainen *et al.* "A single-chip CMOS transceiver for 802.11a/b/g wireless LANs", *IEEE J. Solid-State Circuits*, vol. 39, no. 12, pp. 2250–2258, December 2004.
[43] M. Sida, R. Ahola and D. Wallner, "Bluetooth transceiver design and simulation with VHDL-AMS," *IEEE Circuits and Devices Mag.*, vol. 19, no. 2, pp. 11–14 March 2003.
[44] IEEE, "1076.1-1999 IEEE standard VHDL analog and mixed-signal extensions," 1999. URL, 2008: http://www.ieee.org.
[45] SubstrateStorm of Simplex, URL, 2008: http://www.simplex.com.
[46] M. van Heijningen, J. Compiet, P. Wambacq, S. Donnay, M. Engels and I. Bolsens, "Analysis and experimental verification of digital substrate noise generation for epi-type substrates," *IEEE J. Solid-State Circuits*, vol. 35, no. 7, pp. 1002–1008, July 2000.
[47] M. O'Droma and N. Mgebrishvili, "Signal modelling classes for linearized OFDM SSPA behavioural analysis," *IEEE Communications Lett.*, vol. 9, no. 2, pp. 127–129, February 2005.
[48] K. S. Shanmugam and P. Balaban, "A modified Monte-Carlo simulation technique for the evaluation of error rate in digital communication systems,"*IEEE Trans. Communications*, vol. COM-28, no. 11, pp. 1916–1924, November 1980.
[49] L. Dingqing and Y. Kung, "Improved importance sampling technique for efficient simulation of digital communication systems," *IEEE J. Selected Areas in Communications*, vol. 6, no. 1, pp. 67–75, January 1988.
[50] NEWCOM, URL, 2008: http://newcom.ismb.it.

Appendix A Recent wireless standards

Introduction

Since the early 1980s, the range and variety of wireless communication air interfaces has seen immense growth. This has been driven, and is being driven further, by the need for ever greater information throughput. This requires greater bandwidths, higher output transmitter powers and the more efficient use of handheld battery energy resources and all at ever higher frequencies, although some lower frequencies have been freed up from the traditional broadcast communication services and are becoming available.

One complex modulation scheme that has grown in importance over the last decade is orthogonal frequency-division multiplexing (OFDM). It is a scheme which increases bandwidth efficiency and data capacity by splitting broadband channels into multiple narrowband channels, each using a different frequency, which can then carry different parts of a message simultaneously at bit per hertz capacities that are dynamically adaptable to the wireless channel quality.

This trend is set to continue for the foreseeable future and will result in a wide range of complex wireless communications systems sharing a common physical space. This has major implications for the development of communication-simulation tools.

In this appendix some advanced wireless interface standards are summarised with a view to highlighting the broad range of signal formats and figures of merit that have to be considered when developing system simulators. These are especially relevant when simulators with embedded nonlinear elements such as nonlinear PA behavioural models are being developed.

The three interfaces selected provide wireless coverage from long range to very short range and have been developed by the Institute of Electrical and Electronics Engineers Inc. through the IEEE 802.11, 802.15 and 802.16 Work Groups.

IEEE 802.11a WLAN

The IEEE 802.11a standard was agreed in 1999 and is a set of specifications for implementing high-speed wireless local area networks (WLANs) at 5 GHz [1]. These specifications define an over-the-air interface between a wireless client and a base station or between two or more wireless clients. Its complex modulation scheme is based on a 52-subcarrier OFDM technique with a maximum raw data rate of 54 Mb/s. By using the 5 GHz band, 802.11a WLAN systems have the advantage of less interference (seeing or causing less interference) than if they were located in the heavily used 2.4 GHz band. However, this high carrier frequency means that it is an almost line of sight (LOS) wireless system. Its core air-interface specifications are as follows.

- Data rates up to 54 Mb/s are available using a 20 MHz channel bandwidth in the 5 GHz UNII (unlicensed national information infrastructure) band.
- The data are modulated with BPSK, QPSK, 16QAM or 64QAM to achieve variable data rates and mapped onto the 52 subcarriers of an OFDM signal.
- A highest data rate of 54 Mb/s is achieved with 64QAM modulation.
- There are 52 subcarriers, 48 data and four pilot.
- The subcarriers are spaced 312.5 kHz apart.
- The signal bandwidth is 16.25 MHz per channel with a 20 MHz channel spacing.
- In the US, three bands are specified, each band constituting four channels.
- The maximum allowed transmit power is 16, 23 and 29 dBm for the lowest, mid and highest bands respectively.
- The range is up to 33 m (100 ft) indoors.
- The bit-error rate (BER) is 0.1%.
- The minimum sensitivity at the receiver is −82 dBm at a 6 Mb/s data rate and −65 dBm at a 54 Mb/s data rate.
- The adjacent-channel rejection is 4 dB at 54 Mb/s and 21 dB at 6 Mb/s.
- The non-adjacent-channel rejection is 23 dB at 54 Mb/s and 40 dB at 6 Mb/s.

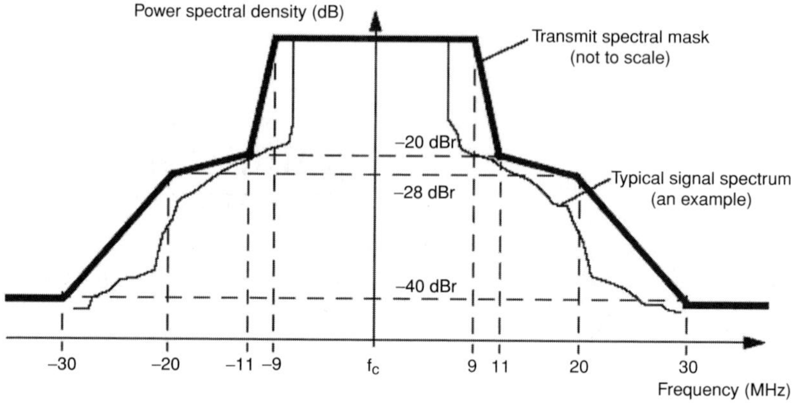

Figure A.1 The allowable transmit spectral mask for the IEEE 802.11a standard. The values are in dBr, i.e. dB relative to the peak power. Reprinted with permission from [1], © 2007, IEEE.

Table A.1 Allowable EVM (dB) values for the IEEE 802.11a standard. Reprinted with permission from [1], © 2007, IEEE

Data transmission speed (Mb/s)	Allowable EVM (dB)
6	−5
9	−8
12	−10
18	−13
24	−16
36	−19
48	−22
54	−25

In Figure A.1 and Table A.1, the transmit spectral mask and the allowable error-vector magnitude (EVM) for the IEEE 802.11a standard are given. Three different frequency bands are allocated for this standard; this frequency allocation is shown in Figure A.2.

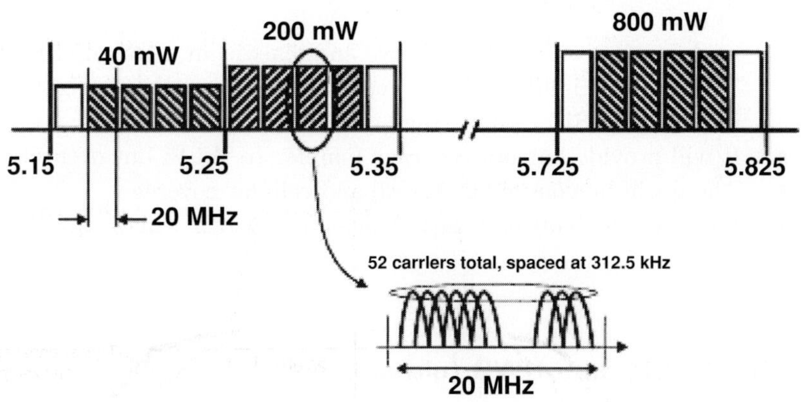

Figure A.2 Frequency-band allocation for the IEEE 802.11a standard. In the upper figure, frequencies are measured along the horizontal axis in GHz. Reprinted with permission from [2], © 2002, IEEE.

IEEE 802.16 (WiMAX)

The IEEE 802.16 WiMAX (worldwide interoperability for microwave access) standard is a wireless technology that provides high-throughput broadband connections over long distances [3]. It is targeted as a wireless replacement for, or alternative to, wired broadband digital subscriber-line (DSL) connections or cabled access networks such as fiber optic links and broadcast television coaxial cable systems using cable modems.

The WiMAX technology is intended to deliver high bit rates, of up to 70 Mb/s per user, over large areas and to have the capacity to handle large numbers of users. The initial application is point-to-multipoint broadband wireless networking for Internet access. Service operators would set up rooftop transceivers as base stations connected to the Internet. Each base station would use WiMAX technology to communicate with fixed externally mounted customer antennas. The following are its core air-interface specifications.

- Its range is up to 50 km (31 miles).
- The initial 802.16 standard operates in the 10 to 66 GHz range; these high frequencies require a direct LOS.
- The WiMAX standard operates in the 2 to 11 GHz frequency range.
- The channel bandwidth is variable (it can be an integer multiple of 1.25 MHz, 1.5 MHz or 1.75 MHz with a maximum of 20 MHz).
- It has a 256-carrier OFDM physical layer (PHY) (192 carriers are used for user data, with 56 nulled as guard bands and eight used as permanent pilot symbols).
- Unlike other wireless standards, which address transmissions over a single frequency range, WiMAX allows data transport over multiple broad frequency ranges.
- Each station will probably serve an area within a 16 km (10 mile) radius.
- There will be infill broadband coverage for areas that a digital subscriber fixed local loop (DSL) cannot reach, primarily rural areas.
- It will provide hotspot coverage (similar to Wi-Fi but over wider areas).
- There will be contiguous broadband cellular coverage.
- Flexible backhaul or targeted high bandwidth will be available to large customers.

IEEE 802.15.3a – UWB (ultrawideband)

Ultrawideband (UWB) wireless personal area networking (WPAN) technology has been conceived to meet the growing needs of high-speed WPANs providing the short-range high-bandwidth multimedia connections required for handling multiple digital, audio and video streams. This will complement and supplement currently deployed WLAN-type networks. Ultrawideband is the WPAN technology targeted for standardisation under the IEEE 802.15.3a 'label' [4]. The USA's FCC have set the defining frequency range that is now understood as UWB as 3.1 GHz to 10.6 GHz, a band 7.5 GHz wide. The spectral mask is shown in Figure A.3 and some of the basic UWB requirements (i.e. for 802.15.3a) are given in Table A.2.

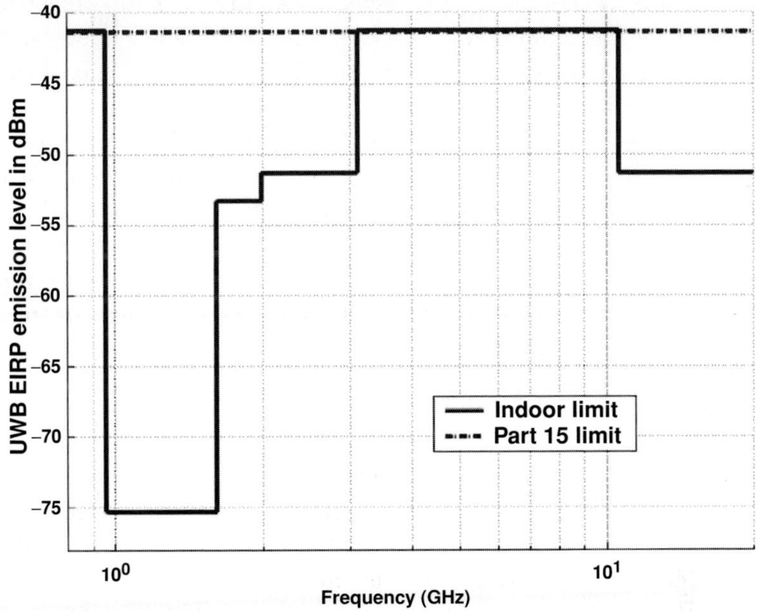

Figure A.3 Emitted UWB power spectral density limitation as defined in the FCC regulations.

Table A.2 Summary of air-interface requirements for 802.15.3a standard

Parameter	Value
bit rate (PHY-SAP)	110, 200, 480 Mb/s
range	10 m (30 ft), 4 m (12 ft)
power consumption	250 mW, 100 mW
bit error rate	10^{-5}
interference capability	robust to IEEE systems
co-existence capability	reduced interference to IEEE systems

Figure A.4 illustrates the generally wide bandwidth occupancy and low output transmitted energy density (the EIRP level must actually be less than −41 dBm/MHz) of UWB when compared with conventional narrowband single-carrier air interfaces or spread-spectrum air interfaces, such as Bluetooth and 802.11a/b/g. In fact the transmitted output power is so low that in most cases the use of transmitter power amplifiers is not envisaged.

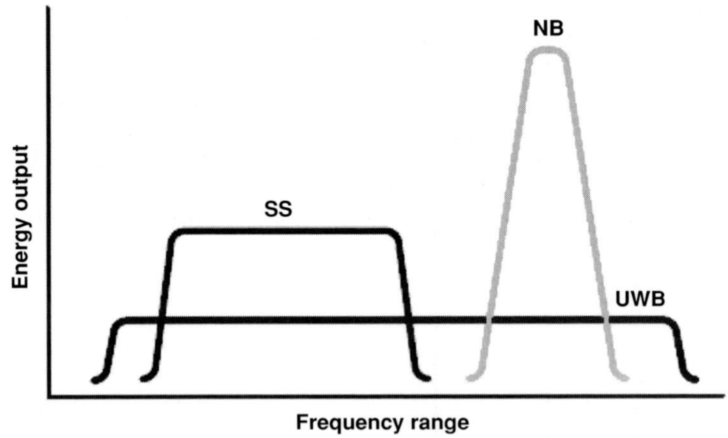

Figure A.4 Ultrawideband (UWB), narrowband (NB) and spread spectrum (SS) signals.

To be regarded as UWB, the transmitted signal must occupy:

- either a fractional bandwidth of more than 20% of the centre frequency
- or an absolute bandwidth in excess of 500 MHz.

Each radio channel can have a bandwidth in excess of 500 MHz. The FCC put in place severe broadcast-power restrictions to allow for the use of such wide signal bandwidths. In this way UWB devices can make use of an extremely wide frequency band while not emitting enough energy to be noticed by any narrower-band devices, such as 802.11a/b/g radios, that are nearby.

Possible applications of UWB systems are:

- replacing cables between portable multimedia devices, such as camcorders, digital cameras, and portable MP3 players, thus providing wireless connectivity;
- enabling high-speed wireless universal serial bus (WUSB) connectivity for PCs and PC peripherals, including printers, scanners and external storage devices;
- replacing short-range wideband communication cables, e.g. in offices;
- enabling short-range very-high-speed broadband access to the Internet;
- enabling precision navigation and asset tracking.

In conclusion, UWB technology complements currently deployed wireless networks in the WLAN environment.

References

[1] IEEE 802.11 Working Group (2007-06-12). IEEE 802.11-2007, *Wireless LAN Medium Access Control (MAC) and Physical Layer (PHY) Specifications.* ISBN 0-7381-5656-9.

[2] T. H. Meng, B. McFarland, D. Su and J. Thomson, "Design and implementation of an All-CMOS 802.11a wireless LAN chipset," *IEEE Communications Mag.*, pp. 160–163, August 2003.

[3] IEEE 802.16 Working Group (2004-10-01). IEEE 802.16-2004, *Air Interface for Fixed Broadband Wireless Access Systems.* ISBN 0-7381-4069-4.

[4] IEEE 802.15.3 Working Group (2003-09-29). IEEE 802.15.3-2003, *Wireless Medium Access Control (MAC) and Physical Layer (PHY) Specifications for High Rate Wireless Personal Area Networks (WPANs).* ISBN 0-7381-3704-9.

Appendix B Authors and contributors

Chapter 1

J. C. Pedro (*Universidade de Aveiro, Portugal*)

Chapter 2

M. Gadringer, D. Silveira (*Technische Universität Wien, Austria*)
D. Schreurs, M. Myslinski (*Katholieke Universiteit Leuven, Belgium*)
M. O'Droma (*University of Limerick, Ireland*)
J. A. García, A. Mediavilla (*Universidad de Cantabria, Spain*)

Chapter 3

M. O'Droma, Y. Lei, S. Meza (*University of Limerick, Ireland*)
V. Camarchia, M. Pirola (*Politecnico di Torino, Italy*)

Chapter 4

M. Gadringer, D. Silveira (*Technische Universtität Wien, Austria*)
A. Zhu, C. Devlin (*University College Dublin, Ireland*)

Chapter 5

E. R. Srinidhi, G. Kompa (*Universität Kassel, Germany*)
D. Schreurs, M. Myslinski (*Katholieke Universiteit Leuven, Belgium*)
P. Gilabert, G. Montoro Lopez, E. Bertran (*Universitat Politecnica de Catalunya, Spain*)
A. Zhu, C. Devlin (*University College Dublin, Ireland*)
J. A. Lonac, C. Florian, I. Melczarsky (*Università di Bologna, Italy*)

Y. Gurbuz, I. Tekin (*Sabanci Üniversitesi, Turkey*)
S. Donati Guerrieri, M. Pirola (*Politecnico di Torino, Italy*)

Chapter 6

I. Melczarsky, A. Santarelli, F. Filicori (*Università di Bologna, Italy*)
M. Gadringer, D. Silveira (*Technische Universtität Wien, Austria*)
D. Schreurs (*Katholieke Universiteit Leuven, Belgium*)

Chapter 7

M. O'Droma, A. A. Goacher (*University of Limerick, Ireland*)
M. Olavsbråten (*NTNU Trondheim, Norway*)

Appendix A

Y. Gurbuz, C. Kavlak, I. Tekin (*Sabanci Üniversitesi, Turkey*)
A. A. Goacher (*University of Limerick, Ireland*)

Index